Progress in Mathematics
Volume 224

Modular Curves and Abelian Varieties

John Cremona
Joan-Carles Lario
Jordi Quer
Kenneth Ribet
Editors

Birkhäuser Verlag
Basel · Boston · Berlin

Editors:

J. E. Cremona
University of Nottingham
School of Mathematical Sciences
University Park
Nottingham NG7 2RD
UK
e-mail: john.cremona@nottingham.ac.uk

K. A. Ribet
Department of Mathematics
M/C 3840
University of California
Berkeley, CA 94720-3840
USA
e-mail: ribet@math.berkeley.edu

J.-C. Lario and J. Quer
Facultat de Matemàtiques i Estadística
Universitat Politècnica de Catalunya
Pau Gargallo, 5
08028 Barcelona
Spain
e-mail: Joan.Carles.Lario@upc.es
Jordi.Quer@upc.es

2000 Mathematics Subject Classification 11Gxx, 14Gxx, 14Kxx

A CIP catalogue record for this book is available from the Library of Congress, Washington D.C., USA

Bibliographic information published by Die Deutsche Bibliothek
Die Deutsche Bibliothek lists this publication in the Deutsche Nationalbibliografie; detailed bibliographic data is available in the Internet at <http://dnb.ddb.de>.

ISBN 3-7643-6586-2 Birkhäuser Verlag, Basel – Boston – Berlin

Printed on acid-free paper produced of chlorine-free pulp. TCF ∞

ISBN 3-7643-6586-2

9 8 7 6 5 4 3 2 1

www.birkhauser.ch

Table of Contents

Preface

It would be difficult to overestimate the influence and importance of modular forms, modular curves, and modular abelian varieties in the development of number theory and arithmetic geometry during the last fifty years. These subjects lie at the heart of many past achievements and future challenges. For example, the theory of complex multiplication, the classification of rational torsion on elliptic curves, the proof of Fermat's Last Theorem, and many results towards the Birch and Swinnerton-Dyer conjecture all make crucial use of modular forms and modular curves.

A conference was held from July 15 to 18, 2002, at the Centre de Recerca Matemàtica (Bellaterra, Barcelona) under the title "Modular Curves and Abelian Varieties". Our conference presented some of the latest achievements in the theory to a diverse audience that included both specialists and young researchers. We emphasized especially the conjectural generalization of the Shimura–Taniyama conjecture to elliptic curves over number fields other than the field of rational numbers (elliptic $\mathbb{Q}$-curves) and abelian varieties of dimension larger than one (abelian varieties of GL_2-type).

This book evolved as a byproduct of that conference: most of the articles here describe the research of their authors as it was presented in their talks. We have also included a few articles that were not presented at the conference but are closely allied with the conference's theme. Among these are three that deserve special mention because they contributed strongly to the current interest in elliptic $\mathbb{Q}$-curves and abelian varieties of GL_2-type; they were written soon after the publication of Barry Mazur's delightful article "Number Theory as a Gadfly" (American Mathematical Monthly, Vol. 98 issue 7, 593–610), which reformulated the Shimura–Taniyama conjecture as a statement over $\overline{\mathbb{Q}}$. The article by Noam Elkies is a revised version of his important unpublished 1993 manuscript on elliptic k-curves; in that manuscript, Elkies explained the precise relation between elliptic k-curves and k-rational points on Atkin-Lehner quotients of the modular curves $X_0(N)$. The editors are grateful to Professor Elkies for revising his paper for this volume. The contribution by Elizabeth Pyle is her doctoral thesis, presented in UCLA in 1995 and realized under the supervision of K. Ribet; it contains a complete characterization of the abelian varieties that should occur as factors of modular abelian varieties over the algebraic closure of the rationals, if the generalized Shimura-Taniyama conjecture holds. Shortly after finishing her thesis, Pyle left academia and did not have the opportunity to publish her thesis. The paper by Kenneth Ribet is a reprint of a paper already published in the proceedings of a conference held in Taejŏn (Korea) in 1993; it contains the first conjectural characterization of $\mathbb{Q}$-curves as the modular elliptic curves over $\overline{\mathbb{Q}}$, and gives evidence for this conjecture by relating it to Serre's conjecture on 2-dimensional modular Galois representations. (At the suggestion of a number of conference participants, we included Ribet's article in this volume, since the original publication is out of print and not available in many university libraries.)

We would like to thank all the participants at the conference for their contribution to the scientific success of the event. Special thanks are due to the contributors to this volume. We are also indebted to the Centre de Recerca Matemàtica: to its director Prof. Manuel Castellet for his continued support since the beginning of the project, and to its staff, Ms. Maria Julià and Ms. Consol Roca, for their superb work on the administrative and organizational tasks, in connection with the local Organising Committee. Thanks are also due to all the administrations whose financial support made the conference possible: the European Community under its program High Level Scientific Conferences (HPCF-CT-2001-00386), the spanish Ministerio de Ciencia y Tecnologia (BFM-2001-4280-E), the Generalitat de Catalunya, the Universitat Autònoma de Barcelona and the Universitat Politècnica de Catalunya.

The editorial committee:

John Cremona
Joan-Carles Lario
Jordi Quer
Kenneth A. Ribet

Progress in Mathematics, Vol. 224, 1–22

Stable Reduction of Modular Curves

Irene I. Bouw and Stefan Wewers

1. Introduction

Modular curves are quotients of the upper half-plane by congruence subgroups of $\mathrm{SL}_2(\mathbb{Z})$. The most prominent examples are $X_0(N)$, $X_1(N)$ and $X(N)$, for $N \geq 1$. Modular curves are also moduli spaces for elliptic curves endowed with a level structure. Therefore, they are defined over small number fields and have a rich arithmetic structure. Deligne and Rapoport [4] determine the reduction behavior of $X_0(p)$ and $X_1(p)$ at the prime p. Using this result, they prove a conjecture of Shimura saying that the quotient of the Jacobian of $X_1(p)$ by the Jacobian of $X_0(p)$ acquires good reduction over $\mathbb{Q}(\zeta_p)$. Both the reduction result and its corollary have been generalized by Katz and Mazur [10] to arbitrary level N and various level structures.

The basic method employed in [4] and [10] is to generalize the moduli problem defining the modular curve in question, in such a way that it makes sense in arbitrary characteristic. By general results on representability of moduli problems one then obtains a model of the modular curve over the ring of integers of a subfield of $\mathbb{Q}(\zeta_N)$. This model has bad reduction at all primes dividing N, and it is far from being semistable, in general. In spite of the very general results of [10], it remains an unsolved problem to describe the stable reduction of $X_0(N)$ at p if p^3 divides N (see [5] for the case $p^2||N$).

In this note we suggest a different approach to study the reduction of modular curves. Our starting point is the observation that the j-invariant

$$X(N) \longrightarrow X(1) \cong \mathbb{P}^1$$

presents the modular curve $X(N)$ as a Galois cover of the projective line (with Galois group $\mathrm{PSL}_2(\mathbb{Z}/N)$) which is branched only at the three rational points 0, 1728 and ∞. Let us call a Galois cover of the projective line branched only at three points a *three point cover*. In [12] Raynaud studies the stable reduction of three point covers under the assumption that the residue characteristic p strictly divides the order of the Galois group. His results have been sharpened in [18] and [17]. In this paper we determine the stable reduction of all three point covers with Galois group $\mathrm{PSL}_2(p)$, using the results of [17]. As a corollary we obtain a new proof of the results of Deligne and Rapoport on the reduction of $X_0(p)$ and $X_1(p)$, and the result of Edixhoven [5] on the reduction of $X_0(p^2)$. Somewhat surprisingly, our proof does not use the modular interpretation of these curves. On the other hand, the determination of the stable reduction of some of the other $\mathrm{PSL}_2(p)$-covers (which are not modular curves) does use the fact that they are moduli spaces of a certain kind.

What is so special about the group $\mathrm{PSL}_2(p)$? From our point of view there are two main aspects. The first is *rigidity.* Let G be a finite group and $\mathbf{C} = (C_1, C_2, C_3)$ a triple of conjugacy classes of G. Suppose that there exists a triple $\mathbf{g} = (g_1, g_2, g_3)$ of generators of G with $g_i \in C_i$ and $g_1 g_2 g_3 = 1$. The triple $\mathbf{g}$ corresponds to a G-cover $Y \to \mathbb{P}^1$ with three branch points. If G equals $\mathrm{PSL}_2(p)$, such a triple $\mathbf{g}$ is (essentially) unique, up to uniform conjugation in $\mathrm{PGL}_2(p)$. Therefore there exists (essentially) at most one three point cover with a given branch cycle description $\mathbf{C}$, up to isomorphism. (In some cases, one needs a further invariant called the *lifting invariant.*)

We use rigidity as follows. By a result of [17] we can construct three point G-covers $Y \to \mathbb{P}^1$ with bad reduction at p by lifting a certain type of 'stable G-cover' $\bar{Y} \to \bar{X}$ from characteristic p to characteristic zero. Here $\bar{Y} \to \bar{X}$ is a finite map between semistable curves in characteristic p together with some extra structure (g, ω) which we call the *special deformation datum.* In the case of the modular curve $X(p)$ (where $G = \mathrm{PSL}_2(p)$) one can construct $\bar{Y} \to \bar{X}$ very explicitly; the special deformation datum (g, ω) corresponds to a solution of the Gauß hypergeometric differential equation. (A similar phenomenon occurs in Ihara's work on congruence relations [7].) The $\mathrm{PSL}_2(p)$-cover $Y \to \mathbb{P}^1$ resulting from the lifting process has branch cycle description $(3A, 2A, pA)$. By rigidity, $Y \to \mathbb{P}^1$ is isomorphic to $X(p) \to X(1)$. In particular, the stable reduction of $X(p)$ is isomorphic to $\bar{Y}$.

The other nice thing about $\mathrm{PSL}_2(p)$ is that p strictly divides its order. The results of [12] and [17] which describe the stable reduction of a given G-cover require that p strictly divides the order of G, whereas the Sylow p-subgroup of $\mathrm{PSL}_2(\mathbb{Z}/p^n)$ is rather big for $n > 1$. This is the main obstruction for extending the method of the present paper to modular curves of higher p-power level. There are partial results of the authors generalizing some of the results of [12] and [17] to groups G with a cyclic or an elementary abelian Sylow p-subgroup (unpublished). It seems hopeless to obtain general results beyond these cases. But maybe a combination of the methods presented in the present paper with the modular approach might shed some light on the stable reduction of modular curves of higher p-power level.

The organization of this paper is as follows. In Section 1 we define special deformation data and explain how to associate a special deformation datum to a three point cover with bad reduction. We recall a result from [17] which essentially says that we can reverse this process. In Section 3 we introduce hypergeometric deformation data. These are special deformation data satisfying an addition condition (Definition 3.1). We classify all hypergeometric deformation data by showing that they correspond to the solution in characteristic p of some hypergeometric differential equation. In Sections 4 and 5 we use these results to give a new proof of the stable reduction of the modular curves $X(p)$ and $X_0(p)$. In Section 6 we generalize these results to all three point covers with Galois group $\mathrm{SL}_2(p)$.

2. The special deformation datum

Let k be an algebraically closed field of characteristic $p > 2$. Let H be a finite group of order prime to p. Fix a character $\chi : H \to \mathbb{F}_p^\times$. Let $g : Z_k \to \mathbb{P}^1_k$ be an H-Galois cover and ω be a meromorphic differential on Z_k. We assume that ω is logarithmic (i.e., can be written as $\omega = du/u$) and

$$\beta^*\omega = \chi(\beta) \cdot \omega, \qquad \text{for all } \beta \in H.$$

Let $\xi \in Z_k$ be a closed point and τ its image in $\mathbb{P}^1_k$. Denote by H_ξ the stabilizer of ξ in H. Define

$$m_\tau := |H_\xi|, \qquad h_\tau := \mathrm{ord}_\xi(\omega) + 1, \qquad \sigma_\tau = h_\tau/m_\tau. \tag{1}$$

Since ω is logarithmic, we have $h_\tau \geq 0$. We say that τ is a *critical point* of the differential ω if $(m_\xi, h_\xi) \neq (1,1)$. Let $(\tau_i)_{i \in \mathbb{B}}$ be the critical points of ω, indexed by a finite set $\mathbb{B}$. For $i \in \mathbb{B}$, we write m_i, h_i, σ_i instead of $m_{\tau_i}, h_{\tau_i}, \sigma_{\tau_i}$. For every i, choose a point $\xi_i \in Z_k$ above τ_i and write $H(\xi_i) \subset H$ for its stabilizer. Define $\mathbb{B}_{\mathrm{wild}} := \{\, i \in \mathbb{B} \mid h_i = 0 \,\}$.

Definition 2.1. *A* special deformation datum *of type (H, χ) is a pair (g, ω), where $g : Z_k \to \mathbb{P}^1_k$ is an H-Galois cover and ω is a logarithmic differential on Z_k such that the following holds.*

(i) *We have*

$$\beta^*\omega = \chi(\beta) \cdot \omega, \qquad \textit{for all } \beta \in H. \tag{2}$$

(ii) *For every $i \in \mathbb{B} - \mathbb{B}_{\mathrm{wild}}$, we have that $0 < \sigma_i \leq 2$.*
(iii) *Define $\mathbb{B}_{\mathrm{prim}} = \{i \in \mathbb{B} \,|\, 0 < \sigma_i \leq 1\}$. Then $|\mathbb{B}_{\mathrm{prim}} \cup \mathbb{B}_{\mathrm{wild}}| = 3$.*

Write $\mathbb{B}_0 = \mathbb{B}_{\mathrm{wild}} \cup \mathbb{B}_{\mathrm{prim}}$ and $\mathbb{B}_{\mathrm{new}} = \mathbb{B} - \mathbb{B}_0$. See [17] for more details and an explanation of the terminology. If $\sigma_i \neq 1, 2$ for all i, the above definition coincides with [17, Definition 2.7].

Definition 2.2. *Let G be a finite group. A G-*tail cover *is a (not necessarily connected) G-Galois cover $f_k : Y_k \to \mathbb{P}^1_k$ such that f_k is wildly branched at ∞ of order pn with n prime to p and tamely branched at no more than one other point. We say that f is a* primitive tail cover *if it is branched at two points. Otherwise, we call f a* new tail cover.

If the group G is understood, we talk about tail covers instead of G-tail covers. To a G-tail cover $f_k : Y_k \to \mathbb{P}^1_k$ we associate its *ramification invariant* $\sigma(f) = h/n$, where h is the conductor of f_k at ∞ and n the order of the prime-to-p ramification, as in Definition 2.2. The ramification invariant is the jump in the filtration of higher ramification groups in the upper numbering.

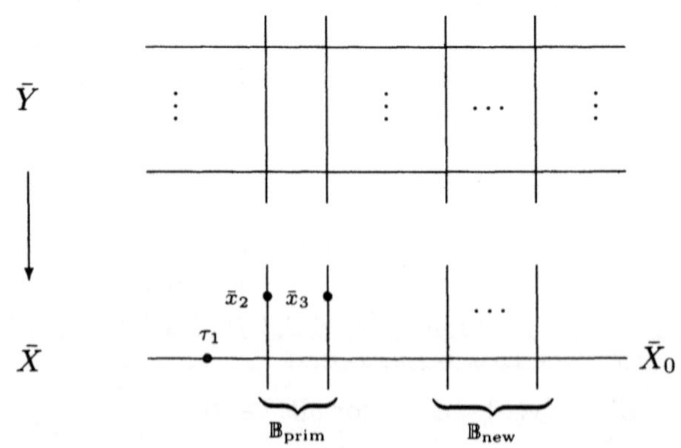

FIGURE 1. The stable reduction of $f : Y \to \mathbb{P}^1$

Three point covers with bad reduction give rise to a special deformation datum and a set of tail covers. Essentially, the stable reduction of the cover is determined by these data. Proposition 2.3 states that given a special deformation datum and a set of tail covers satisfying some compatibility conditions, there exists a three point cover in characteristic zero which gives rise to the given datum.

To be more precise, let R be a complete discrete valuation ring with fraction field K of characteristic zero and residue field an algebraically closed field k of characteristic p. Let G be a finite group whose order is strictly divisible by p and let $f : Y \to X = \mathbb{P}^1_K$ be a G-Galois cover branched at three points x_1, x_2, x_3. We assume that the points x_1, x_2, x_3 specialize to pairwise distinct points on the special fiber $\mathbb{P}^1_k$. Denote the ramification points of f by $y_1, \ldots, y_s$. We consider the y_i as markings on Y. After replacing K by a finite extension, there exists a unique extension $(Y_R; y_i)$ of $(Y; y_i)$ to a stably marked curve over R. The action of G extends to Y_R; write X_R for the quotient of Y_R by G. The map $f_R : Y_R \to X_R$ is called the *stable model* of f; its special fiber $\bar{f} : \bar{Y} \to \bar{X}$ is called the *stable reduction* of f, [17, Definition 1.1].

We say that f has *good reduction* if $\bar{f}$ is separable. This is equivalent to $\bar{X}$ being smooth. If f does not have good reduction, we say it has *bad reduction*.

Suppose that f has bad reduction. Then the (semistable) curve $\bar{X}$ is a 'comb', see [17, Theorem 2.6] and Figure 1. The central component $\bar{X}_0 \subset \bar{X}$ is canonically isomorphic to $\mathbb{P}^1_k$. Choose a component $\bar{Y}_0$ of $\bar{Y}$ above $\bar{X}_0$. Let $G_0 \subset G$ denote the decomposition group of the component $\bar{Y}_0$ and $I_0 \lhd G_0$ the inertia group. Then $I_0 \cong \mathbb{Z}/p$. Therefore, the restriction of $\bar{f}$ to $\bar{Y}_0$ factors as $\bar{Y}_0 \to \bar{Z}_0 \to \bar{X}_0$, with $g : \bar{Z}_0 \to \bar{X}_0$ a separable Galois cover of prime-to-p order and $\bar{Y}_0 \to \bar{Z}_0$ purely inseparable of degree p. The map $\bar{Y}_0 \to \bar{Z}_0$ is generically endowed with the structure of a $\boldsymbol{\mu}_p$-torsor. This structure is encoded in a logarithmic differential ω.

Define $H := \mathrm{Gal}(\bar{Z}_0/\bar{X}_0)$; then G_0 is an extension of H by I_0. The action of H on I_0 by conjugation gives rise to a character $\chi : H \to \mathbb{F}_p^\times$. It is proved in [17, Proposition 2.15] that (g, ω) is a special deformation datum of type (H, χ).

The irreducible components of $\bar{X}$ different from the central component $\bar{X}_0$ are called the *tails* of $\bar{X}$. The tails are parameterized by $i \in \mathbb{B} - \mathbb{B}_{\text{wild}}$, [17, Proposition 1.7]. For $i \in \mathbb{B} - \mathbb{B}_{\text{wild}}$, write $\bar{X}_i$ for the corresponding tail of $\bar{X}$ and write $\bar{f}_i : \bar{Y}_i \to \bar{X}_i$ for the restriction of $\bar{f}$ to $\bar{X}_i$. The map $\bar{f}_i$ is a primitive tail cover if $i \in \mathbb{B}_{\text{prim}}$ and a new tail cover if $i \in \mathbb{B}_{\text{new}}$, [17, Lemma 1.4]. We say that $\bar{X}_i$ is a *primitive* (resp. *new*) tail of $\bar{X}$ if $i \in \mathbb{B}_{\text{prim}}$ (resp. $i \in \mathbb{B}_{\text{new}}$).

The three branch points specialize as follows. If p divides the ramification index of x_j, then x_j specializes to $\tau_i \in \bar{X}_0$ for a unique $i \in \mathbb{B}_{\text{wild}}$. Otherwise, x_j specializes to one of the primitive tails $\bar{X}_i$, for $i \in \mathbb{B}_{\text{prim}}$. From now on, we will identify the set $\mathbb{B}_0 = \mathbb{B}_{\text{prim}} \cup \mathbb{B}_{\text{wild}}$ with $\{1, 2, 3\}$ such that $i \in \mathbb{B}_{\text{wild}}$ if and only if p divides the order of the ramification index at x_i. For $i \in \mathbb{B}_{\text{prim}}$ we denote by $\bar{x}_i \in \bar{X}_i$ the specialization of x_i.

The special deformation datum (g, ω) and the tail covers $\bar{f}_i$ satisfy the following compatibility condition. For every $i \in \mathbb{B} - \mathbb{B}_{\text{wild}}$, there exists a point $\xi_i \in \bar{Z}_0$ above τ_i and a point $\eta_i \in \bar{Y}_i$ which is wildly ramified in $\bar{f}_i$ such that

$$I(\eta_i) \subset G_0, \qquad I(\eta_i) \cap H = H(\xi_i). \tag{3}$$

Here $I(\eta_i) \subset G$ (resp. $H(\xi_i) \subset H$) denotes the stabilizer of η_i (resp. of ξ_i). Furthermore, the ramification invariant $\sigma(\bar{f}_i)$ of the tail cover $\bar{f}_i$ is equal to σ_i as in Definition 2.1.

Proposition 2.3. *Let G be a finite group. Fix a subgroup $G_0 \simeq \mathbb{Z}/p \rtimes_\chi H$ of G, where H has order prime-to-p and acts on $\mathbb{Z}/p$ via a character $\chi : H \to \mathbb{F}_p^\times$. Let (g, ω) be a special deformation datum of type (H, χ). Let $\mathbb{B}$, τ_i and $\sigma_i = h_i/m_i$ be as in Definition* 2.1.

For $i \in \mathbb{B}_{\text{prim}}$ (resp. $i \in \mathbb{B}_{\text{new}}$), suppose we are given a primitive (resp. new) G-tail cover $\bar{f}_i : \bar{Y}_i \to \bar{X}_i = \mathbb{P}^1_k$ with ramification invariant σ_i. Suppose furthermore that there exists a point $\xi_i \in Z_k$ above $\tau_i \in \mathbb{P}^1_k$ and a point $\eta_i \in \bar{Y}_i$ which is wildly ramified in $\bar{f}_i$ such that (3) *holds.*

Let K_0 be the fraction field of $W(k)$ (the Witt vectors over k) and let K/K_0 be the (unique) tame extension of degree

$$N := (p-1) \cdot \operatorname*{lcm}_{i \notin \mathbb{B}_{\text{wild}}} (h_i). \tag{4}$$

There exists a G-Galois cover $f : Y \to X = \mathbb{P}^1_K$ over K, branched at 0, 1 *and* ∞, *which gives rise to the special deformation datum (g, ω) and the tail covers $\bar{f}_i$.*

Proof. The datum $(g, \omega, \bar{f}_i)$ together with the choice of the points ξ_i and η_i is called a *special G-deformation datum* in [17, Definition 2.13]. Hence the proposition follows from [17, Corollary 4.6]. □

In this paper, the group G is either $\mathrm{SL}_2(p)$ or $\mathrm{PSL}_2(p)$. Therefore, the group H is cyclic and $H(\xi_i)$ only depends on i and not on the choice of ξ_i.

The G-cover $f : Y \to \mathbb{P}^1_K$ in Proposition 2.3 which gives rise to the datum $(g, \omega, \bar{f}_i)$ is not unique. In [17, Theorem 4.5] one finds an explicit parameterization of the set of isomorphism classes of all such G-covers.

3. Hypergeometric deformation data

In this section we classify a certain type of special deformation data which we call *hypergeometric*. A hypergeometric (special) deformation datum corresponds to a polynomial solution of a hypergeometric differential equation. For this reason it is possible to find all such deformation data. The definition of hypergeometricity looks very restrictive, but we show in Section 6 that the special deformation datum corresponding to a three point cover with Galois group $\mathrm{PSL}_2(p)$ or $\mathrm{SL}_2(p)$ is hypergeometric. In particular, this holds for the $\mathrm{PSL}_2(p)$-cover $X(p) \to X(1) \simeq \mathbb{P}^1_j$.

Definition 3.1. *A* hypergeometric deformation datum *is a special deformation datum* (g, ω) *with* $\sigma_i = (p+1)/(p-1)$ *for all* $i \in \mathbb{B}_{\mathrm{new}}$.

Let (g, ω) be a hypergeometric deformation datum. Recall that $\mathbb{B}_0 = \{1, 2, 3\}$, by assumption. In this section we assume $x_1 = 0, x_2 = 1, x_3 = \infty$. In particular, $\tau_i \neq \infty$ for all $i \in \mathbb{B}_{\mathrm{new}}$. For $i \in \mathbb{B}$, we define integers $0 \le a_i < p-1$ such that

$$a_i \equiv (p-1)\sigma_i \mod p-1.$$

In particular, $a_i = 2$ for $i \in \mathbb{B}_{\mathrm{new}}$.

To $g : Z_k \to \mathbb{P}^1_k$ we associate the (not necessarily connected) $\mathbb{F}_p^\times$-Galois cover

$$g' : Z'_k := \mathrm{Ind}_{\mathrm{Im}(\chi)}^{\mathbb{F}_p^\times}(Z_k/\mathrm{Ker}\,\chi) \to \mathbb{P}^1_k.$$

It follows easily from Definition 3.1 and the expansion of ω into local coordinates that Z'_k is the complete nonsingular curve defined by

$$z^{p-1} = x^{a_1}(x-1)^{a_2} \prod_{i \in \mathbb{B}_{\mathrm{new}}} (x - \tau_i)^2, \tag{5}$$

where $\beta^* z = \chi(\beta) \cdot z$. We call (a_1, a_2, a_3) the *signature* of the hypergeometric deformation datum (g, ω).

We may identify the differential ω on Z_k with a differential on Z'_k which we also denote by ω. Definition 2.1(i) and (1) imply that

$$\omega = \epsilon \frac{z\,dx}{x(x-1)}, \tag{6}$$

for some $\epsilon \in k^\times$. The condition that ω is logarithmic imposes a strong condition on $u := \prod_{i \in \mathbb{B}_{\mathrm{new}}} (x - \tau_i)$. Proposition 3.2 below states that the differential (6) is logarithmic if and only if u is the solution to a certain hypergeometric differential equation.

It follows from (1) and the Riemann–Roch Theorem applied to ω that

$$|\mathbb{B}_{\mathrm{new}}| = (p - 1 - a_1 - a_2 - a_3)/2. \tag{7}$$

In particular, $a_1 + a_2 + a_3$ is an even integer between 0 and $p - 1$. Note that if $a_1 + a_2 + a_3 = p - 1$ then $\deg(u) = |\mathbb{B}_{\text{new}}| = 0$. Therefore, we exclude this case in Proposition 3.2.(i). However, Proposition 3.2.(ii) applies also if $a_1+a_2+a_3 = p-1$.

Proposition 3.2.

(i) *Let (g, ω) be a hypergeometric deformation datum. Let (a_1, a_2, a_3) be its signature, and suppose that $a_1 + a_2 + a_3 < p - 1$. Then $u(x) := \prod_{i\in\mathbb{B}_{\text{new}}}(x - \tau_i)$ is a solution to the hypergeometric differential equation*

$$x(x-1)u'' + [(A+B+1)x - C]u' + ABu = 0,$$

where $A = (1 + a_1 + a_2 + a_3)/2$, $B = (1 + a_1 + a_2 - a_3)/2$ and $C = 1 + a_1$. The degree of u in x is $d := (p - 1 - a_1 - a_2 - a_3)/2$.

(ii) *Let $0 \le a_1, a_2, a_3 < p - 1$ be integers with $a_1 + a_2 + a_3 \le p - 1$. We assume that $a_1 + a_2 + a_3$ is even. Let $H = \mathbb{Z}/m$, where $m = (p - 1)/2$ if a_1, a_2, a_3 are all even and $m = p-1$ otherwise. Choose an injective character $\chi : H \to \mathbb{F}_p^\times$. There exists a hypergeometric deformation datum (g, ω) of signature (a_1, a_2, a_3) and type (H, χ). The deformation datum (g, ω) is uniquely determined by (a_1, a_2, a_3), up to multiplying the differential ω by an element of $\mathbb{F}_p^\times$.*

Proof. The following proof is inspired by [1, Lemma 3]. A similar argument can be found in [7, Theorem 5]. Write

$$Q = x^{1+a_1}(x-1)^{1+a_2}, \qquad F = \epsilon\frac{z}{x(x-1)} = \frac{\epsilon z^p}{Qu^2}.$$

It is well known that the fact that $\omega = F\,\mathrm{d}x$ is logarithmic is equivalent to $D^{p-1}F = -F^p$, where $D := \mathrm{d}/\mathrm{d}x$. Since $D^{p-1}F = \epsilon z^p D^{p-1}[1/(Qu^2)]$, we find

$$D^{p-1}\frac{1}{Qu^2} = -\epsilon^{p-1}\frac{1}{x^p(x-1)^p}. \tag{8}$$

Choose $i \in \mathbb{B}_{\text{new}}$ and write

$$\frac{1}{Qu^2} = \sum_{n\ge -2} c_n(x-\tau_i)^n.$$

Then

$$D^{p-1}\frac{1}{Qu^2} = -\left[\frac{c_{-1}}{(x-\tau_i)^p} + c_{p-1} + \cdots\right] = -\epsilon^{p-1}\frac{1}{x^p(x-1)^p}.$$

We conclude that $c_{-1} = 0$.

Write

$$\begin{aligned}Qu^2 &= [Q(\tau_i) + Q'(\tau_i)(x-\tau_i)\cdots][u'(\tau_i)(x-\tau_i) + \frac{1}{2}u''(\tau_i)(x-\tau_i)^2 + \cdots]^2\\ &= (x-\tau_i)^2u'(\tau_i)\,[\,Q(\tau_i)u'(\tau_i) + (\,Q'(\tau_i)u'(\tau_i) + Q(\tau_i)u''(\tau_i)\,)\,(x-\tau_i) + \cdots].\end{aligned}$$

We see that $-c_{-1} = u'(\tau_i)[\,Q'(\tau_i)u'(\tau_i) + Q(\tau_i)u''(\tau_i)\,]$. Since $u'(\tau_i) \ne 0$, we have that $Q'(\tau_i)u'(\tau_i) + Q(\tau_i)u''(\tau_i) = 0$ for all $i \in \mathbb{B}_{\text{new}}$.

Define $G = Q'u' + Qu''$. This is a polynomial of degree less than or equal to $e := \deg(Q) + \deg(u) - 2 = |\mathbb{B}_{\rm new}| + a_1 + a_2$. The coefficient of x^e in G is $g_e = \deg(u)(\deg(Q) + \deg(u) - 1)$. The condition $0 \leq a_1 + a_2 + a_3 < p - 1$ implies that the coefficient g_e is nonzero modulo p.

The polynomial G is obviously divisible by $x^{a_1}(x-1)^{a_2}$, and $G(\tau_i) = 0$, for all $i \in \mathbb{B}_{\rm new}$. Therefore $G = g_e x^{a_1}(x-1)^{a_2}u$. Dividing by $x^{a_1}(x-1)^{a_2}$, we find that u is a solution to the hypergeometric differential equation

$$P_0 u'' + P_1 u' + P_2 u = 0, \tag{9}$$

where $P_0 = x(x-1)$, $P_1 = P_0 Q'/Q = (2 + a_1 + a_2)x - (1 + a_1)$, and $P_2 = -g_e = ((1 + a_1 + a_2)^2 - a_3^2)/4$. Part (i) of the proposition follows.

To prove (ii), let $0 \leq a_1, a_2, a_3 < p-1$ be integers such that $a_1 + a_2 + a_3$ is even and less than or equal to $p - 1$. Define $d = (p - 1 - a_1 - a_2 - a_3)/2$. We claim that the differential equation (9) has a unique polynomial solution $u = \sum_i u_i x^i \in k[x]$ of degree d with $u_d = 1$.

Suppose that $u = \sum_{i \in \mathbb{Z}} u_i x^i$ is a solution of (9). Put $A_i = (i+1)(i + a_1 + 1)$ and $B_i = (i + (1 + a_1 + a_2 + a_3)/2)(i + (1 + a_1 + a_2 - a_3)/2)$. The differential equation gives a recursion

$$A_i u_{i+1} = B_i u_i \tag{10}$$

for the coefficients of u. One checks that A_i and B_i are nonzero for all $0 \leq i < d$ and that $A_{-1} = B_d = 0$ and $A_d \cdot B_{-1} \neq 0$. Therefore there is a unique polynomial solution u of degree d with $u_d = 1$. This implies the claim.

Since u is the solution to a hypergeometric differential equation, it has at most simple zeros outside 0 and 1. We know that $u(0) = u_0 = \prod_{i=0}^{d-1}(B_i/A_i) \neq 0$. Therefore u does not have a zero at $x = 0$. We claim that u does not have a zero at 1, also. To see this, let $t = 1 - x$ be a new coordinate on $\mathbb{P}^1_k$ and write $\tilde{u}(t) = u(1-t)$. The coefficients of $\tilde{u}$ satisfy a recursion similar to (10), but with a_1 and a_2 interchanged. The same argument as above applied to $\tilde{u}$ shows now that $\tilde{u}(0) = u(1) \neq 0$. We conclude that u has exactly d zeros different from 0 and 1.

Let m be as in the statement of the proposition. Let Z'_k be the complete nonsingular curve defined by $z^{p-1} = x^{a_1}(x-1)^{a_2}u^2$ and let Z_k be a connected component of Z'_k. Write $g : Z_k \to \mathbb{P}^1_k$ for the m-cyclic cover defined by $(x, z) \mapsto x$. We may identify H with $\mathrm{Gal}(Z_k/\mathbb{P}^1_k)$ such that H acts on z via some chosen character $\chi : H \to \mathbb{F}_p^\times$. Let

$$\omega' = \frac{z\,\mathrm{d}x}{x(x-1)} = \frac{z^p\,\mathrm{d}x}{Qu^2}.$$

We consider the partial fractions expansion of $1/Qu^2$. Write

$$\frac{1}{Qu^2} = \frac{\gamma_{1+a_1}}{x^{1+a_1}} + \cdots + \frac{\gamma_1}{x} + \frac{\delta_{1+a_2}}{(x-1)^{1+a_2}} + \cdots + \frac{\delta_1}{x-1} + \sum_{i \in \mathbb{B}_{\rm new}} \left(\frac{\epsilon_i}{(x-\tau_i)^2} + \frac{\nu_i}{x - \tau_i} \right).$$

The reverse direction of the argument in (i) shows that $1/Qu^2$ has no residues at τ_i, i.e., $\nu_i = 0$, for all i.

Denote by $\mathcal{C}$ the Cartier operator. Recall that a differential ω is logarithmic if and only if $\mathcal{C}\omega = \omega$. For the definition and properties of the Cartier operator, see for example [13]. The previous computation implies that

$$\mathcal{C}\omega' = \left(\frac{\gamma_1^{1/p}}{x^p} + \frac{\delta_1^{1/p}}{(x-1)^p}\right) z\,\mathrm{d}x = \left(\frac{-(\gamma_1^{1/p} + \delta_1^{1/p})\,x^p + \gamma_1^{1/p}}{x^p\,(x-1)^p}\right) z\,\mathrm{d}x.$$

Note that ω' has at most logarithmic poles above $x = 0, 1, \infty$ and is holomorphic elsewhere. It is well known that this implies that $\mathcal{C}\omega'$ has also at most logarithmic poles above $x = 0, 1, \infty$ and is holomorphic elsewhere. We conclude therefore that $\gamma_1 + \delta_1 = 0$, and that $\mathcal{C}\omega' = \gamma_1^{1/p}\omega$. It follows from [18, Lemma 1.4] that $\gamma_1 \neq 0$. (This lemma is stated for holomorphic differentials, but it can easily be extended to differentials with logarithmic poles above $x = 0, 1, \infty$.) Define $\omega = \gamma_1^{1/(p-1)}\omega'$. The fact that $\mathcal{C}$ is $(1/p)$-semilinear implies that $\mathcal{C}\omega = \omega$. We conclude that ω is a logarithmic differential and that (g, ω) is a special deformation datum. The uniqueness statement is obvious. □

4. The reduction of $X(p)$

In the rest of the paper we suppose that $p \geq 5$. Choose a primitive $(p-1)$th root of unity $\zeta \in \mathbb{F}_p$ and a primitive $(p+1)$th root of unity $\tilde{\zeta} \in \mathbb{F}_{p^2}$. Define

$$C(l) = \{A \in \mathrm{SL}_2(p) \mid \mathrm{tr}(A) = \zeta^l + \zeta^{-l}\}, \qquad \text{for } 0 < l \leq (p-1)/2$$

and

$$\tilde{C}(l) = \{A \in \mathrm{SL}_2(p) \mid \mathrm{tr}(A) = \tilde{\zeta}^l + \tilde{\zeta}^{-l}\}, \qquad \text{for } 0 < l < (p+1)/2.$$

These are the conjugacy classes of $\mathrm{SL}_2(p)$ of non-trivial elements of order prime to p. We write pA and pB for the two conjugacy classes of elements of order p. Note that the elements of order p have trace 2.

Let $\mathbf{C} = (C_1, C_2, C_3)$ be a triple of conjugacy classes of $\mathrm{SL}_2(p)$. In the rest of this section, we suppose that the C_i do not contain $\pm I$. We write $\mathrm{Ni}_3^{\mathrm{in}}(\mathbf{C})$ for the set of isomorphism classes of $\mathrm{SL}_2(p)$-covers $Y \to \mathbb{P}^1$ branched at $0, 1, \infty$ with class vector $\mathbf{C}$. This means that the canonical generator of some point of Y above $x_i \in \{0, 1, \infty\}$ with respect to the chosen roots of unity is contained in the conjugacy class C_i. In this paper, we call two G-covers $f_i : Y_i \to X$ isomorphic if there exists a G-equivariant automorphism $\phi : Y_1 \to Y_2$ such that $f_1 = f_2 \circ \phi$.

Proposition 4.1 (Linear rigidity). *Let $\mathbf{C} = (C_1, C_2, C_3)$ be a triple of conjugacy classes of $\mathrm{SL}_2(p)$.*

(i) *Suppose that the elements of C_i have prime-to-p order. Then $\# \mathrm{Ni}_3^{\mathrm{in}}(\mathbf{C}) \in \{0, 2\}$.*

(ii) *Otherwise, $\# \mathrm{Ni}_3^{\mathrm{in}}(\mathbf{C}) \in \{0, 1\}$.*

Proof. See [15]. The difference between the two cases comes from the fact that the outer automorphism group of G has order two and interchanges the two conjugacy classes of order p but fixes all other conjugacy classes. □

In the following we denote by $2A$ (resp. $3A$) the unique conjugacy classes in $\mathrm{PSL}_2(p)$ of elements of order 2 (resp. 3). Note that elements of $2A$ (resp. elements of $3A$) lift to elements of order 4 (resp. of order 3 or 6) in $\mathrm{SL}_2(p)$. Furthermore, we denote by pA and pB the images in $\mathrm{PSL}_2(p)$ of the conjugacy classes of $\mathrm{SL}_2(p)$ with the same name.

Let $X(N)$ be the modular curve parameterizing (generalized) elliptic curves with full level N-structure [4]. Consider the cover $X(2p) \to X(2) \simeq \mathbb{P}^1_\lambda$. This is a Galois cover with Galois group $\mathrm{PSL}_2(p)$ which is branched at $0, 1, \infty$ of order p and unbranched elsewhere. One easily checks as in [16, Lemma 3.27] that there are no $\mathrm{PSL}_2(p)$-covers with class vector (pA, pA, pB). After renaming the conjugacy classes, the class vector of $X(2p) \to X(2)$ is therefore $\mathbf{C} = (pA, pA, pA)$.

Proposition 4.1 implies that the triple $\mathbf{C} = (pA, pA, pA)$ is rigid, i.e., there is a unique $\mathrm{PSL}_2(p)$-cover of $\mathbb{P}^1$ whose class vector is $\mathbf{C}$, up to automorphism. Since $\mathrm{PSL}_2(p)$ has trivial center, it follows that the $\mathrm{PSL}_2(p)$-cover $X(2p) \to X(2)$ has a unique model $X(2p)_K \to X(2)_K = \mathbb{P}^1_K$ over any field K of characteristic zero containing $\sqrt{p^*}$, [14, Chapter 7]. Let K_0 be the fraction field of the ring of Witt vectors over $k = \bar{\mathbb{F}}_p$, and let K/K_0 be a sufficiently large finite extension. We define the stable reduction of $X(2p) \to X(2)$ at p as the stable reduction of $X(2p)_K \to \mathbb{P}^1_K$.

The cover $X(2p) \to X(2)$ may be considered as a variant with ordered branch points of the cover $X(p) \to X(1) \simeq \mathbb{P}^1_j$. We find it easier to first compute the stable reduction of the former and then deduce the stable reduction of the latter.

Proposition 4.2. *Let $\bar{Y} \to \bar{X}$ be the stable reduction of $X(2p) \to X(2)$ at p. Then $\bar{X}$ has $(p-1)/2$ new tails with ramification invariant $\sigma = (p+1)/(p-1)$. The points $\tau_i \in \mathbb{P}^1_k$, $i \in \mathbb{B}_{\mathrm{new}}$ (where the new tails occur) are precisely the zeros of the polynomial*

$$u = \sum_{j=0}^{(p-1)/2} \binom{(p-1)/2}{j}^2 x^j. \tag{11}$$

Moreover, the stable reduction occurs over the (unique) tame extension K/K_0 of degree $(p^2-1)/2$.

The polynomial u is known as the *Hasse invariant.* It has the property that the elliptic curve given by $y^2 = x(x-1)(x-\lambda)$ is supersingular in characteristic p if and only if λ is a zero of u. It is a solution to the Gauss hypergeometric differential equation, with parameters $A = 1/2, B = 1/2, C = 1$. The relation between this differential equation and the reduction of elliptic curve is well understood, see for example [9].

Proof. Let $a_1 = a_2 = a_3 = 0$ and $m = (p-1)/2$. Choose an injective character $\chi : \mathbb{Z}/m \to \mathbb{F}_p^\times$. Let (g, ω) be the special deformation datum of signature

$(0,0,0)$ constructed in Proposition 3.2.(ii). Recall that we associate to the triple $(a_1, a_2, a_3) = (0,0,0)$ the hypergeometric differential equation with parameters $A = B = 1/2$ and $C = 1$. Using the recursion (10) one checks that the polynomial u defined by (11) is a solution in characteristic p to this differential equation. This differential equation has a unique monic solution of degree $(p-1)/2$. Therefore, u is the same as in Proposition 3.2 (ii).

Let $Y_k' \to \mathbb{P}^1_k$ be a connected and new $\mathrm{PSL}_2(p)$-tail cover with ramification invariant $\sigma = (p+1)/(p-1)$. The existence of such a tail cover is proved in Lemma 4.3 below. The inertia group of a point above ∞ has order $p(p-1)/2$. For every $i \in \mathbb{B}_{\mathrm{new}}$, we let $\bar{f}_i : \bar{Y}_i \to \bar{X}_i$ be a copy of $Y_k' \to \mathbb{P}^1_k$. Let $\eta_i \in \bar{Y}_i$ be the point whose inertia group consists of the upper triangular matrices in $\mathrm{PSL}_2(p)$. Then the condition of Proposition 2.3 is satisfied. Therefore, there exists a $\mathrm{PSL}_2(p)$-cover $f_K : Y_K \to \mathbb{P}^1_K$ branched at three points of order p whose stable reduction gives rise to the special deformation datum (g, ω) and the tail covers $\bar{f}_i$. This cover is defined and has stable reduction over the tame extension K/K_0 of degree $(p^2-1)/2$. It follows from rigidity that f_K is isomorphic to $X(2p)_K \to X(2)_K$. This proves the proposition. □

Lemma 4.3. *There exists a connected new* $\mathrm{SL}_2(p)$*-tail cover* $h : Y_k \to X_k = \mathbb{P}^1_k$, *with inertia group of order* $p(p-1)$ *and ramification invariant* $\sigma = (p+1)/(p-1)$.

Proof. Let Y_k be the complete nonsingular curve defined by

$$y^{p+1} = x^p - x.$$

The group $\mathrm{SL}_2(p)$ acts on Y_k as follows. For

$$M = \begin{pmatrix} a & b \\ c & d \end{pmatrix} \in \mathrm{SL}_2(p), \qquad \text{define } Mx = \frac{ax+b}{cx+d}, \quad My = \frac{y}{cx+d}.$$

It is easy to check that the quotient map $h : Y_k \to X_k \simeq \mathbb{P}^1_k$ is branched at exactly one point which we may assume to be ∞. The inertia group of some point above ∞ has order $p(p-1)$ and $\sigma = (p+1)/(p-1)$. □

Let $X(p) \to X(1) \simeq \mathbb{P}^1_j$ be the projection of the modular curve $X(p)$ to the j-line. This is a $\mathrm{PSL}_2(p)$-cover with branch cycle description $(3A, 2A, pA)$, which we may assume to be branched at $x_1 = 0, x_2 = 1728, x_3 = \infty$. By Proposition 4.1, it has a unique model $X(p)_K \to \mathbb{P}^1_K$ over any field K of characteristic 0 containing $\sqrt{p^*}$. As before, we take K/K_0 a sufficiently large finite extension and define the stable reduction of $X(p) \to X(1)$ at p as the stable reduction of the model $X(p)_K \to \mathbb{P}^1_K$.

Corollary 4.4. *Define* $\alpha = \lfloor p/12 \rfloor$. *Let* $\bar{Y}' \to \bar{X}'$ *be the stable reduction of* $X(p) \to X(1)$. *Then* $\bar{X}'$ *has two primitive tails and* α *new tails. For* $i = 1, 2$, *denote by* $\bar{X}_i$

the primitive tail to which x_i specializes. Then

$$\sigma_i = \begin{cases} \dfrac{p+1}{p-1} & \text{if } i \in \mathbb{B}_{\text{new}}, \\ \dfrac{\lceil (p-1)/3 \rceil}{p-1} & \text{if } i = 1, \\ \dfrac{2\lceil (p-1)/4 \rceil}{p-1} & \text{if } i = 2. \end{cases}$$

Moreover, the stable reduction occurs over the tame extension K/K_0 of degree $(p^2-1)/2$.

Proof. Consider the commutative diagram

$$\begin{array}{ccc} X(2p)_K & \longrightarrow & X(p)_K \\ \downarrow & & \downarrow \\ X(2)_K & \longrightarrow & X(1)_K. \end{array}$$

The horizontal arrows are Galois covers whose Galois group is the symmetric group S_3 on three letters. Let Y_R be the stable model of $X(2p)_K$. The action of S_3 extends to Y_R and $Y_R/S_3 := Y_R^*$ is a semistable curve, [11, Appendice].

Since S_3 acts on Y_R it follows that the set of new tails of $\bar{X}_0$ is stable under the action of S_3. We claim that the action of S_3 on the set of new tails of $\bar{X}$ has α orbits of length six, one orbit of length three if $p \equiv 3 \bmod 4$ and one orbit of length two if $p \equiv 2 \bmod 3$. Since $\bar{X}_0$ has genus zero, an element of order three in S_3 has two fixed points in $\bar{X}_0 - \{0, 1, \infty\}$. An element of order two in S_3 fixes one of the points $0, 1, \infty$ and has therefore one fixed point in $\bar{X}_0 - \{0, 1, \infty\}$. The claim now follows by distinguishing the four possibilities for $p \bmod 12$.

The curve $X_R^* := Y_R^*/G$ is almost equal to X_R'. The special fiber $\bar{X}^*$ has a tail at zero (resp. 1728) if and only if $p \equiv -1 \bmod 3$ (resp. $p \equiv -1 \bmod 4$), i.e., exactly when zero (resp. 1728) is supersingular. To obtain X_R' one has to perform a suitable blow up centered at the point $j = 0$ (resp. $j = 1728$) on $\bar{X}^*$ if $p \equiv 1 \bmod 3$ (resp. $p \equiv 1 \bmod 4$.) The corollary follows, by analyzing the ramification of $\bar{Y}_0^* \to \bar{X}_0^*$. For future reference, we note that a connected component of $\bar{Y}_1'$ has decomposition group $\mathrm{PSL}_2(p)$ if $p \equiv -1 \bmod 3$ and the nonabelian group of order $3p$ if $p \equiv 1 \bmod 3$. A connected component of $\bar{Y}_2'$ has decomposition group $\mathrm{PSL}_2(p)$ if $p \equiv -1 \bmod 4$ and the dihedral group of order $2p$ if $p \equiv 1 \bmod 4$. □

The hypergeometric differential equation corresponding to $X(p) \to X(1)$ is

$$x(x-1728)u'' + [(2+a_1+a_2)x - 1728(1+a_1)]u' + (1+a_1+a_2)^2 u/4,$$

where $a_1 = \lceil (p-1)/3 \rceil$ and $a_2 = 2\lceil (p-1)/4 \rceil$. The polynomial

$$u = (1728)^\alpha \sum_{n=0}^{\alpha} \binom{a_1+\alpha}{\alpha-n} \binom{\alpha}{n} \left(\frac{x}{1728}\right)^n \in k[x]$$

is a solution to this differential equation. Recall that $j \in k$ is *supersingular* if and only if the corresponding elliptic curve over k is supersingular. It follows that $j \neq 0, 1728$ is supersingular if and only if $u(j) = 0$. We refer the interested reader to [8] for more formulae for this polynomial.

5. The reduction of $X_0(p)$

Let $X_0(p)$ denote the modular curve parameterizing (generalized) elliptic curves with a $\Gamma_0(p)$-structure [4]. We may and will identify $X_0(p)$ with the quotient curve $X(p)/G_0$, where $G_0 \subset G := \mathrm{PSL}_2(p)$ is the standard Borel group, modulo $\pm I$. The G-cover $X(p) \to X(1) = \mathbb{P}^1_j$ gives rise to a non-Galois cover $X_0(p) \to \mathbb{P}^1_j$ which we call the *j-invariant*. We claim that the curve $X_0(p)$ has a *unique* $\mathbb{Q}$-model $X_0(p)_{\mathbb{Q}}$ such that the j-invariant descends to a cover $X_0(p)_{\mathbb{Q}} \to \mathbb{P}^1_{\mathbb{Q}}$. (Of course, the existence of such a $\mathbb{Q}$-model is also a consequence of the modular interpretation of $X_0(p)$.)

To prove the claim, let $G \hookrightarrow S_{p+1}$ denote the standard permutation representation of G coming from the natural action on $\mathbb{P}^1(\mathbb{F}_p)$. Note that G_0 is the stabilizer of ∞. It is well known that the normalizer of G in S_{p+1} is equal to $\mathrm{PGL}_2(p) = \mathrm{Aut}(G)$. In particular, the centralizer of G in S_{p+1} is trivial. Let $\mathbf{C} = (3A, 2A, pA)$ denote the triple of conjugacy classes of G corresponding to the G-cover $X(p) \to X(1)$, as in the previous section. Let $\mathbf{C}'$ denote the image of $\mathbf{C}$ in $\mathrm{PGL}_2(p)$. Then $\mathbf{C}'$ is rigid and rational. Hence the claim follows from standard Galois theory.

For any field K of characteristic zero we denote by $X_0(p)_K$ the K-model of $X_0(p)$ obtained from $X_0(p)_{\mathbb{Q}}$ by base change. In [4, VI.6.16] Deligne and Rapoport prove the following result.

Theorem 5.1. *Let $X_0(p)_{\mathbb{Z}_p}$ denote the normalization of $\mathbb{P}^1_{\mathbb{Z}_p}$ inside $X_0(p)_{\mathbb{Q}_p}$. Set $k := \bar{\mathbb{F}}_p$.*

(i) *The $\mathbb{Z}_p$-curve $X_0(p)_{\mathbb{Z}_p}$ is semistable (and stable if $g(X_0(p)) > 1$).*
(ii) *The geometric special fiber $X_0(p)_k := X_0(p)_{\mathbb{Z}_p} \otimes k$ is the union of two smooth curves $\bar{W}_0'$, $\bar{W}_0''$ of genus 0. The induced map $\bar{W}_0' \to \mathbb{P}^1_k$ (resp. $\bar{W}_0'' \to \mathbb{P}^1_k$) is an isomorphism (resp. purely inseparable of degree p). The components $\bar{W}_0'$ and $\bar{W}_0''$ intersect precisely in the* supersingular points, *i.e., the points of $X_0(p)_k$ with supersingular j-invariant.*
(iii) *Let $x \in X_0(p)_k$ be a supersingular point, with j-invariant $j_x \in k$. If $j_x \equiv 0$ then $X_0(p)_{\mathbb{Z}_p}$ has a singularity of type A_3 at x. If $j_x \equiv 1728$ then x presents a singularity of type A_2. Otherwise, x is a regular point of $X_0(p)_{\mathbb{Z}_p}$.*

We give a proof of this theorem using the results of the previous section.

Proof. Let $R_0 := W(k)$ denote the ring of Witt vectors and K_0 the fraction field of R_0. Note that

$$X_0(p)_{R_0} := X_0(p)_{\mathbb{Z}_p} \otimes_{\mathbb{Z}_p} R_0$$

is equal to the normalization of $\mathbb{P}^1_{R_0}$ in the function field of $X_0(p)_{K_0}$. Hence it suffices to prove the theorem with the ring $\mathbb{Z}_p$ replaced by R_0.

Let K/K_0 be the tame extension of degree $(p^2-1)/2$. By Corollary 4.4 the $\mathrm{PSL}_2(p)$-cover $Y_K := X(p)_K \to X_K := \mathbb{P}^1_K$ extends to a stable model $Y_R \to X_R$ over the ring of integers R of K. As before, we denote by $\bar{Y} \to \bar{X}$ the special fiber of $Y_R \to X_R$.

Set $W_R := Y_R/G_0$. By [11, Appendice], W_R is a semistable curve over R with generic fiber $W_K = X_0(p)_K$. By construction, we have a finite map $W_R \to X_R$. Let $\bar{W}$ denote the special fiber of W_R. The formation of the quotient Y_R/G_0 does not commute with base change, in general. However, the canonical map $\bar{Y}/G_0 \to \bar{W}$ is a homeomorphism. Let V be an irreducible component of $\bar{Y}$ and U its image in $\bar{W}$. Let $D(V) \subset G$ (resp. $I(V) \subset G$) denote the decomposition group (resp. the inertia group) of V. Then the natural map $V/(D(V)\cap G_0) \to U$ is purely inseparable of degree $|I(V) \cap G_0|$. Using Corollary 4.4 and the description of the stable reduction in Section 2, we see that $\bar{W}$ has four different types of components (see Figure 2 for the case $p = 29$). In the description below we call an irreducible component of $\bar{W}$ *horizontal* if it is mapped to the central component $\bar{X}_0$ of $\bar{X}$ and *vertical* otherwise, cf. Figure 1.

(a) A horizontal component $\bar{W}_0'$ which is the image of the (unique) horizontal component $\bar{Y}_0 \subset \bar{Y}$ with decomposition group G_0. The natural map $\bar{W}_0' \to \bar{X}_0 = \mathbb{P}^1_k$ is an isomorphism.

(b) Another horizontal component $\bar{W}_0''$ which is the image of the set of horizontal components of $\bar{Y}$ different from $\bar{Y}_0$. Note that if $\bar{Y}'' \neq \bar{Y}_0$ is such a horizontal component, with decomposition group D'', then $D'' \cap G_0$ is a cyclic group of order $(p-1)/2$ (a split torus). Therefore, the natural map $\bar{W}_0'' \to \bar{X}_0$ is purely inseparable of degree p.

(c) For $i \in \mathbb{B}_{\mathrm{new}} \cup \mathbb{B}_{\mathrm{prim}}$ we let $\bar{W}_i$ be the image of the tail cover $\bar{Y}_i$. Suppose that i is supersingular, i.e., that the tail $\bar{X}_i$ intersects $\bar{X}_0 = \mathbb{P}^1_k$ in a supersingular point. Then $\bar{Y}_i$ and – hence $\bar{W}_i$ as well – is connected. The curve $\bar{W}_i$ intersects each of the horizontal components $\bar{W}_0$ and $\bar{W}_0'$ in a unique point.

(d) With notation as in (c), suppose that i is not supersingular (in particular, $i \in \mathbb{B}_{\mathrm{prim}}$). Then $\bar{Y}_i$ is not connected, and the decomposition group of any connected component is isomorphic to $\mathbb{Z}/p \rtimes \mathbb{Z}/2$ or $\mathbb{Z}/p \rtimes \mathbb{Z}/3$ (depending on whether $x_i \equiv 1728$ or $x_i \equiv 0$). It follows that $\bar{W}_i$ has exactly two connected components $\bar{W}_i'$ and $\bar{W}_i''$. The curve $\bar{W}_i'$ (resp. $\bar{W}_i''$) intersects $\bar{W}_0'$ (resp. $\bar{W}_0''$) in a unique point. (See the end of the proof of Corollary 4.4.)

In the modular picture, the component $\bar{W}_0'$ (resp. $\bar{W}_0''$) corresponds generically to ordinary elliptic curves together with a subgroup scheme isomorphic to μ_p (resp. to $\mathbb{Z}/p$), see [4, VI.6]. Note that every component of $\bar{W}$ has genus 0.

Let W_R' be the curve obtained from W_R by contracting all vertical components of $\bar{W}$. Since these vertical components have genus zero and self-intersection number -1 or -2, the curve W_R' is semistable. In particular, W_R' is a normal scheme. Recall that the curve obtained from X_R by contracting the tails $\bar{X}_i$

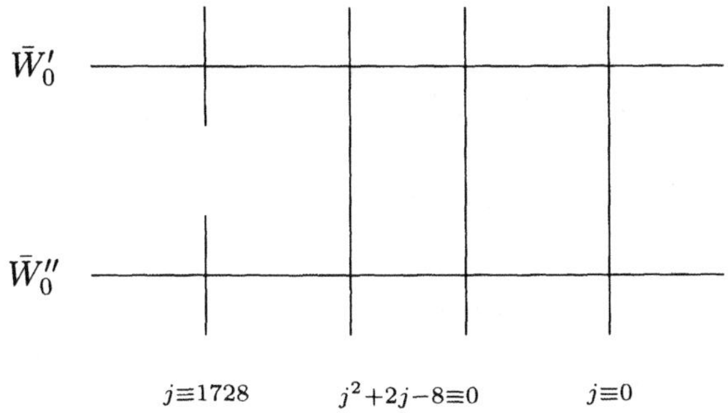

FIGURE 2. The curve $\bar{W}$ for $p = 29$

is equal to $\mathbb{P}^1_R$. Therefore the finite map $W_R \to X_R$ gives rise to a finite map $W'_R \to \mathbb{P}^1_R$. We conclude that W'_R is the normalization of $\mathbb{P}^1_R$ inside the function field of $W_K = X_0(p)_K$.

Let $\Gamma := \mathrm{Gal}(K/K_0) \cong \mathbb{Z}/N$, and recall that $N = (p^2-1)/2$ is prime to p. Since $W_K = X_0(p)_{K_0} \otimes_{K_0} K$, the group Γ acts naturally on W_K. Since W'_R is the normalization of $\mathbb{P}^1_R$ in W_K this action extends to the R-model W'_R and

$$X_0(p)_{R_0} \;=\; W'_R/\Gamma. \tag{12}$$

Since the order of Γ is prime to p, formation of the quotient commutes with base change. In particular, we have

$$X_0(p)_k \;=\; \bar{W}'/\Gamma. \tag{13}$$

It is clear that the action of Γ commutes with the map $\bar{W}' \to \mathbb{P}^1_k$ coming from the j-invariant. Since this map is an isomorphism on one and purely inseparable of degree p on the other component, Γ acts trivially on $\bar{W}'$. Now Parts (i) and (ii) of the theorem follow from (13) and the description of the components of $\bar{W}$ given above.

It remains to prove Part (iii) of the theorem. Let $x \in X_0(p)_k = \bar{W}'$ be a supersingular point, with j-invariant $j_x \in k$. Let v denote the valuation on K, normalized such that $v(p) = 1$. Since x is an ordinary double point of $\bar{W}'$, the local ring of $X_0(p)_{R_0}$ at x is of the form $R_0[[u,v \mid uv = \pi]]$, with $\pi \in R_0$. We define the *thickness* of $X_0(p)_{R_0}$ at x as the rational number $e(x) \;:=\; v(\pi)$. We have to show that $e(x)$ is equal to 3 if $j_x \equiv 1728$, equal to 2 if $j_x \equiv 0$ and equal to 1 otherwise.

We assume that $j_x \not\equiv 0, 1728$. The other cases follow in the same manner. Let $x' \in \bar{W}'$ be the unique point lying above x. Also, let x'' and x''' be the two double points on $\bar{W}$ lying above x'. We assume that $x'' \in \bar{W}_0'$ and $x''' \in \bar{W}_0''$. It is easy to see that

$$e(x) \;=\; e(x') \;=\; e(x'') + e(x'''). \tag{14}$$

Let $y'', y''' \in \bar{Y}$ be points above x'', x'''. Recall that $W_R = Y_R/G_0$. By [11, Appendice], we have $e(x'') = |\mathrm{Stab}_{G_0}(y'')| \cdot e(y'')$ and $e(x''') = |\mathrm{Stab}_{G_0}(y''')| \cdot e(y''')$. Using (14) and our knowledge of the G-action on $\bar{Y}$, we see that

$$e(x) \;=\; \frac{p(p-1)}{2} \cdot e(y'') + \frac{p-1}{2} \cdot e(y'''). \tag{15}$$

It is proved in [17] that the thickness of any double point $y \in \bar{Y}$ is equal to $(h_i(p-1))^{-1}$ if y lies on the tail $\bar{Y}_i$ and h_i denotes the conductor of the tail cover $\bar{Y}_i \to \bar{X}_i$ at ∞. Since we assumed that $j_x \not\equiv 0, 1728$ we have $y'', y''' \in \bar{Y}_i$ for some $i \in \mathbb{B}_{\mathrm{new}}$. By Corollary 4.4 we have $h_i = (p+1)/2$ for $i \in \mathbb{B}_{\mathrm{new}}$. With (15) we conclude that

$$e(x) \;=\; \frac{p}{p+1} + \frac{1}{p+1} \;=\; 1.$$

This finishes the proof of the theorem. □

One can analyze the natural R_0-model of $X_1(p)$ in an entirely analogous manner. This would give a new proof of the result of [4] that $J_1(p)/J_0(p)$ has good reduction over $\mathbb{Q}(\zeta_p)$. One can also derive this result directly from our knowledge of the stable reduction of $X(p)$. The curve $X_0(p^2)$ is also a quotient of $X(p)$. Therefore one can use a similar method to deduce the stable reduction of $X_0(p^2)$, giving a new proof of the result of [5].

6. Three point covers with Galois group $\mathrm{SL}_2(p)$

In this section we compute the stable reduction of all $\mathrm{SL}_2(p)$-covers of $\mathbb{P}^1$ branched at three points. In particular, we obtain an explicit formula for the number of such covers with good reduction (Theorem 6.6). To prove this we use the results of the previous sections. In addition, we use the fact that some $\mathrm{SL}_2(p)$-covers have a modular interpretation as a *Hurwitz space*.

Let us give an outline of the proof of Theorem 6.6. We first show that the $\mathrm{SL}_2(p)$-covers of $\mathbb{P}^1$ branched at three points of order p, p, n are essentially Hurwitz spaces. These covers have bad reduction to characteristic p; their stable reduction is described in [2]. By considering the (unique) primitive tail of the stable reduction, we obtain one primitive tail cover for every noncentral conjugacy class of $\mathrm{SL}_2(p)$ whose order is prime to p. Proposition 6.5 states that every primitive $\mathrm{SL}_2(p)$-tail cover with group $\mathrm{SL}_2(p)$ arises in this way. Then we show (Proposition 6.3) that the only new tail cover that occurs is the cover constructed in Lemma 4.3. This implies that the special deformation datum of an $\mathrm{SL}_2(p)$-cover branched at three points is hypergeometric (Definition 3.1). Therefore, Proposition 2.3 allows to construct all $\mathrm{SL}_2(p)$-covers with bad reduction at p, by lifting. This also gives a formula for the number of covers with good reduction. Passing to the quotient modulo $-I$, one obtains a similar result for three point covers with Galois group $\mathrm{PSL}_2(p)$.

In this section, R is a complete mixed characteristic discrete valuation ring which contains $\sqrt{p^*}$ and whose residue field is an algebraically closed field k of characteristic p. We let K denote the fraction field of R.

Lemma 6.1. *Let $D \in \{C(l), \tilde{C}(l)\}$. Let $f: Y \to \mathbb{P}^1_K$ be an $\mathrm{SL}_2(p)$-Galois cover with class vector $\mathbf{C}_1 = (pA, pA, D)$ or $\mathbf{C}_2 = (pA, pB, D)$. Let $\bar{Y} \to \bar{X}$ be the stable reduction of f. The curve $\bar{X}$ has exactly one primitive tail, with ramification invariant*

$$\sigma = \begin{cases} \dfrac{p-1-2l}{p-1} & \text{if } D = C(l), \\ \dfrac{p+1-2l}{p-1} & \text{if } D = \tilde{C}(l). \end{cases}$$

The inertia group of the primitive tail has order $p(p-1)/\gcd(p-1,l)$ if $D = C(l)$ and $p(p-1)/\gcd(p-1,l-1)$ otherwise. The new tails have ramification invariant $\sigma = (p+1)/(p-1)$ and an inertia group of order $p(p-1)$.

For $\mathrm{SL}_2(p)$-covers with ramification of order n, one can prove a strengthening of Proposition 4.1, cf. [16, Lemma 3.29]. Either there exists a cover with class vector $\mathbf{C}_1$ or there exists a cover with class vector $\mathbf{C}_2$. In both cases, there is a unique such cover. The outer automorphism of $\mathrm{SL}_2(p)$ conjugates the cover with class vector $\mathbf{C}_1$ (resp. $\mathbf{C}_2$) to a cover with class vector (pB, pB, D) (resp. (pB, pA, D)). Therefore, in both cases, there are exactly two covers with ramification (p, p, D), up to isomorphism.

Proof. The case $D = C(l)$ is treated in [2, Section 8]. The idea of the proof is as follows. Define $m = (p-1)/\gcd(p-1,l)$. For $\lambda \in \mathbb{P}^1 - \{0,1,\infty\}$, let W_λ be a connected component of the nonsingular projective curve corresponding to

$$w^{p-1} = x^l(x-1)^l(x-\lambda)^{p-1-l}.$$

This defines a family of m-cyclic covers $g_\lambda : W_\lambda \to \mathbb{P}^1$, branched at four points. Fix an injective character $\chi : \mathbb{Z}/m \to \mathbb{F}_p^\times$ and let $N = \mathbb{Z}/p \rtimes_\chi \mathbb{Z}/m$. Define the Hurwitz space $\mathbb{H}$ parameterizing N-Galois covers branched at $0, 1, \lambda, \infty$ of order prime to p which factor via g_λ. Define $\mathbb{H} \to \mathbb{P}^1_\lambda$ by mapping an N-cover to the branch point λ and let $f : \tilde{\mathbb{H}} \to \mathbb{P}^1_\lambda$ its Galois closure. The Galois group of f is either $\mathrm{SL}_2(p)$ or $\mathrm{PSL}_2(p)$. If it is $\mathrm{SL}_2(p)$, the class vector of f is (D_1, D_2, D), with $D_1, D_2 \in \{pA, pB\}$. Otherwise, there exists a unique lift of f to a cover with such a class vector. Therefore, f is isomorphic to the cover of the statement of the proposition. In [2], the stable reduction of f is computed, by using the moduli interpretation of $\mathbb{H}$.

The proof for $D = \tilde{C}(l)$ is analogous. (Take $m|(p+1)$ and $N \simeq (\mathbb{Z}/p)^2 \rtimes \mathbb{Z}/m$.) □

The following lemma is due to R. Pries (unpublished).

Lemma 6.2. *Let G be a finite group. Let $\varphi : W \to \mathbb{P}^1_k$ be a new G-tail cover with ramification invariant σ and inertia group I of order pn. Suppose $1 < \sigma < 2$. Then* *there exists a G-cover $f : Y \to \mathbb{P}^1_K$ with signature (p, p, n) such that φ occurs as the unique new tail of the stable reduction of f.*

If $\sigma = 2$ then φ can be lifted to a cover in characteristic zero which is branched at three points of order p, [3, Section 3.2].

Proof. Write $n = m_1 m_2$, where m_2 is the order of the prime-to-p part of the center of I. Then m_1 divides $p-1$ and $\sigma - 1 = a/m_1$ for some $0 < a < m_1$. Write $\alpha = (p-1)/m_1$. Choose $\lambda \in k - \{0,1\}$ such that $(1-\lambda)^{\alpha a} = 1$. Note that $\sigma \neq p/(p-1)$, since conductors are prime to p. Therefore, $\alpha a \neq 1$ and it is possible to choose λ as asserted.

Let $g_0 : \bar{Z}_0 \to \bar{X}_0$ be the m_1-cyclic cover where $\bar{Z}_0$ is defined by $z^{m_1} = x^{-a}(x-\lambda)^a$. Let $g : \bar{Z} \to \mathbb{P}^1_k$ be the n-cyclic cover branched only at 0 and λ which has g_0 as a quotient. Define the differential

$$\tilde{\omega} = \frac{z\,\mathrm{d}x}{x(x-1)}$$

on $\bar{Z}$. A straightforward computation, as in the proof of Proposition 3.2, shows that one can choose $\epsilon \in k^\times$ such that $\omega := \epsilon\tilde{\omega}$ is a logarithmic differential. It follows that (g, ω) is a special deformation datum. There is one primitive critical point at zero and one new critical point at λ.

Define a primitive I-tail cover $\psi' : \bar{Y}' \to \mathbb{P}^1_k$ which is totally ramified at ∞ with ramification invariant $\sigma = (m_1 - a)/m_1$ and branched of order n at 0. Let $\psi := \mathrm{Ind}_I^G \psi'$. One checks that the datum $(g, \omega, \psi, \varphi)$ satisfies the conditions of Proposition 2.3. We conclude that there exists a G-Galois cover $f : Y \to \mathbb{P}^1_K$ which is branched at three points with ramification of order (p, p, n) such that its stable reduction gives rise to $(g, \omega, \psi, \varphi)$. This proves the lemma. □

It follows from rigidity that the special deformation datum associated to an $\mathrm{SL}_2(p)$-cover with bad reduction is uniquely determined by the invariants h_i and m_i. On the other hand, the special deformation datum (g, ω) in the proof of Lemma 6.2 depends also on the choice of $\lambda \in k - \{0,1\}$ with $(1-\lambda)^{\alpha a} = 1$. It follows that we have $\alpha a = 2$ in case $G = \mathrm{SL}_2(p)$. This statement is also a consequence of the next proposition.

Proposition 6.3. *Suppose that G is either $\mathrm{SL}_2(p)$ or $\mathrm{PSL}_2(p)$. Let $\varphi : W \to \mathbb{P}^1_k$ be a new G-tail cover with ramification invariant σ. Suppose that $1 < \sigma \leq 2$. Then $\sigma = (p+1)/(p-1)$. The order of the inertia group is $p(p-1)$ if $G = \mathrm{SL}_2(p)$ and $p(p-1)/2$ if $G = \mathrm{PSL}_2(p)$. Furthermore, the cover φ is unique, up to isomorphism.*

Proof. If there exists a new $\mathrm{SL}_2(p)$-tail cover φ with $\sigma \neq 2$, Lemma 6.2 implies that φ occurs as the restriction to a new tail of the stable reduction of a $\mathrm{SL}_2(p)$-cover with signature (p, p, n) for some n prime-to-p. Therefore Lemma 6.1 implies that $\sigma = (p+1)/(p-1)$.

Similarly, one shows that every $\mathrm{SL}_2(p)$-cover of $\mathbb{P}^1_k$ branched at one point, with $\sigma = 2$, can be lifted to a cover of $\mathbb{P}^1_K$ which is branched at three points of order p. The proposition follows in this case from Proposition 4.2 and Proposition 4.1, cf. [3, Corollary 3.2.2].

Since every $\mathrm{PSL}_2(p)$-cover of $\mathbb{P}^1_K$ with ramification of order p, p, n' can be lifted to an $\mathrm{SL}_2(p)$-cover with ramification of order p, p, n (where n is n' or $2n'$), the proposition also follows for $G = \mathrm{PSL}_2(p)$. □

Corollary 6.4. *Let G be $\mathrm{SL}_2(p)$ or $\mathrm{PSL}_2(p)$. Let $f : Y \to \mathbb{P}^1_K$ be a G-Galois cover branched at three points. Suppose that f has bad reduction. Then the special deformation datum corresponding to f is hypergeometric.*

Proof. $f : Y \to \mathbb{P}^1_K$ be a G-Galois cover branched at three points. Suppose that f has bad reduction. Proposition 6.3 implies that $\sigma_i = (p+1)/(p-1)$ for $i \in \mathbb{B}_{\mathrm{new}}$. □

Proposition 6.5.

(i) *Let $\varphi : W \to \mathbb{P}^1_k$ be a (possibly nonconnected) primitive $\mathrm{SL}_2(p)$-tail cover defined over k, with ramification invariant σ. Suppose that $0 < \sigma < 1$. Denote by D the conjugacy class in $\mathrm{SL}_2(p)$ of the canonical generator of a point of W above 0, with respect to the fixed roots of unity. Then*

$$\sigma = \begin{cases} \dfrac{p-1-2l}{p-1} & \text{if } D = C(l), \\[2ex] \dfrac{p+1-2l}{p-1} & \text{if } D = \tilde{C}(l). \end{cases}$$

(ii) *The decomposition group of an irreducible component of W is $\mathrm{SL}_2(p)$ if $D = \tilde{C}(l)$ and is a semidirect product of order $p(p-1)/gcd(p-1,l)$ if $C = C(l)$.*

Proof. Write $\sigma = a/(p-1)$. The normalizer $N_G(P)$ of a Sylow p-subgroup P in $\mathrm{SL}_2(p)$ has order $p(p-1)$. Since the center of $N_G(P)$ has order two, it follows that a is even. Therefore $(p-1-a)/2$ is a positive integer.

Proposition 3.2 (ii) shows that there exists a hypergeometric deformation datum (g', ω') of signature $(0,0,a)$, where $g' : Z'_k \to \mathbb{P}^1_k$ is a $(p-1)/2$-cyclic cover, given by

$$z'^{(p-1)/2} = \prod_{i \in \mathbb{B}_{\mathrm{new}}} (x - \tau_i).$$

We define $g : Z_k \to \mathbb{P}^1_k$ to be the $(p-1)$-cyclic cover given by

$$z^{p-1} = \prod_{i \in \mathbb{B}_{\mathrm{new}}} (x - \tau_i).$$

Let ω be the pull-back of ω' to Z_k. Then (g, ω) is a special deformation datum. Note that $|\mathbb{B}_{\mathrm{prim}}| = 1$ and $|\mathbb{B}_{\mathrm{new}}| = (p-1-a)/2$, with $a = p-1-2l$ if $D = C(l)$ and $a = p+1-2l$ if $D = \tilde{C}(l)$. The primitive critical point is at ∞, where the cover g is branched of order $(p-1)/(p-1,l)$ if $D = C(l)$ and of order $(p-1)/(p-1,l-1)$ otherwise.

Let h be the $\mathrm{SL}_2(p)$-tail cover constructed in Lemma 4.3. For every $i \in \mathbb{B}_{\mathrm{new}}$, let $\bar{f}_i$ be a copy of h. Proposition 2.3 shows that there exists an $\mathrm{SL}_2(p)$-cover $f : Y \to \mathbb{P}^1_K$ whose stable reduction gives rise to the special deformation datum (g, ω), the new tail covers $\bar{f}_i$ and the primitive tail cover φ. By construction, the class vector of f is (D_1, D_2, D) with $D_1, D_2 \in \{pA, pB\}$. Therefore f is isomorphic to a cover described in Lemma 6.1. Part (i) of the proposition follows from this.

Let N denote the decomposition group of an irreducible component of W. Suppose $D = \tilde{C}(l)$, for some $l < (p-1)/2$. Then N contains an element of order p and an element of order $(p+1)/\gcd(p+1, l) > 2$. Therefore, $N = \mathrm{SL}_2(p)$, [6, Section II.8]. Alternatively, it is also easy to write down equations for these covers, analogous to Lemma 4.3.

Next suppose that $D = C(l)$. Put $n_l = (p-1)/\gcd(p-1, l)$. It is easy to construct a primitive tail cover with Galois group $\mathbb{Z}/p \rtimes \mathbb{Z}/n_l$ which is branched at zero of order n_l and has ramification invariant $\sigma = (p-1-2l)/(p-1)$. Let φ' be the primitive $\mathrm{SL}_2(p)$-tail cover obtained by induction. Using again Proposition 2.3, we see that there exists an $\mathrm{SL}_2(p)$-cover $f' : Y' \to \mathbb{P}^1_K$ whose stable reduction gives rise to the special deformation datum (g, ω), the new tail covers $\bar{f}_i$ and the primitive tail cover φ'. Note that f' has the same branch cycle description as f. It follows from Proposition 4.1 that f and f' are outer isomorphic, i.e., become isomorphic as $\mathrm{SL}_2(p)$-covers after twisting by an outer automorphism of $\mathrm{SL}_2(p)$. We conclude that N is an extension of $\mathbb{Z}/n_l$ by $\mathbb{Z}/p$. □

Theorem 6.6. *Let $\mathbf{C} = (C_1, C_2, C_3)$ be a triple of conjugacy classes of $\mathrm{SL}_2(p)$, where we suppose that the elements of the C_i are not contained in the center.*

(a) *If C_i contains an element of order p for some i, all covers with class vector $\mathbf{C}$ have bad reduction.*

(b) *Suppose $C_i \in \{C(l), \tilde{C}(l)\}$ and put $a_i = p - 1 - 2l$ if $C_i = C(l)$ and $a_i = p + 1 - 2l$ otherwise. Write*

$$\mathrm{Ni}_3^{\mathrm{bad}}(\mathbf{C}) = \{ f \in \mathrm{Ni}_2^{\mathrm{in}}(\mathbf{C}) \mid f \textit{ has bad reduction.} \}.$$

Then

$$|\,\mathrm{Ni}_3^{\mathrm{bad}}(\mathbf{C})| = \begin{cases} 2 & \textit{if } a_1 + a_2 + a_3 < p - 1, \\ 2 & \textit{if } a_1 + a_2 + a_3 = p - 1 \textit{ and } C_i = \tilde{C}(l) \textit{ for some } i, \\ 0 & \textit{otherwise.} \end{cases}$$

Proof. Let $f : Y \to \mathbb{P}^1_K$ be an $\mathrm{SL}_2(p)$-cover with class vector $\mathbf{C} = (C_1, C_2, C_3)$. If p divides one of the ramification indices of f, then f has bad reduction, so (a) is obvious.

Suppose that p does not divide the ramification indices of f. If f has bad reduction, then (7) implies that $a_1 + a_2 + a_3 \le p - 1$. If $a_1 + a_2 + a_3 = p - 1$, there are no new tails. Since $\bar{Y}$ is connected, at least one of the primitive tails has decomposition group $\mathrm{SL}_2(p)$. Therefore Proposition 6.5 implies that $C_i = \tilde{C}(l)$, for some i. This shows that if there exists a cover with bad reduction, then either $a_1 + a_2 + a_3 < p - 1$ or $a_1 + a_2 + a_3 = p - 1$ and $C_i = \tilde{C}(l)$ for some i.

Now suppose that C_1, C_2, C_3 are conjugacy classes of noncentral elements of order prime to p such that the condition of the theorem is satisfied. Let a_1, a_2, a_3 be as in the statement of the theorem. Recall that a_i is even and different from $p-1$. Let (g', ω') be the hypergeometric deformation datum of signature (a_1, a_2, a_3) constructed in Proposition 3.2 (ii). The cover $g' : Z'_k \to \mathbb{P}^1_k$ is the $(p-1)/2$-cyclic cover defined by

$$z'^{(p-1)/2} = x^{a_1/2}(x-1)^{a_2/2} \prod_{i \in \mathbb{B}_{\mathrm{new}}} (x - \tau_i), \quad (z, x) \mapsto x.$$

We define a $(p-1)$-cyclic cover $g : Z_k \to \mathbb{P}^1_k$, factoring through g' by

$$z^{p-1} = x^{b_1}(x-1)^{b_2} \prod_{i \in \mathbb{B}_{\mathrm{new}}} (x - \tau_i), \quad (z, x) \mapsto x,$$

where $b_i = a_i/2 + (p-1)/2$, i.e., $b_i = p-1-l$ if $C_i = C(l)$ and $b_i = p-l$ otherwise. Then (g, ω) is a special deformation datum.

We define a set of tail covers, as follows. For $i \in \mathbb{B}_{\mathrm{new}}$, we define $\bar{f}_i \simeq h$, where h is the new $\mathrm{SL}_2(p)$-tail cover of $\mathbb{P}^1_k$ with $\sigma = (p+1)/(p-1)$ from Lemma 4.3. For $i \in \mathbb{B}_{\mathrm{prim}} = \{1, 2, 3\}$, we let $\bar{f}_i$ be the primitive $\mathrm{SL}_2(p)$-tail cover with $\sigma = a_i/(p-1)$ whose existence is guaranteed by Lemma 6.1. As in the proof of Lemma 6.2, one checks the datum $(g, \omega; \bar{f}_i)$ satisfies the conditions of Proposition 2.3. Hence there exists an $\mathrm{SL}_2(p)$-cover $f : Y \to \mathbb{P}^1_K$ with class vector $\mathbf{C} = (C_1, C_2, C_3)$ whose stable reduction gives rise to the datum $(g, \omega; \bar{f}_i)$.

Since the outer automorphism group of $\mathrm{SL}_2(p)$ has order two, the number of liftings, up to isomorphism, is at least two. Proposition 4.1 implies that the number of elements of $\mathrm{Ni}_3^{\mathrm{in}}(\mathbf{C})$ is at most two. We conclude that the number of liftings is exactly two. This proves that all $\mathrm{SL}_2(p)$-covers over K with class vector $\mathbf{C}$ have bad reduction to characteristic p. □

References

[1] F. Beukers. On Dwork's accessory parameter problem. *Math. Z.*, 241:425–444, 2002.

[2] I.I. Bouw. Reduction of the Hurwitz space of metacyclic covers. arXiv:math.AG/020403, to appear in *Duke J. Math.*

[3] I.I. Bouw and R.J. Pries. Rigidity, reduction, and ramification. *Math. Ann.*, 326:803–824, 2003.

[4] P. Deligne and M. Rapoport. Les schémas de modules de courbes elliptiques. In *Modular functions of one variable II*, number 349 in LNM, pages 143–316. Springer-Verlag, 1972.

[5] B. Edixhoven. Minimal resolution and stable reduction of $X_0(N)$. *Ann. Inst. Fourier*, 40(1):31–67, 1990.

[6] B. Huppert. *Endliche Gruppen I.* Number 134 in Grundlehren. Springer-Verlag, 1967.

[7] Y. Ihara. On the differentials associated to congruence relations and Schwarzian equations defining uniformizations. *J. Fac. Sci. Univ. Tokyo, Sect. IA Math.*, 21:309–332, 1974.

[8] M. Kaneko and D. Zagier. Supersingular j-invariants, hypergeometric series, and Atkin's orthogonal polynomials. In *Computational perspectives on number theory (Chicago, IL, 1995)*, volume 7 of *AMS/IP Stud. Adv. Math.*, pages 97–126. Amer. Math. Soc., 1998.

[9] N.M. Katz. Expansion-coefficients as approximate solutions of differential equations. In *p-adic cohomology*, Astérisque, 119-120, pages 183–189, 1984.

[10] N.M. Katz and B. Mazur. *Arithmetic Moduli of Elliptic Curves.* Number 108 in Annals of Mathematical Studies. Princeton Univ. Press, 1985.

[11] M. Raynaud. p-groupes et réduction semi-stable des courbes. In P. Cartier, editor, *Grothendieck Festschrift III*, number 88 in Progress in Math., pages 179–197. Birkhäuser, 1990.

[12] M. Raynaud. Spécialisation des revêtements en caractéristique $p > 0$. *Ann. Sci. École Norm. Sup.*, 32(1):87–126, 1999.

[13] J.-P. Serre. Sur la topologie des variétés algébriques en caractéristique p. In *Symp. Int. Top. Alg.*, pages 24–53, 1958.

[14] J.-P. Serre. *Topics in Galois Theory.* Research notes in mathematics, 1. Jones and Bartlett Publishers, 1992. Lecture notes prepared by Henri Darmon.

[15] K. Strambach and H. Völklein. On linear rigid triples. *J. Reine Angew. Math.*, 510:57–62, 1999.

[16] H. Völklein. *Groups as Galois Groups.* Number 53 in Cambridge Studies in Adv. Math. Cambridge Univ. Press, 1996.

[17] S. Wewers. Three point covers with bad reduction. arXiv:math.AG/0205026, *J. Amer. Math. Soc.*, 16:991–1032, 2003.

[18] S. Wewers. Reduction and lifting of special metacyclic covers. *Ann. Sci. École Norm. Sup.*, 36:1113–138, 2003.

Irene I. Bouw
Institut für Experimentelle Mathematik
Universität Duisburg–Essen
Ellernstr. 29
D-45326 Essen, Germany
e-mail: `bouw@exp-math.uni-essen.de`

Stefan Wewers
Mathematisches Institut
Beringstr. 1
D-53115 Bonn, Germany
e-mail: `wewers@math.uni-bonn.de`

Progress in Mathematics, Vol. 224, 23–44

On p-adic Families of Automorphic Forms

Kevin Buzzard

Abstract. Coleman and Mazur have constructed "eigencurves", geometric objects parametrising certain overconvergent p-adic modular forms. We formulate definitions of overconvergent p-adic automorphic forms for two more classes of reductive groups – firstly for GL_1 over a number field, and secondly for $D^\times$, D a definite quaternion algebra over the rationals. We give several reasons why we believe the objects we construct to be the correct analogue of an overconvergent p-adic modular form in this setting.

1. Introduction

The definition of an overconvergent modular form was formulated in [14] and used in a very powerful way in work of Coleman (see [5], [6], [7] etc.) as a key tool for constructing families of modular forms. Using these arguments Coleman was able to resolve some questions of Gouvêa and Mazur about the existence of analytic families of modular forms. In fact, Coleman's ideas gave more: using them, Coleman and Mazur constructed "eigencurves", geometric objects which parametrise finite slope normalised overconvergent modular eigenforms.

There is now evidence that these parameter spaces are just the tip of an iceberg, and that there should exist parameter spaces, or "eigenvarieties", parametrising systems of eigenvalues occurring in p-adic Fréchet spaces of "overconvergent automorphic forms" for a wide class of reductive groups. One way of looking at the formalism is as follows: firstly there should be some appropriate set of classical infinity types (for example, the regular algebraic ones), which naturally form a subset of a rigid analytic space of "p-adic infinity types", or "weights" for short. We call such a space a *weight space*, we call points of this space *weights*, and we call the subset mentioned above the set of *classical weights*. If κ is a weight, then the space of overconvergent automorphic forms of some fixed level U and weight κ should be a (typically infinite-dimensional) p-adic vector space that, if κ is classical, will naturally contain the finite-dimensional space of classical automorphic forms of this weight (after one has chosen isomorphisms $\mathbf{C} \cong \mathbf{C}_p$ and so on). Furthermore, the spaces should naturally p-adically interpolate the classical spaces of forms as the infinity type varies in weight space. Note that in this context, as was observed in the 1970s by Serre and Katz, the notion of a weight should include both an infinity type and also a level structure at the prime p.

Now fix a level structure U prime to p. Then there should be a second rigid space, the eigenvariety associated to this data – this will be a rigid-analytic space equipped with a map to weight space, such that the fibre above a weight will be equal to the spectrum of a Hecke algebra acting on the space of "finite slope" overconvergent forms of that weight, and "tame level" U. In this perspective, a Hecke operator becomes a map from the eigenvariety to the affine line, sending a point to the eigenvalue of the corresponding eigenform.

There should be a family of p-adic Galois representations living on the eigenvariety, and the eigenvariety should in general be strongly connected to the deformation theory of such Galois representations. Moreover, the eigenvariety should perhaps be the natural domain for the special values of p-adic L-functions, in analogy with the role of the collection of Riemann surfaces that Tate uses to great effect in the study of classical L-functions in his thesis (see p. 314 of [3]). These sides of the story will not be treated in this paper, however.

One interesting application of the existence of these eigenvarieties is that one can re-formulate cases of Langlands functoriality in this setting as simply predicting the existence of a morphism between two such parameter spaces that intertwines Hecke operators appropriately.

It is not currently clear to the author in what generality one would expect these parameter spaces to exist, but this may be due to a large extent to the fact that the author is certainly no expert in the theory of automorphic forms. On the other hand, there do seem to be some cases where this philosophy does not apply. For example, we know of no evidence to suggest that Maass forms can be p-adically interpolated in such a way, although perhaps Mazur's observation that it seems harder to deform even Galois representations than odd ones might be evidence to suggest that the associated rigid spaces may all be 0-dimensional. On the other hand, there is a growing body of evidence to suggest that in many other common cases, non-trivial parameter spaces do exist. The author is optimistic that, for example, if the reductive group in question is compact mod centre at infinity then almost all the techniques now exist for constructing such parameter spaces. See the thesis of G. Chenevier [4], forthcoming work [2] of the author and also future joint work of the author and Chenevier, where the constructions will be given in many cases. Unfortunately, the works cited above are perhaps daunting to read, relying on many technical computations in representation theory, as well as a generalisation of much of Coleman's theory of families of p-adic Banach spaces to families of dimension greater than one.

The purpose of this paper however is to give an explicit construction of the parameter spaces in two of the simplest cases: firstly in the case of GL_1 over a number field, and secondly in the case of $D^\times$, D a definite quaternion algebra over the rationals. The motivation for explaining these cases explicitly is several-fold. Firstly, the theory for GL_1 drops out from class field theory and known results about p-adic Galois representations, and is excellent motivation for believing that these parameter spaces exist in some generality. One avoids having to generalise Coleman's theorems because the Banach spaces in question are finite-dimensional and everything is easy. Note that already the dimension of the parameter spaces in this case depends on the truth of Leopoldt's conjecture, but not knowing the dimension of the spaces constructed here does not seem to lead to any problems in the theory. Secondly, the representation theory involved in the GL_1 case is trivial, so one sees a concrete example of the theory without having to get too bogged down in technicalities.

One motivation for the construction in the definite quaternion algebra case is that it is easy to make all the representation theory very explicit, thus avoiding the technical computations in [4] where analogous results are proved for compact forms of GL_n. Another reason is that by the Jacquet-Langlands theorem, classical automorphic forms in this setting will correspond to classical modular forms, and overconvergent automorphic forms in this setting should correspond to overconvergent p-adic modular forms (in fact this has recently been proved by Chenevier: see [4]). The families one constructs are one-dimensional, so one does not need to generalise Coleman's work either. We refer to Theorem 1 below for some of the results proved by Coleman and others about classical and overconvergent modular forms. On the other hand, we show in this paper that the analogues of these theorems in this setting are, for the most part, elementary consequences of the definitions. This gives weight to our hope that our definitions are the right ones in this setting.

Some of this work in this paper was done during the semestre automorphe at the IHP in Paris in 2000, and some was done during a visit to Paris 13 funded by the Arithmetic Algebraic Geometry RTN network. The author thanks all these institutions. The results for GL_1 in this paper were presented at the conference on modular curves at the CRM in Barcelona, and the author thanks the organisers of this conference for giving him the opportunity to speak there. The author also thanks the referee for several useful comments and observations.

Note added in proof: recent preprints of Emerton seem to give the construction of eigenvarieties in a very wide class of situations, using more representation-theoretic techniques.

2. The case of GL_1 over a number field

Throughout this paper, $p > 0$ will be a prime, and K will denote a field which is complete with respect to a non-trivial non-Archimedean valuation, and whose residue field has characteristic p. The rigid spaces we construct will be rigid spaces over K. We will consider K as being fixed – the interested reader can verify that all rigid spaces in this paper are quasi-separated and their definitions commute with base change if K' is a complete extension of K. Up to and including Lemma 2, K may have characteristic 0 or p, but after Lemma 2 and for the rest of the paper we assume that the characteristic of K is zero. From Section 3 onwards, we assume furthermore that K is a complete subfield of $\mathbf{C}_p$, but this is only for notational convenience; one has to do little more than to replace p by $|p|^{-1}$ on many occasions to deal with the general case.

Let $L/\mathbf{Q}$ be a number field of degree d, and let $\mathcal{O}_L$ denote the integers of L. Let L_∞ denote $L \otimes_{\mathbf{Q}} \mathbf{R}$ and $(L_\infty^\times)^\circ$ the identity component of the topological group $L_\infty^\times$. Similarly, let L_p denote $L \otimes_{\mathbf{Q}} \mathbf{Q}_p$, let $\mathcal{O}_{L,p}$ denote $\mathcal{O}_L \otimes_{\mathbf{Z}} \mathbf{Z}_p$ and let $\mathcal{O}_{L,p}^\times$ denote the topological group of units in $\mathcal{O}_{L,p}$.

We would like to formulate a definition of an overconvergent p-adic automorphic form for the group $\mathrm{Res}_{L/\mathbf{Q}} \mathrm{GL}_1$, defined over K. Associated to an overconvergent eigenform we should expect a one-dimensional p-adic Galois representation. On the other hand, any Hodge-Tate one-dimensional p-adic representation of $\mathrm{Gal}(\overline{L}/L)$ comes from an algebraic Grössencharacter, and hence one should suspect that for classical weights, the overconvergent eigenforms should be in bijection with the classical eigenforms. In fact the main problem for GL_1 is simply that of constructing weight space and we present the natural definition below.

Lemma 1. *Let R be an affinoid algebra over K. For there to exist a continuous group homomorphism $\mathbf{Z}_p \to R^\times$ sending 1 to $1 + r \in R$, it is necessary and sufficient that r should be topologically nilpotent, and when this is the case, the group homomorphism sending 1 to $1 + r$ is unique.*

Proof. Fix a complete K-algebra norm on R inducing the topology on R. If there is a continuous group homomorphism $\mathbf{Z}_p \to R^\times$ sending 1 to $1+r$ then $(1+r)^{p^n} \to 1$ as $n \to \infty$ in R, and the same is true in $R/\mathfrak{m}$ for $\mathfrak{m}$ any maximal ideal of R. This means that modulo $\mathfrak{m}$, the residue norm of r must be less than 1 (look at the residue field). Hence r is topologically nilpotent by Proposition 6.2.3.2 of [1]. Conversely, if r is topologically nilpotent, then $|r^n| \to 0$ as $n \to \infty$ and hence an easy argument using the binomial theorem shows that $(1+r)^{p^N}$ tends to 1 as $N \to \infty$, which suffices to construct the group homomorphism $\mathbf{Z}_p \to R^\times$, and uniqueness follows because of continuity. □

Now let X be an arbitrary rigid space, and let $\mathcal{O}(X)$ denote the global sections of the structure sheaf on X. If H is a topological group, then we say that a group homomorphism $H \to \mathcal{O}(X)^\times$ is *continuous* if for all admissible affinoid subdomains U of X, the resulting map $H \to \mathcal{O}(U)^\times$ is continuous. The lemma above shows that there is a canonical functorial bijection between the continuous homomorphisms $\mathbf{Z}_p \to \mathcal{O}(X)^\times$ and the maps of rigid spaces $X \to \Delta$, where Δ denotes the open disc of radius 1 and centre 1, considered as a subspace of rigid affine 1-space.

Let $\mathbf{G}_m$ denote the rigid space over K associated to the affine scheme $\mathbf{A}^1 \backslash \{0\}$. We now explain an elementary but key construction. Let H be an abelian profinite group which contains an open subgroup isomorphic to $(\mathbf{Z}_p)^n$ for some $n \geq 0$. Let $\mathrm{Hom}(H, \mathbf{G}_m)$ denote the functor on rigid spaces over K sending a rigid space X to the set of continuous group homomorphisms $H \to \mathcal{O}(X)^\times$.

Lemma 2.

(i) *The functor* $\mathrm{Hom}(H, \mathbf{G}_m)$ *is represented by a quasi-separated rigid space over K, also denoted* $\mathrm{Hom}(H, \mathbf{G}_m)$.

(ii) *The space* $\mathrm{Hom}(H, \mathbf{G}_m)$ *is a union of finitely many open* n*-balls and naturally has the structure of a group object in the category of rigid spaces over* K.

(iii) *If* $H \to J$ *is a surjective morphism of profinite groups with finite kernel, then the canonical map* $\mathrm{Hom}(J, \mathbf{G}_m) \to \mathrm{Hom}(H, \mathbf{G}_m)$ *is a closed immersion. If furthermore the size of the kernel of* $H \to J$ *is non-zero in* K *(for example, if the characteristic of* K *is 0), then the canonical map* $\mathrm{Hom}(J, \mathbf{G}_m) \to \mathrm{Hom}(H, \mathbf{G}_m)$ *is a closed and open immersion, identifying* $\mathrm{Hom}(J, \mathbf{G}_m)$ *with a union of components of* $\mathrm{Hom}(H, \mathbf{G}_m)$.

(iv) *If* $H \to J$ *is an injective morphism of profinite groups with finite cokernel, then the canonical map* $\mathrm{Hom}(J, \mathbf{G}_m) \to \mathrm{Hom}(H, \mathbf{G}_m)$ *is finite and flat, of degree equal to* $[J : H]$*. If furthermore* $[J : H] \neq 0$ *in* K *then the map is etale.*

Proof. Note firstly that if H_1 and H_2 satisfy the hypotheses put on H above, and $\mathrm{Hom}(H_1, \mathbf{G}_m)$ and $\mathrm{Hom}(H_2, \mathbf{G}_m)$ are representable, then one can represent $\mathrm{Hom}(H_1 \times H_2, \mathbf{G}_m)$ by the product $\mathrm{Hom}(H_1, \mathbf{G}_m) \times \mathrm{Hom}(H_2, \mathbf{G}_m)$. Furthermore, if H is as in the lemma, then by the structure theorem for topologically finitely-generated abelian profinite groups, H is isomorphic to the product of a finite group and a group isomorphic to $(\mathbf{Z}_p)^n$.

(i) By the remarks above, we are reduced to representing the functor in the following two cases: the case of finite cyclic H, and the case $H \cong \mathbf{Z}_p$. If H is finite cyclic of order n then the rigid space associated to the affinoid algebra $K\langle T\rangle/(T^n - 1)$ is easily shown to do the trick. If on the other hand $H \cong \mathbf{Z}_p$, then by Lemma 1 and the remarks following it, the open disc with centre 1 and radius 1 over K represents the functor in question.

(ii) By the construction of the representing space above, we see it is a product of rigid spaces each of which is either finite or an open unit ball, and the results follow easily.

(iii) Here one only need observe that H is isomorphic to the product of a free $\mathbf{Z}_p$-module and a finite group, and the kernel of the map $H \to J$ is contained in this finite group. One now easily reduces to the case where H is finite, where the result is easy.

(iv) One reduces easily to the case where J/H is finite cyclic of degree n for some $n \geq 1$. In this case the cover $\mathrm{Hom}(J, \mathbf{G}_m)$ of $\mathrm{Hom}(H, \mathbf{G}_m)$ is defined by taking the nth root of some function f on $\mathrm{Hom}(H, \mathbf{G}_m)$, and the fact that $|f - 1| < 1$ for any residue norm means that f is a unit. Hence the covering is finite and flat, and moreover it is etale if n is prime to the characteristic of K. □

We now use this lemma to construct a "weight space". Assume for the rest of this paper that the characteristic of K is zero. If Γ is a subgroup of $\mathcal{O}_L^\times$ of finite index, and $\overline{\Gamma}$ denotes the closure of Γ in $\mathcal{O}_{L,p}^\times$, then the quotient $H_\Gamma := \mathcal{O}_{L,p}^\times/\overline{\Gamma}$ satisfies the hypotheses of the lemma. Hence one has a rigid space $\mathcal{W}_\Gamma := \mathrm{Hom}(H_\Gamma, \mathbf{G}_m)$ defined over K. If $\Delta \subseteq \Gamma$ is a subgroup of finite index, then the corresponding surjection $H_\Delta \to H_\Gamma$ has kernel of finite order $[\overline{\Gamma} : \overline{\Delta}]$ and hence by part (iii) of Lemma 2 there is a closed and open immersion $\mathcal{W}_\Gamma \to \mathcal{W}_\Delta$, identifying the former with a union of components of the latter.

Definition. *We define* weight-space $\mathcal{W}$ *to be the direct limit* $\lim_{\to} \mathcal{W}_\Gamma$, *as* Γ *varies over the set of subgroups of finite index in* $\mathcal{O}_L^\times$, *partially ordered by inclusion.*

The space $\mathcal{W}$ is a rigid space defined over K. It has a group structure, which we in fact shall never use. By definition, a K-point in $\mathcal{W}$ corresponds to a continuous group homomorphism $\mathcal{O}_{L,p}^\times \to K^\times$ which vanishes on some subgroup of finite index in $\mathcal{O}_L^\times$.

Let $\sigma_1, \ldots, \sigma_d$ denote the d embeddings $L \to \mathbf{C}_p$. Each such embedding gives a continuous group homomorphism $\mathcal{O}_{L,p}^\times \to \mathbf{C}_p^\times$, which we also denote by σ_i.

Definition. *We say that a weight in* $\mathcal{W}(K)$ *is* classical *if there are a collection* $(n_1, n_2, \ldots, n_d)$ *of integers, and a subgroup* H *of finite index of* $\mathcal{O}_{L,p}^\times$, *such that the corresponding group homomorphism* κ, *when restricted to* H, *equals*

$$(\sigma_1)^{n_1}(\sigma_2)^{n_2}\ldots(\sigma_d)^{n_d}.$$

Choose a K-point of $\mathcal{W}_K$ and let κ denote the corresponding map $\mathcal{O}_{L,p}^\times \to K^\times$. If S is a finite set of places of $\mathbf{Q}$, let $\mathbf{A}_L^S$ denote the adeles of L away from the places in L above the places in S.

Set $S = \{p, \infty\}$ once and for all, and let U be a compact open subgroup of $(\mathbf{A}_L^S)^\times$. We refer to such a subgroup U as a *tame level.*

Let G denote the group $\mathrm{Res}_{L/\mathbf{Q}}\,\mathrm{GL}_1$.

Definition. *An* overconvergent automorphic eigenform *for* G, *of weight* κ *and tame level* U, *defined over* K, *is a continuous group homomorphism* $L^\times\backslash\mathbf{A}_L^\times \to K^\times$ *which contains* $U.(L_\infty^\times)^\circ$ *in its kernel, and whose restriction to* $\mathcal{O}_{L,p}^\times$ *equals* κ. *An* overconvergent automorphic form *for* G, *of weight* κ *and tame level* U, *defined over* K, *is a continuous function* $f : L^\times\backslash\mathbf{A}_L^\times \to K$ *such that* $f(gu) = f(g)$ *for all* $u \in U.(L_\infty^\times)^\circ$, *the* K*-vector space generated by all the* $\mathbf{A}_L^\times$*-translates of* f *is finite-dimensional, and such that, after a finite base extension of* K *if necessary, there is a basis of this finite-dimensional space consisting of overconvergent automorphic eigenforms for* G *of weight* κ *and tame level* U.

Note that any continuous group homomorphism $L^\times\backslash\mathbf{A}_L^\times \to K$ will automatically contain $(L_\infty^\times)^\circ$ and some compact open subgroup of $(\mathbf{A}_L^S)^\times$ in its kernel, and hence its restriction to $\mathcal{O}_{L,p}^\times$ will automatically be a weight, in the sense that it must vanish on a subgroup of $\mathcal{O}_L^\times$ of finite index.

Note that there are only finitely many overconvergent automorphic eigenforms of a given weight and tame level, because $L^\times\backslash\mathbf{A}_L^\times/\mathcal{O}_{L,p}^\times U(L_\infty^\times)^\circ$ is finite. By the main theorem of global class field theory, to an overconvergent eigenform defined over K there is associated a continuous one-dimensional representation $\mathrm{Gal}(\overline{L}/L) \to \mathrm{GL}_1(K)$. If K is a finite extension of $\mathbf{Q}_p$ and $\kappa \in \mathcal{W}(K)$, then this associated Galois representation will be Hodge-Tate if and only if κ is classical. In fact, for classical κ, the space we have constructed perhaps deserves to be thought of as either the space of p-adic, overconvergent p-adic, or classical forms defined over K.

Note that Weil associated a Hodge-Tate p-adic Galois representation to any algebraic Grössencharacter $L^\times \backslash \mathbf{A}_L^\times \to \mathbf{C}^\times$ and this p-adic Galois representation gives rise to an overconvergent automorphic eigenform (although of course one does not need class field theory here – Weil's argument directly constructs an overconvergent automorphic eigenform from a classical one). The weight of such an eigenform is classical, and the corresponding integers $(n_1, n_2, \ldots, n_d)$ are the Hodge-Tate weights of the associated Galois representation, if one normalises the isomorphism of global class field theory appropriately.

It is easy to see, given what we have done, that these definitions interpolate naturally. More precisely, if V is an affinoid subdomain of $\mathcal{W}$ with $\mathcal{O}(V) = R$, then the inclusion $V \to \mathcal{W}_K$ gives us, from the functorial property of $\mathcal{W}_K$, a continuous group homomorphism $\mathcal{O}_{L,p}^\times \to R^\times$. Now for a tame level U one can define an overconvergent automorphic eigenform of weight V as simply being a continuous group homomorphism $f : L^\times \backslash \mathbf{A}_L^\times \to R^\times$ containing $U.(L_\infty^\times)^\circ$ in its kernel, such that the restriction of f to $\mathcal{O}_{L,p}^\times$ is the group homomorphism $\mathcal{O}_{L,p}^\times \to R^\times$ corresponding to the map $V \to \mathcal{W}_K$, and one can give a similar definition of an overconvergent automorphic form of weight V. These spaces should of course be thought of as interpolating the spaces of overconvergent forms of weight κ introduced above, as κ varies through V.

Let U be a fixed tame level, and let Γ denote the intersection

$$L^\times \cap U.\mathcal{O}_{L,p}^\times (L_\infty^\times)^\circ.$$

Then Γ is a subgroup of $\mathcal{O}_L^\times$ of finite index. We will now define an "eigenvariety" of tame level U, which will be a finite cover of the weight space $\mathcal{W}_\Gamma$ introduced earlier. Its definition is simple, given what we have already. Let H denote the quotient of the group $L^\times \backslash \mathbf{A}_L^\times$ by the closure of the image of $U.(L_\infty^\times)^\circ$. Then the map $\mathcal{O}_{L,p}^\times \to H$ is continuous with finite cokernel, and hence H satisfies the hypotheses of Lemma 2.

Definition. *We define the eigenvariety of tame level U to be the rigid space*

$$\mathrm{Hom}(H, \mathbf{G}_m).$$

Let $\overline{\Gamma}$ denote the closure of Γ in $\mathcal{O}_{L,p}^\times$. Then the natural map $\mathcal{O}_{L,p}^\times / \overline{\Gamma} \to H$ is injective and the cokernel is finite. The corresponding map of rigid spaces $\mathrm{Hom}(H, \mathbf{G}_m) \to \mathcal{W}_\Gamma$ is finite flat of degree equal to the order of this cokernel, by part (iv) of Lemma 2. Note that it also follows from this lemma that the eigenvariety is a finite union of open balls, and that the fibre of a point κ in weight space is canonically the set of overconvergent automorphic forms of weight κ and tame level U. The eigenvariety itself can be thought of as a parameter space interpolating overconvergent eigenforms of varying weights. Finally, the Hecke operator corresponding to a uniformiser at a place $\mathfrak{q}$ of L manifests itself as the evaluation homomorphism from the eigenvariety to $\mathbf{G}_m$ corresponding to evaluation at this uniformiser.

In this setting, the eigenvariety is smooth and has finitely many components. It still appears to be an open question as to whether the Coleman-Mazur eigencurve has finitely or infinitely many components (indeed, almost every natural question about the Coleman-Mazur eigencurve is still open).

3. Classical and overconvergent modular forms: Results of Coleman and others

In this section we briefly summarise (see Theorem 1 below) some of the main results from the theory of classical and overconvergent modular forms. Our goal in the rest of this paper is to propose analogous definitions in the setting of automorphic forms for $D^\times$, D a definite quaternion algebra over $\mathbf{Q}$, and then to prove analogues of the results in Theorem 1, although we shall only sketch the construction of the eigencurve in this setting, and the construction will be made under a minor but unnecessary technical restriction, which we shall remove in forthcoming work.

Let p be prime, and let N be a positive integer prime to p. Let K be a complete subfield of $\mathbf{C}_p$ (this is just for notational convenience; all of this works for a general K of characteristic zero, complete with respect to a non-trivial non-archimedian valuation and with residue field of characteristic p), and for $\Gamma \subseteq \mathrm{SL}_2(\mathbf{Z})$ let $S_k(\Gamma)$ denote the space of classical cusp forms of level Γ and weight k defined over K. If $\Gamma = \Gamma_1(N)$ then we refer to this space as the space of classical cusp forms of level N and weight k defined over K. For $r = p^{-t}$ with $t \in \mathbf{Q}$ and $0 \leq t < p/(p+1)$, let $\mathbf{S}_k(\Gamma_1(N); r)$ denote the p-adic Banach space of r-overconvergent cusp forms of tame level N and weight k, that is, at least for $N \geq 5$, the cuspidal sections of ω^k on the rigid subspace of $X_1(N)$ obtained by removing open discs of radius r above every supersingular point in characteristic p (a more precise definition may be found in, for example, [7] and [5]. We omit the more precise definitions as we shall never be using them). Let $\mathbf{S}_k(\Gamma_1(N))$ denote the space of forms which overconverge as far as one can reasonably ask, that is, sections of ω^k on the wide-open subspace of $X_1(N)$ obtained by removing closed discs of radius $p^{-p/(p+1)}$ in each supersingular disc. By definition, $S_k(\Gamma_1(N)) \subset \mathbf{S}_k(\Gamma_1(N)) \subset \mathbf{S}_k(\Gamma_1(N); r)$ and if $r < s$ then $\mathbf{S}_k(\Gamma_1(N); r) \subset \mathbf{S}_k(\Gamma_1(N); s)$. We remind the reader of some theorems about these objects (again we shall be slightly sketchy, and refer the reader to the original papers for more precise statements)

Theorem 1.

1. *If $f \in S_k(\Gamma_1(N) \cap \Gamma_0(p^n))$ then $f \in \mathbf{S}_k(\Gamma_1(N); r)$ for some appropriate r.*
2. *If $f \in \mathbf{S}_k(\Gamma_1(N); r)$ and $U_p f = \lambda f$ for some non-zero λ, then $f \in \mathbf{S}_k(\Gamma_1(N))$.*
3. *For $k \geq 1$, there is a map $\theta^{k-1} : \mathbf{S}_{2-k}(\Gamma_1(N)) \to \mathbf{S}_k(\Gamma_1(N))$ which on q-expansions is $\left(q\frac{d}{dq}\right)^{k-1}$, and whose cokernel is finite-dimensional.*

4. *Say* $0 \neq f \in \mathbf{S}_k(\Gamma_1(N);r)$ *and* $U_p f = \lambda f$ *for some non-zero* λ. *If* $v_p(\lambda) < k-1$, *then* $f \in S_k(\Gamma_1(N))$. *On the other hand, if* $v_p(\lambda) > k-1$ *then* $f \notin S_k(\Gamma_1(N))$.
5. *Any finite slope overconvergent eigenform is part of a family of constant slope, in some precise sense.*
6. *Finite slope normalised overconvergent eigenforms of tame level* N *are parametrised by a rigid-analytic geometric object, the eigencurve of tame level* N *(at least if* $p > 2$ *and* $N = 1$*).*

Proof. Parts 1 and 2 are classical. Part 1 is proved in [14] for $n = 1$ and in [10] for general n. Part 2 is also proved in [14]. Parts 3 and 4 are related and deeper, and are both proved in [5]. Part 5 is proved in [6] and part 6 in [7]. Note also that the constructions in [7] should also go through for general p and N, although nothing is published in this generality as yet (see [2] and also forthcoming work of Emerton, however). □

As already explained, we will prove analogues of all these results in the setting of automorphic forms on definite quaternion algebras over the rationals. The main question to be solved here is that of finding the correct definition of an overconvergent automorphic form in this setting. The problem with the classical definition is that it is inherently geometric, and relies (amongst other things) on the fact that modular curves are one-dimensional. This causes problems when trying to construct analogues of the Coleman-Mazur construction for other reductive groups, although some progress has been made by Goren and others in the case of Hilbert modular forms. See also forthcoming work of Nevens, and recent preprints of Abbes-Mokrane and Kisin-Lai, which should hopefully culminate in a geometric construction of p-adic parameter spaces for Hilbert modular forms.

In the setting of groups which are compact mod centre at infinity, the problem is rather different: now one cannot remove canonical small regions from the moduli space because it is 0-dimensional, and it appears that a completely different approach is necessary. Our approach is very much inspired by a preprint [9] of Stevens and involves using an "Eichler-Shimura" philosophy which re-interprets the theory of modular forms in a much more combinatorial way, which seems to be more amenable to generalisation. One consequence of this is that the proofs of the analogues of the results above in this setting are much easier, requiring no hard algebraic geometry.

One natural question that can be raised is what relation there is between the eigencurves constructed here and those of Coleman-Mazur. Is there a map between them "explaining" the Jacquet-Langlands correspondence? In fact Chenevier in [4] has recently constructed such a map, although his construction assumes the classical Jacquet-Langlands result and hence does not give another proof of it.

4. Definitions and classical forms

Let p be a fixed prime. For $\alpha \geq 1$ an integer, let $\mathbf{M}_\alpha$ be the monoid consisting of 2 by 2 matrices $\left(\begin{smallmatrix} a & b \\ c & d \end{smallmatrix}\right)$ with $a, b, c, d \in \mathbf{Z}_p$ and such that $p^\alpha \mid c$, $p \nmid d$ and $ad - bc \neq 0$. Let D be a definite quaternion algebra over $\mathbf{Q}$ with discriminant δ prime to p. Fix once and for all a maximal order $\mathcal{O}_D$ of D and an isomorphism $\mathcal{O}_D \otimes B \cong \mathrm{M}_2(B)$, where $B = \lim_{\leftarrow}(\mathbf{Z}/M\mathbf{Z})$, the limit being taken over all integers M prime to δ. This isomorphism induces isomorphisms $\mathcal{O}_D \otimes \mathbf{Z}_l \cong \mathrm{M}_2(\mathbf{Z}_l)$ and $\mathcal{O}_D \otimes \mathbf{Q}_l \cong \mathrm{M}_2(\mathbf{Q}_l)$ for all primes $l \nmid \delta$, and we will henceforth identify these rings with one another. Let $\mathbf{A}_f$ denote the finite adeles over $\mathbf{Q}$ and define $D_f = D \otimes_{\mathbf{Q}} \mathbf{A}_f$. Then D_f can be thought of as the restricted product over all primes l of $D \otimes \mathbf{Q}_l$ and in particular if $g \in D_f$ then the component g_p of g at p can be regarded as an element of $\mathrm{M}_2(\mathbf{Q}_p)$.

Definition. *If U is an open compact subgroup of $D_f^\times$ and $\alpha \geq 1$ then we say that U has* wild level $\geq p^\alpha$ *if the projection of U onto $\mathrm{GL}_2(\mathbf{Q}_p)$ is contained in $\mathbf{M}_\alpha$.*

We offer the following as examples. If M is any integer prime to δ, then define $U_0(M)$ (resp. $U_1(M)$) to be the compact open in $D_f^\times$ formed as a product $\prod_l U_l$ where $U_l = (\mathcal{O}_D \otimes \mathbf{Z}_l)^\times$ for $l \mid \delta$, and U_l is the matrices which are of the form $\left(\begin{smallmatrix} * & * \\ 0 & * \end{smallmatrix}\right)$ (resp. $\left(\begin{smallmatrix} * & * \\ 0 & 1 \end{smallmatrix}\right)$) mod $l^{v_l(M)}$ for all other l. If p^α divides M then the projection of $U_1(M)$ to the factor at p is contained in $\mathbf{M}_\alpha$ and hence U has wild level $\geq p^\alpha$. At the other extreme, we say that a compact open U is "prime to p" if U can be written as $U' \times \mathrm{GL}_2(\mathbf{Z}_p)$ where U' is a compact open of $(D \otimes \mathbf{A}_{f,p})^\times$, where $\mathbf{A}_{f,p}$ denotes the adeles with trivial components at infinity and p. As examples, if N is a positive integer prime to $p\delta$ then $U_0(N)$ and $U_1(N)$ have level prime to p. We refer to the subgroups U' above as "tame levels".

Let $\alpha \geq 1$ be an integer, and let U be a compact open of wild level $\geq p^\alpha$. If R is any commutative ring and A is an R-module with an R-linear right action of $\mathbf{M}_\alpha$, then define the R-module $\mathcal{L}(U, A)$ by

$$\mathcal{L}(U, A) = \left\{ f : D_f^\times \to A \,:\, f(dgu) = f(g)u_p \right\},$$

where $d \in D^\times$ (embedded diagonally in $D_f^\times$), $g \in D_f^\times$ and $u \in U$. This object looks rather terrifying if one is not familiar with this kind of thing, but in fact it is well known that if one writes $D_f^\times$ as a disjoint union of double cosets $D_f^\times = \coprod_i D^\times d_i U$ then this union is finite, of size n say, and hence any $f \in \mathcal{L}(U, A)$ is determined uniquely by the finite amount of data $f(d_1), f(d_2), \ldots, f(d_n)$. More precisely it can be shown that if $\Gamma_i = d_i^{-1} D^\times d_i \cap U$ then Γ_i is a finite group, and then an easy formal argument shows that the map

$$\mathcal{L}(U; A) \to \bigoplus_{i=1}^{n} A^{\Gamma_i} \tag{1}$$

sending f to $\big(f(d_i)\big)_{1 \leq i \leq n}$ is an isomorphism of R-modules. In particular, if U is chosen such that all the groups Γ_i are trivial, then the map $\mathcal{L}(U; A) \to A^n$ sending f to $\big(f(d_1), f(d_2), \ldots\big)$ is an isomorphism of R-modules. Triviality of all the Γ_i is a common occurrence – if $\gamma \in \Gamma_i$ is non-trivial then the characteristic polynomial of γ must be of the form $X^2 - aX + 1$ where $a \in \{-2, -1, 0, 1\}$, with $a = -2$ occurring iff $\gamma = -1$. Hence if $U \subseteq U_1(M)$ for some $M \geq 4$, for example, then all the Γ_i are trivial. The non-triviality of the groups Γ_i can be dealt with if one extends some of Coleman's theory on families of compact operators – see forthcoming work [2] of the author. For simplicity we shall here assume that all Γ_i are trivial whenever this eases the exposition. The more sophisticated reader might be happier thinking of $\mathcal{L}(U, A)$ in the general case as being the global sections of the sheaf corresponding to A on the stack $D^\times \backslash D_f / U$.

We recall the classical definition of the space of automorphic forms of level U and weight k. Strictly speaking these are not classical forms because we have shifted the "weight" action from ∞ to p, so one should perhaps think of the spaces below as being analogues of modular forms with coefficients in K. It is an elementary exercise (see for example the argument around equation (2) on p. 443 of [8]) to verify, however, that if there is an embedding of fields $K \to \mathbf{C}$ then the K-spaces we are about to define, when tensored up to $\mathbf{C}$, become isomorphic as Hecke modules to spaces of classical forms.

Recall that K is a complete subfield of $\mathbf{C}_p$. Let L_k denote the space of polynomials in one variable z, of degree at most $k-2$, and with coefficients in K. Define an action of $\mathbf{M}_1$ (and hence of $\mathbf{M}_\alpha$ for all $\alpha \geq 1$) by, for $\gamma = \left(\begin{smallmatrix} a & b \\ c & d \end{smallmatrix}\right) \in \mathbf{M}_1$ and $h \in L_k$,

$$(h|\gamma)(z) = (cz+d)^{k-2} h\left(\frac{az+b}{cz+d}\right).$$

If U is a compact open subgroup of $D_f^\times$ of wild level p, we shall refer to the K-vector space $S_k^D(U) := \mathcal{L}(U, L_k)$ as the space of classical automorphic forms of weight k and level U for D, defined over K. If $U = U_1(M)$ for some integer M which is a multiple of p and prime to δ as usual, then we refer to $S_k^D(U_1(M))$ as the space of classical automorphic forms for D of level M and weight k, defined over K. If K contains all the $\phi(M)$th roots of unity then this space decomposes as a direct sum

$$S_k^D(U_1(M)) = \bigoplus_{\varepsilon} S_k^D(U_1(M))(\varepsilon)$$

where the sum is over all characters $\varepsilon : (\mathbf{Z}/M\mathbf{Z})^\times \to K^\times$, and $S_k^D(U_1(M))(\varepsilon)$ is defined to be

$$\left\{ f : D_f^\times \to L_k \, : \, f(dgu) = \varepsilon(u) f(g) u_p \text{ for all } d \in D^\times, u \in U_0(M) \right\}.$$

Here ε is considered as a character of $U_0(M)$ via the map $U_0(M) \to (\mathbf{Z}/M\mathbf{Z})^\times$ sending $\left(\begin{smallmatrix} a & b \\ c & d \end{smallmatrix}\right)$ to $d \bmod M$.

It is more convenient to split up a character into its p-power part and its p-primary part. If $M = Np^n$ with $n \geq 1$ and N prime to p, and we continue to assume that K contains all $\phi(M)$th roots of unity, then one can write $\varepsilon : (\mathbf{Z}/M\mathbf{Z})^\times \to K^\times$ as a product $\varepsilon^p \varepsilon_p$, where ε^p has level N and ε_p has level p^n. Let $\alpha \geq 1$ be minimal such that ε_p factors through $(\mathbf{Z}/p^\alpha\mathbf{Z})^\times$. One can then define a "weight-character (k,ε_p)"-action of $\mathbf{M}_\alpha$ on L_k by, for $h\in L_k$ and $\gamma=\left(\begin{smallmatrix}a&b\\c&d\end{smallmatrix}\right)\in\mathbf{M}_\alpha$,

$$(h|\gamma)(z)=(cz+d)^{k-2}\varepsilon_p(d)h\left(\frac{az+b}{cz+d}\right).$$

Let L_{k,ε_p} denote L_k equipped with this action. If U is any compact open of level prime to p, then it makes sense to define $S_k^D(U\cap U_1(p^n))(\varepsilon_p):=\mathcal{L}(U\cap U_0(p^n),L_{k,\varepsilon_p})$ and we see that

$$\bigoplus_{\varepsilon^p:(\mathbf{Z}/N\mathbf{Z})^\times\to K^\times} S_k^D(U_1(M))(\varepsilon^p\varepsilon_p)=S_k^D(U_1(M))(\varepsilon_p),$$

$$\bigoplus_{\varepsilon_p:(\mathbf{Z}/p^n\mathbf{Z})^\times\to K^\times} S_k^D(U_1(M))(\varepsilon_p)=S_k^D(U_1(M)),$$

and more generally

$$\bigoplus_{\varepsilon_p:(\mathbf{Z}/p^n\mathbf{Z})^\times\to K^\times} S_k^D(U\cap U_1(p^n))(\varepsilon_p)=S_k^D(U\cap U_1(p^n))$$

if U has level prime to p. This philosophy, of treating characters at p differently to characters of level prime to p, by amalgamating the character at p with the weight, is already present in the "classical" theory of p-adic modular forms, and we also saw it in the GL_1 case.

Now let us return to the setting of a compact open U of wild level p^α, and a general right $\mathbf{M}_\alpha$-module A. The space $\mathcal{L}(U;A)$ comes equipped with certain standard operators, which we now define. Firstly say $\eta\in D_f^\times$ such that $\eta_p\in\mathbf{M}_\alpha$. If $f:D_f^\times\to A$ then define $f|\eta:D_f^\times\to A$ by

$$(f|\eta)(g)=f(g\eta^{-1})\eta_p.$$

Note that using this definition we have

$$\mathcal{L}(U,A)=\{f:D^\times\backslash D_f^\times\to A\ :\ f|\eta=f\text{ for all }\eta\in U\}.$$

Now say η is as above. We define $A^{[\eta_p]}$ to be the abelian group A equipped with the action, denoted $\cdot$, of $\eta_p^{-1}\mathbf{M}_\alpha\eta_p\cap\mathbf{M}_\alpha$, defined by $a\cdot(\eta_p^{-1}\gamma\eta_p):=a\gamma$. The map $f\mapsto f|\eta$ is easily shown to be an isomorphism $\mathcal{L}(U,A)\to\mathcal{L}(\eta^{-1}U\eta,A^{[\eta_p]})$, which we shall denote $|\eta$. If the map $a\mapsto a\eta_p$ induces an isomorphism of abelian groups $A\to A$, then one can think of this map as an isomorphism $A^{[\eta_p]}\to A$ and in this case the map $|\eta$ sends $\mathcal{L}(U,A)$ to $\mathcal{L}(\eta^{-1}U\eta,A)$ isomorphically.

Next we define Hecke operators. Let η be as above, write $U\eta U=\coprod_i U\eta_i$ (a finite union) and define the Hecke operator $[U\eta U]:\mathcal{L}(U,A)\to\mathcal{L}(U,A)$ by $[U\eta U]f=\sum_i f|\eta_i$. Of particular interest are the standard Hecke operators, defined as follows. If l is a prime not dividing δ, then define $\varpi_l\in\mathbf{A}_f$ to be the finite adele which is l at l and 1 at all other finite places. We sometimes think of ϖ_l as an element of $D_f^\times$ via the diagonal embedding. By a mild abuse of notation, we define $\eta_l=\left(\begin{smallmatrix}\varpi_l&0\\0&1\end{smallmatrix}\right)$ to mean the element of $D_f^\times$ which is $\left(\begin{smallmatrix}l&0\\0&1\end{smallmatrix}\right)$ at l and is the identity at all other places. We define $T_l = [U\eta_l U]$ and $S_l = [U\varpi_l U]$. An easy computation verifies for example that if $f \in S_k^D(U_1(M))(\varepsilon)$, then for all $l \nmid M$ we have $S_l(f) = \varepsilon(l)l^{k-2}f$. We conclude this section by reminding the reader of the explicit form that the Jacquet-Langlands theorem takes in this setting:

Theorem 2 (Jacquet-Langlands, Shimizu, Arthur). *If $k \geq 3$ then the space $S_k^D(U_1(M))$ is isomorphic to the space $S_k^{\delta\text{-new}}(\Gamma_1(M) \cap \Gamma_0(\delta))$ of classical δ-new forms, and this isomorphism commutes with the action of the standard Hecke operators defined above. If $k = 2$ then $S_k^{\delta\text{-new}}(\Gamma_1(M) \cap \Gamma_0(\delta))$ is isomorphic to the quotient of $S_k^D(U_1(M))$ by the subspace of forms which factor through the norm map, and again the isomorphism commutes with the action of the standard Hecke operators defined above.*

Proof. This is a "concrete" realisation of the Jacquet-Langlands theorem; a good reference for how to deduce it from Theorem 16.1 of [13] is the first few pages of Section 5 of [8] (with $t = 1$ in their notation). □

As a consequence we see that if $f \in S_k^D(U_1(M))(\varepsilon)$ is an eigenform then there is a continuous Galois representation $\rho_f : \mathrm{Gal}(\overline{\mathbf{Q}}/\mathbf{Q}) \to \mathrm{GL}_2(K)$ associated to f satisfying the usual properties.

5. Overconvergent automorphic forms of a fixed weight

Let K be a complete subfield of $\mathbf{C}_p$ as usual.

Definition. *A real number r is said to be a* radius of convergence *if $r = p^{-n}$ for some integer $n \geq 0$.*

Definition. *If r is a radius of convergence then we let $\mathbf{B}_r$ be the rigid subspace of affine 1-space over K whose $\mathbf{C}_p$-points are*

$$\mathbf{B}_r(\mathbf{C}_p) = \{z \in \mathbf{C}_p : \text{there exists } y \text{ in } \mathbf{Z}_p \text{ such that } |z - y| \leq r\}.$$

If furthermore $r < 1$ then we define $\mathbf{B}_r^\times$ as the rigid subspace of affine 1-space whose $\mathbf{C}_p$-points are

$$\mathbf{B}_r^\times(\mathbf{C}_p) = \{z \in \mathbf{C}_p : \text{there exists } y \text{ in } \mathbf{Z}_p^\times \text{ such that } |z - y| \leq r\}.$$

In fact $\mathbf{B}_r$ and $\mathbf{B}_r^\times$ are just disjoint unions of finitely many closed affinoid discs. One can think of $\mathbf{B}_r$ and $\mathbf{B}_r^\times$ as being a sequence of rigid neighbourhoods of $\mathbf{Z}_p$ and $\mathbf{Z}_p^\times$ respectively in the analytification of affine 1-space over K. If $s < r$ are both radii of convergence then $\mathbf{B}_s \subset \mathbf{B}_r$ and $\mathbf{B}_s^\times \subset \mathbf{B}_r^\times$.

We say that a map between K-Banach spaces is *compact* (some authors use the term "completely continuous") if it is a limit of maps having finite rank. If r is a radius of convergence then let $\mathcal{A}_r$ denote the ring of functions on $\mathbf{B}_r$, equipped with the supremum semi-norm (which is a norm in this case, as $\mathcal{A}_r$ is reduced). A basic result is

Lemma 3. *If $s < r$ are radii of convergence, then the inclusion $\mathbf{B}_s \subset \mathbf{B}_r$ induces a compact map $\mathcal{A}_r \to \mathcal{A}_s$.*

Proof. This is an immediate consequence of the fact that the inclusion $\mathbf{B}_s \subset \mathbf{B}_r$ is inner. Alternatively, one can give a direct proof, as follows: the space $\mathbf{B}_r$ is a finite disjoint union of closed discs, and hence $\mathcal{A}_r$ is a finite direct sum of rings topologically isomorphic to the affinoid $K\langle x\rangle$. The inclusion $\mathbf{B}_s \to \mathbf{B}_r$ induces, on each affinoid corresponding to a component of $\mathbf{B}_r$, a map $K\langle x\rangle \to K\langle y\rangle$ which, if one chooses appropriate parameters on the discs, is equal to the ring homomorphism defined by $x \mapsto (r/s)y$. But this map is compact because for any $\varepsilon > 0$, the image of the basis $\{1, x, x^2, \ldots\}$ of $K\langle x\rangle$ only contains finitely many terms with norm greater than ε. □

Note that we think of $\mathbf{B}_r$ and $\mathbf{B}_r^\times$ not as being abstract rigid spaces but as being explicit rigid subspaces of rigid affine 1-space. Hence these spaces come with with a fixed parameter z, upon which several constructions below will depend.

Let $\mathcal{W}$ denote "weight space" in this setting, that is, the rigid space

$$\mathrm{Hom}(\mathbf{Z}_p^\times, \mathbf{G}_m)$$

over K (see Lemma 2). Then the L-points (L any finite extension of K) of $\mathcal{W}$ are the continuous group homomorphisms $\kappa : \mathbf{Z}_p^\times \to L^\times$. We refer to points of $\mathcal{W}$ as weights, or as weight-characters. Points of $\mathcal{W}$ corresponding to weight-characters of the form $\kappa(x) = x^k \varepsilon_p(x)$, where k is an integer and ε_p is a finite order character, will be called classical weight-characters.

We have seen that $\mathcal{W}$ is a union of finitely many open discs. Say $\kappa \in \mathcal{W}(K)$. Then it is an easy check (use log and exp) that κ extends to a morphism of rigid spaces $\mathbf{B}_r^\times \to \mathbf{G}_m^{\mathrm{an}}$ for some radius of convergence $r < 1$, (and hence to a morphism $\mathbf{B}_s^\times \to \mathbf{G}_{m,K}$ for any radius of convergence $s < r$). For example, one can take $r = p^{-1}$ if $\kappa(x) = x^k$ for some $k \in \mathbf{Z}$, and $r = p^{-n}$ if κ is of finite order and factors through $(\mathbf{Z}/p^n\mathbf{Z})^\times$.

If $\left(\begin{smallmatrix} a & b \\ c & d \end{smallmatrix}\right) \in \mathbf{M}_\alpha$ then the map $z \mapsto cz + d$ defines a morphism $\mathbf{B}_r \to \mathbf{B}^\times_{rp^{-\alpha}}$, and the map $z \mapsto \frac{az+b}{cz+d}$ defines a morphism $\mathbf{B}_r \to \mathbf{B}_{r|\Delta|}$, where $\Delta = ad - bc$. We frequently use the weaker statement that $z \mapsto \frac{az+b}{cz+d}$ defines a map $\mathbf{B}_r \to \mathbf{B}_r$. Now say $\kappa \in \mathcal{W}(K)$ and r is a radius of convergence.

Definition. *We say an integer $\alpha \geq 1$ is* good *for (κ, r) if $\kappa : \mathbf{Z}_p^\times \to K^\times$ extends to a morphism of rigid spaces $\mathbf{B}^\times_{rp^{-\alpha}} \to \mathbf{G}_m^{an}$.*

For any κ and r, a good integer α will exist, by the remarks above. Of course, given α which is good for (κ, r), the extension of κ to $\mathbf{B}^\times_{rp^{-\alpha}}$ is unique. Finally, if α is good for (κ, r) and $\beta > \alpha$ then β is also good for (κ, r) and the extension of κ to $\mathbf{B}^\times_{rp^{-\beta}}$ is just the restriction of the extension of κ to $\mathbf{B}_{rp^{-\alpha}}$.

Now choose κ and r as above, and let α be any integer which is good for (κ, r). Let κ also denote the extension of κ to $\mathbf{B}^\times_{rp^{-\alpha}}$.

Definition. *Let $\mathcal{A}_{\kappa,r}$ be the ring $\mathcal{A}_r$ of rigid-analytic functions on $\mathbf{B}_r$, equipped with the following right action of $\mathbf{M}_\alpha$: for $h \in \mathcal{A}_{\kappa,r}$ and $\gamma = \left(\begin{smallmatrix} a & b \\ c & d \end{smallmatrix}\right) \in \mathbf{M}_\alpha$,*

$$(h|\gamma)(z) = \frac{\kappa(cz+d)}{(cz+d)^2} h\left(\frac{az+b}{cz+d}\right).$$

Note that α has to be good for (κ, r) for this definition to make sense. Note also that the simplest example of these rings is just $\mathcal{A}_{\kappa,1}$ which is the ring $K\langle z\rangle$ of power series $\sum a_n z^n$ such that $a_n \to 0$ as $n \to \infty$.

We say that a map f between Banach spaces is *norm-decreasing* if $|f(x)| \leq |x|$ for all x. We have equipped $\mathcal{A}_r$ with the structure of a Banach space and we think of $\mathcal{A}_{\kappa,r}$ also as having this Banach space structure. The map $\mathcal{A}_{\kappa,r} \to \mathcal{A}_{\kappa,r}$ defined by $h \mapsto h|\gamma$ is norm-decreasing: this is an immediate consequence of the fact that $\kappa(cz+d)$ and $cz+d$ both have norm 1. If $|\det(\gamma)| < 1$ then one can say much more. In this case, the endomorphism $f \mapsto f|\gamma$ of $\mathcal{A}_{\kappa,r}$ is compact, as it factors as $\mathcal{A}_{\kappa,r} \xrightarrow{\text{res}} \mathcal{A}_{\kappa,r|\Delta|} \xrightarrow{|\gamma} \mathcal{A}_{\kappa,r}$, where $\Delta = \det(\gamma)$, and the restriction map $\mathcal{A}_{\kappa,r} \to \mathcal{A}_{\kappa,r|\Delta|}$ is compact by Lemma 3.

On the other hand if $|\det(\gamma)| = 1$ then γ has an inverse in $\mathbf{M}_\alpha$, and so the induced endomorphism of $\mathcal{A}_{\kappa,r}$ is norm-decreasing with a norm-decreasing inverse and is hence an isometry.

For κ a weight and $r = p^{-n}$, choose $\alpha \geq 1$ which is good for (κ, r), and let U be a compact open subgroup of $D_f^\times$ of wild level $\geq p^\alpha$. The key definition is the following:

Definition. *The space of r-overconvergent automorphic forms of weight-character κ and level U is the space $\mathbf{S}_\kappa^D(U; r) := \mathcal{L}(U, \mathcal{A}_{\kappa,r})$.*

We now make some elementary observations about this space. Firstly, note that this space is independent of the choice of α, in the sense that if β is also good for (κ, r) and U also has wild level $\geq p^\beta$, then the two induced actions of U on $\mathcal{A}_{\kappa,r}$ are the same and hence the definition is independent of the choice of α. Next note that using the "explicit" description of equation (1) in Section 4, one sees that $\mathbf{S}_\kappa^D(U; r)$ is naturally a Banach space over K (note that the groups Γ_i are finite and K has characteristic 0 and hence the subspace of $\mathcal{A}_{\kappa,r}$ where Γ_i acts trivially is a closed direct summand). In fact one can write down an explicit norm on $\mathbf{S}_\kappa^D(U; r)$, useful for computational purposes – if $f \in \mathbf{S}_\kappa^D(U; r)$ then define $|f| = \max_{g \in D_f^\times} |f(g)|$. This max exists, is finite, and is attained, because if $u \in U$ and $h \in \mathcal{A}_{\kappa,r}$ then $\det(u_p) \in \mathbf{Z}_p^\times$, so h and $h|u$ have the same norm, and hence $|f(g)|$ is constant on double cosets $D^\times gU$. One can also view the norm on $\mathbf{S}_\kappa^D(U, r)$ in the following manner: fix once and for all $\{d_1, \ldots, d_n\}$ such that $D_f^\times = \coprod D^\times d_i U$; then there is an induced isomorphism $\mathbf{S}_\kappa^D(U; r) \cong \oplus_i \mathcal{A}_{\kappa,r}^{\Gamma_i}$ as above, and the right-hand side is a closed subspace of the affinoid $\oplus_i \mathcal{A}_{\kappa,r}$ (which can be thought of as being equipped with its usual supremum semi-norm).

If $s < r$ is another weight, then the restriction map $\mathcal{A}_{\kappa,r} \to \mathcal{A}_{\kappa,s}$ is compact and norm-decreasing, and hence induces a compact norm-decreasing map $\mathbf{S}^D_\kappa(U;r) \to \mathbf{S}^D_\kappa(U;s)$.

For $\eta \in D_f^\times$ such that $\eta_p \in \mathbf{M}_\alpha$, the endomorphism $[U\eta U]$ of $\mathbf{S}^D_\kappa(U;r)$ is continuous and norm-decreasing. Moreover if also η_p has determinant Δ a non-unit, then $[U\eta U]$ factors as

$$[U\eta U] : \mathbf{S}^D_\kappa(U;r) \xrightarrow{\text{res}} \mathbf{S}^D_\kappa(U;r|\Delta|) \to \mathbf{S}^D_k(U;r)$$

and hence is norm-decreasing and compact, as the restriction morphism is compact. Recall that a compact map has a characteristic power series, or a Fredholm determinant, which is the natural generalisation of the characteristic polynomial of a finite rank map: see Section 5 of [16] for a beautifully-written introduction to these notions. The fact that $[U\eta U]$ is compact and norm-decreasing means that its characteristic power series is in $\mathcal{O}_K\{\{T\}\}$, that is, in the space of power series that converge on the whole of affine 1-space. The case we are most interested in is the Hecke operator $U_p := [U\eta_p U]$ where we recall $\eta_p = \left(\begin{smallmatrix} \varpi_p & 0 \\ 0 & 1 \end{smallmatrix}\right)$.

If κ is a classical weight-character, that is, $\kappa(x) = x^k\varepsilon_p(x)$ where k is an integer and ε_p is a finite order character, then choose $\alpha \geq 1$ minimal such that ε_p factors through $(\mathbf{Z}/p^\alpha\mathbf{Z})^\times$. We have a natural inclusion $L_{k,\varepsilon_p} \to \mathcal{A}_{\kappa,1}$ of $\mathbf{M}_\alpha$-modules. This induces a map from classical automorphic forms (with p-adic coefficients) to 1-overconvergent automorphic forms. More precisely, if we fix a level U_0 prime to p and any $n \geq \alpha$ then we get a continuous embedding from $S^D_k(U_0 \cap U_1(p^n))(\varepsilon_p) = \mathcal{L}(U_0 \cap U_0(p^n), L_{k,\varepsilon_p})$ to $\mathbf{S}_\kappa(U_0 \cap U_1(p^n), 1)$. This is the analogue in this setting of the statement that classical forms are overconvergent. In the next section we will discuss the slightly more subtle question of letting the level drop at p, and prove an analogue of the result that classical forms of level Np^n with trivial character at p are overconvergent of level Np.

6. The operator U_p

We record a few useful facts about the Hecke operator U_p. Let κ be a weight and let $r = p^{-t}$ be a radius of convergence. Choose any $\alpha \geq 1$ which is good for (κ, r) and choose any $n \geq \alpha$. We will be concerned with forms of level $U = U_0 \cap U_1(p^\alpha) \cap U_0(p^n)$, where U_0 is a compact open of level prime to p.

Lemma 4.

1. *There exist maps* $\alpha_s : \mathbf{S}^D_\kappa(U;s/p) \to \mathbf{S}^D_\kappa(U;s)$, *for all radii of convergence* s, *such that* $U_p : \mathbf{S}^D_\kappa(U;r) \to \mathbf{S}^D_\kappa(U;r)$ *factors as*

$$\mathbf{S}^D_\kappa(U;r) \xrightarrow{res} \mathbf{S}^D_\kappa(U;r/p) \xrightarrow{\alpha_r} \mathbf{S}^D_\kappa(U;r)$$

and if $r < 1$ *it also factors as*

$$\mathbf{S}^D_\kappa(U;r) \xrightarrow{\alpha_{pr}} \mathbf{S}^D_\kappa(U;pr) \xrightarrow{res} \mathbf{S}^D_\kappa(U;r).$$

2. U_p *is a compact operator on* $\mathbf{S}^D_\kappa(U;r)$.

3. *If $s < r$ is another radius of convergence then the characteristic power series of U_p on $\mathbf{S}^D_\kappa(U;s)$ and on $\mathbf{S}^D_\kappa(U;r)$ coincide.*
4. *There is a canonical isomorphism of Hecke modules*

$$\iota : \mathbf{S}^D_\kappa(U;r/p) \cong \mathbf{S}^D_\kappa(U \cap U_0(p^{n+1});r)$$

5. *The characteristic power series of U_p on $\mathbf{S}^D_\kappa(U;r)$ equals the characteristic power series of U_p on $\mathbf{S}^D_\kappa(U_0 \cap U_1(p^\alpha);r)$. That is, "the characteristic power series does not see the higher $U_0(p^n)$ structure".*

Proof. 1. By definition, U_p is the sum of various maps $f \mapsto f|\gamma$ where $\gamma \in \mathbf{M}_\alpha$ has determinant p. Hence the results follow from the fact that $|\gamma : \mathcal{A}_{\kappa,r} \to \mathcal{A}_{\kappa,r}$ factors as $\mathcal{A}_{\kappa,r} \xrightarrow{\text{res}} \mathcal{A}_{\kappa,r/p} \xrightarrow{|\gamma} \mathcal{A}_{\kappa,r}$, and also as $\mathcal{A}_{\kappa,r} \xrightarrow{|\gamma} \mathcal{A}_{\kappa,pr} \xrightarrow{\text{res}} \mathcal{A}_{\kappa,r}$ if $r < 1$. Hence $\alpha_r(f) := \sum_i f|\eta_i$ will do the job, if $U\left(\begin{smallmatrix} \varpi_p & 0 \\ 0 & 1 \end{smallmatrix}\right)U = \coprod_i U\eta_i$.

2. The restriction maps are compact and compactness of U_p follows.

3. Let $\alpha_{ps} : \mathbf{S}^D_\kappa(U;s) \to \mathbf{S}^D_\kappa(U;ps)$ be as in part 1. The characteristic power series of $\alpha_s \circ \text{res}$ and $\text{res} \circ \alpha_s$ are equal, by [16], Corollaire 2 to Proposition 7, and the result follows easily by induction on $-\log_p(s)$.

4. One constructs ι by composing the following sequence of canonical isomorphisms. Firstly define $U^0(p)$ to be the matrices $\gamma \in U_0(1)$ such that γ_p is lower triangular mod p. Then one observes that $\mathbf{S}^D_\kappa(U;r/p) = \mathcal{L}(U, \mathcal{A}_{\kappa,r/p})$ is the subspace of $\mathcal{L}(U \cap U^0(p), \mathcal{A}_{\kappa,r/p})$ where the matrices $\left(\begin{smallmatrix} 1 & i \\ 0 & 1 \end{smallmatrix}\right)$ all act as the identity. This space is naturally isomorphic to the space $\mathcal{L}(U \cap U^0(p), \mathcal{O}(p\mathbf{B}_r))$, where $p\mathbf{B}_r$ is the rigid subspace of $\mathbf{A}^{\text{an}}_1$ whose $\mathbf{C}_p$-points are $\{z : z/p \in \mathbf{B}_r(\mathbf{C}_p)\}$. Note that $p\mathbf{B}_r \subset \mathbf{B}_{r/p}$. Next we observe that the map $h \mapsto h|\left(\begin{smallmatrix} \varpi_p & 0 \\ 0 & 1 \end{smallmatrix}\right)$ induces an isomorphism $\mathcal{L}(U \cap U^0(p), \mathcal{O}(p\mathbf{B}_r)) \to \mathcal{L}(U \cap U_0(p^{n+1}), \mathcal{O}(p\mathbf{B}_r)^{[\left(\begin{smallmatrix} p & 0 \\ 0 & 1 \end{smallmatrix}\right)]})$. Finally, one checks that the map $f \mapsto f|\left(\begin{smallmatrix} p & 0 \\ 0 & 1 \end{smallmatrix}\right)$ induces an isomorphism $\mathcal{O}(p\mathbf{B}_r)^{[\left(\begin{smallmatrix} p & 0 \\ 0 & 1 \end{smallmatrix}\right)]} \to \mathcal{A}_{\kappa,r}$ and hence an isomorphism $\mathcal{L}(U \cap U_0(p^{n+1}), \mathcal{O}(p\mathbf{B}_r)^{[\left(\begin{smallmatrix} p & 0 \\ 0 & 1 \end{smallmatrix}\right)]}) = \mathcal{L}(U \cap U_0(p^{n+1}), \mathcal{A}_{\kappa,r}) = \mathbf{S}^D_\kappa(U \cap U_0(p^{n+1});r)$. If we define ι to be the composite of these isomorphisms then it is elementary to check that ι commutes with the Hecke operators away from p, and an explicit calculation shows that ι also commutes with the Hecke operator U_p. We remark that the way we have normalised things, the operator U_p at level p^n is $\sum_{i=0}^{p-1} |\left(\begin{smallmatrix} p & 0 \\ ip^n & 1 \end{smallmatrix}\right)$, and in particular it depends on n. This annoyance could have been avoided at the expense of introducing other annoyances.

5. It suffices to prove that the characteristic power series of U_p on $\mathbf{S}^D_\kappa(U_0 \cap U_1(p^\alpha) \cap U_0(p^n);r/p)$ and $\mathbf{S}^D_\kappa(U_0 \cap U_1(p^\alpha) \cap U_0(p^{n+1});r)$ are equal, if $n \geq \alpha$. Let T denote the composition

$$\mathbf{S}^D_\kappa(U_0 \cap U_1(p^\alpha) \cap U_0(p^{n+1});r) \xrightarrow{\text{res}} \mathbf{S}^D_\kappa(U_0 \cap U_1(p^\alpha) \cap U_0(p^{n+1});r/p)$$
$$\to \mathbf{S}^D_\kappa(U_0 \cap U_1(p^\alpha) \cap U_0(p^n);r/p),$$

where this latter morphism is given by the trace map $f \mapsto \sum_{i=0}^{p-1} f|\left(\begin{smallmatrix} 1 & 0 \\ p^n i & 1 \end{smallmatrix}\right)$. Then it is an elementary exercise to check that T is compact, and that both $\iota \circ T$ and $T \circ \iota$

are equal to U_p on their respective spaces. The result follows by [16], Corollaire 2 to Proposition 7. □

We can now easily prove analogues of the first two parts of Theorem 1. As before, let U_0 be some compact open of level prime to p, let κ be a weight-character and let r be a radius of convergence. Let α be any positive integer which is good for (κ, r), and let $n \geq \alpha$ be any integer. Set $U = U_0 \cap U_1(p^\alpha) \cap U_0(p^n)$.

Proposition 3.

1. *If $f \in \mathbf{S}_\kappa(U; r)$ then $f|\left(\begin{smallmatrix} \varpi_p^{\alpha-n} & 0 \\ 0 & 1 \end{smallmatrix}\right)$ (appropriately interpreted) is an element of $\mathbf{S}_\kappa(U_0 \cap U_1(p^\alpha); r/p^{n-\alpha})$.*
2. *If $f \in \mathbf{S}_\kappa(U; r)$ and $U_p f = \lambda f$ for some non-zero λ then $f \in \mathbf{S}_\kappa(U; s)$ for any radius of convergence $s \geq r$ such that α is good for (κ, s).*

Proof. 1. One continually applies the inverse of the map ι defined in part 4 of Lemma 4.

2. By part 1 of Lemma 4, $U_p f \in \mathbf{S}_\kappa(U; pr)$ and hence $f \in \mathbf{S}_\kappa(U; pr)$. So we are finished by induction on $-\log_p(r)$. □

Note that the map $f \mapsto f|\left(\begin{smallmatrix} \varpi_p^{n-\alpha} & 0 \\ 0 & 1 \end{smallmatrix}\right)$, the inverse of the map in part 1 of this proposition, is in fact the natural inclusion $\mathbf{S}_\kappa(U_0 \cap U_1(p^\alpha); r/p^{n-\alpha}) \to \mathbf{S}_\kappa(U; r)$, in the sense that it is the one which commutes with the Hecke operator U_p. The fact that it is this morphism which is natural, rather than the identity map, is a mildly annoying consequence of the way we have set things up – more specifically, it is because the matrices defining U_p depend on the level that we are working with.

7. Classical and overconvergent forms

We have just seen how classical forms embed into overconvergent ones. Now we show how to pull them out. Let $\kappa = (k, \varepsilon_p)$ be a classical weight-character, where $k \geq 1$, and as usual choose $\alpha \geq 1$ minimal such that ε_p factors through $(\mathbf{Z}/p^\alpha\mathbf{Z})^\times$. Then $\mathbf{M}_\alpha$ acts on L_{k,ε_p}. Let U_0 be a compact open in $D_f^\times$ of level prime to p, and let $U = U_0 \cap U_0(p^n)$ for any $n \geq \alpha$. Let κ' be the character $(2-k, \varepsilon_p)$. We define a map

$$\theta^{1-k} : \mathbf{S}_\kappa^D(U, 1) \to \mathbf{S}_{\kappa'}^D(U, 1)$$

by

$$\left(\theta^{1-k}(f)\right)(g) = \left(|\nu(g)| \det(g_p)\right)^{1-k} \frac{d^{k-1} f(g)}{dz^{k-1}}.$$

Here ν is the norm map $D_f^\times \to \mathbf{A}_f^\times$, and $|.|$ is the usual absolute value map $\mathbf{A}_f \to \mathbf{Q}$. Note that the superscript $1-k$ is notational and merely there to indicate that θ^{1-k} looks like the $(1-k)$th power of some kind of Tate twisting operator.

One has to verify that the map θ^{1-k} is well defined, which boils down to checking that for $\left(\begin{smallmatrix} a & b \\ c & d \end{smallmatrix}\right) \in \mathbf{M}_\alpha$ and F a power series in z, we have the identity

$$\left(\frac{d^{k-1}}{dz^{k-1}}\right)\left((cz+d)^{k-2}F\left(\frac{az+b}{cz+d}\right)\right) = (ad-bc)^{k-1}(cz+d)^{-k}\left(\frac{d^{k-1}F}{dz^{k-1}}\right)\left(\frac{az+b}{cz+d}\right).$$

This identity is trivial for $k = 1$ and the general case is easily established by induction on k. Next one can analyse the relationship between θ^{1-k} and Hecke operators. Again it is elementary to check that if $f \in \mathbf{S}^D_\kappa(U,1)$ and $\eta \in D_f^\times$ with $\eta_p \in \mathbf{M}_\alpha$, then $(\theta^{1-k}f)|\eta = |\nu(\eta)|^{k-1}\theta^{1-k}(f|\eta)$ and hence that $[U\eta U]\theta^{1-k} = |\nu(\eta)|^{k-1}\theta^{1-k}[U\eta U]$. Applying this with $\eta = \eta_l$ one sees that $T_l\theta^{1-k} = l^{1-k}\theta^{1-k}T_l$ and hence that if $f \in \mathbf{S}^D_\kappa(U,1)$ is an eigenform for T_l with eigenvalue a_l then $\theta^{1-k}f$ is an eigenform for T_l with eigenvalue $a_l l^{1-k}$. We now prove the analogue of the third and fourth results of Theorem 1 in this setting.

Proposition 4. *The kernel of θ^{1-k} is precisely the classical forms*

$$\mathbf{S}^D_k(U_0 \cap U_1(p^n))(\varepsilon_p).$$

Let $0 \neq f \in \mathbf{S}^D_\kappa(U;r)$ be an eigenvector for U_p with non-zero eigenvalue λ. Then $f \in \mathbf{S}^D_\kappa(U;1)$. Moreover, if $v_p(\lambda) < k-1$ then f is classical, and if $v_p(\lambda) > k-1$ then f is not classical.

Proof. The kernel of θ^{1-k} is the function $f \in \mathbf{S}^D_\kappa(U;1)$ whose image is contained within the space of polynomials of degree at most $k-2$, which is precisely the space of classical forms $\mathcal{L}(U, L_{k,\varepsilon_p}) = S^D_k(U_0 \cap U_1(p^n))(\varepsilon_p)$. Next, let $0 \neq f \in \mathbf{S}^D_\kappa(U;r)$ be an eigenvector for U_p with non-zero eigenvalue λ. By the proposition in the previous section, $f \in \mathbf{S}^D_\kappa(U,1)$.

If f is classical then one can easily deduce from the classical theory (see for example Theorem 4.6.17 of [15], the fact that λ is an algebraic integer, and the Jacquet-Langlands theorem) that $v_p(\lambda) \leq k-1$. On the other hand, if $\lambda < k-1$ then $\theta^{1-k}f$ is in $\mathbf{S}^D_{\kappa'}(U;1)$ and if it is non-zero then it is an eigenvector for U_p with eigenvalue λ/p^{k-1}, which has negative valuation. On the other hand, U_p is an operator with norm at most 1, and hence $\theta^{1-k}f = 0$. Hence f is classical. □

In the classical theory, one can find both classical and non-classical forms with $\lambda = k-1$, and it would be an interesting computational exercise to search for examples of this phenomenon in this situation.

We finish by remarking that if we had defined our spaces of overconvergent automorphic forms as $\mathcal{L}(U, (\mathcal{A}_{\kappa,r})^*)$, using the dual of $\mathcal{A}_{\kappa,r}$, then one would see the theta operator going in the other direction, which would be more analogous to the classical theory. On the other hand, using this convention gives a surjection from overconvergent forms to classical ones, rather than an injection from classical forms to overconvergent ones. Using the duals would give a theory more analogous to the one set up by Stevens in the classical case.

8. Families

Let R be a reduced affinoid over K. Then R is a Banach algebra in the sense of Section A1 of [6]. Let A be a Banach module over R. Briefly, the key points are that R is a ring complete with respect to a non-trivial ultrametric norm and A is a complete normed R-module satisfying various natural axioms such as $|ra| \leq |r||a|$ for $r \in R$ and $a \in A$. In fact the only cases of interest to us in this paper are when R is an affinoid disc over K. Assume furthermore that we have an R-linear action of $\mathbf{M}_\alpha$ on A, and that all elements of $\mathbf{M}_\alpha$ act continuously. Then for $U \subset D_f^\times$ open and compact and of wild level $\geq p^\alpha$, we see that $\mathcal{L}(U;A)$ also inherits the structure of a Banach module over R. Let us for simplicity assume in this section that all the finite groups Γ_i are trivial.

We now specialise to the case that we are interested in. Let R be the ring of functions on an affinoid disk V defined over K in $\mathcal{W}$. Define $\mathcal{A}_{V,r}$ to be the ring of functions on the rigid space $V \times_K \mathbf{B}_r$. We say that an integer $\alpha \geq 1$ is *good* for the pair (V,r) if for all $\kappa \in V(\mathbf{C}_p)$, the map $\kappa : \mathbf{Z}_p^\times \to \mathbf{C}_p^\times$ extends to a map $\mathbf{B}_r^\times \to \mathbf{G}_m^{\mathrm{an}}$. Because V is affinoid, one can check that good integers α do exist. Define a right action of $\mathbf{M}_\alpha$ on $\mathbf{A}_{V,r}$ by, for $h \in \mathbf{A}_{V,r}$ and $\gamma = \left(\begin{smallmatrix} a & b \\ c & d \end{smallmatrix}\right) \in \mathbf{M}_\alpha$,

$$(h|\gamma)(\kappa, z) = \frac{\kappa(cz+d)}{(cz+d)^2} h\left(\kappa, \frac{az+b}{cz+d}\right).$$

It is easily verified that $\mathbf{M}_\alpha$ acts by continuous R-linear maps. If U is a compact open of wild level $\geq p^\alpha$ then we define $\mathbf{S}_V^D(U;r) := \mathcal{L}(U, \mathcal{A}_{V,r})$. This is a Banach module over R and it enjoys various base change properties. For example, if $V_1 \subseteq V_2$ are affinoids in $\mathcal{W}$, with associated algebras R_1 and R_2, then $\mathbf{S}_{V_2}^D(U;r)\widehat{\otimes}_{R_2} R_1 = \mathbf{S}_{V_1}^D(U;r)$, and if $\kappa \in V(K)$ inducing a map $R \to K$ then $\mathbf{S}_V^D(U;r) \otimes_R K = \mathbf{S}_\kappa^D(U;r)$. The same formalism as we have seen already in the case of fixed weight shows that we have continuous Banach R-module endomorphisms T_l of $\mathbf{S}_V^D(U;r)$ for l any prime not dividing δ, and furthermore that U_p is a compact morphism. In particular U_p on $\mathbf{S}_V^D(U,r)$ has a characteristic power series in $R\{\{T\}\}$ whose restriction to a weight κ is the characteristic power series of U_p on forms of weight κ.

The reader who has read the construction of the classical eigencurve in [7] will now see that we have all the ingredients to construct the analogous object in this case. The reason we have restricted to the case where the Γ_i are all trivial is to ensure that all the Banach modules $\mathbf{S}_V^D(U,r)$ are orthonormizable, although in future work we shall show that this assumption is unnecessary. We point out here that we only define the "D" eigencurve in this setting, although no doubt one can also construct a "C" eigencurve by mimicking the construction in [7].

By the argument at the beginning of Chapter 7 of [7], for every Hecke operator α we can construct a spectral curve Z_α representing the spectrum of the compact operator αU_p. All of Coleman's Riesz theory applies in this setting, and the construction of the curve D_α in Section 7.2 of [7] goes through word for word in our setting. Similarly the arguments of Section 7.3 go over unchanged, to give us an eigencurve D. Once the existence of D is established, the usual corollaries, such as local constancy of dimension of spaces of overconvergent forms of a given slope follow, although of course one can prove these corollaries directly just by an analysis of the spaces $\mathbf{S}_V^D(U,r)$. Finally, the fact that forms of small slope are classical enables one to deduce that the dimensions of spaces of classical forms of a given slope are also locally constant.

We remark that Dan Jacobs in his thesis ([12]) has make some explicit computations which indicate that, just as in the classical case, these new eigencurves exhibit some surprising and unexplained regularity in their structure.

We end with a remark about what is lacking in this theory. One can prove slightly less about these eigencurves than the classical eigencurves, because the arguments in [7] that rely on the existence of q-expansions seem to have no analogue in this setting. The fact that points on the eigencurve correspond to overconvergent eigenforms follows from some elementary commutative algebra, but on the other hand, the lack of a natural pairing between spaces of forms and Hecke algebras in this setting means that the author cannot currently prove that the rank of the Hecke algebra acting on a space of overconvergent forms of some fixed slope is as big as the rank of the space of the forms. We hope to resolve this problem in the future.

References

[1] S. Bosch, U. Güntzer and R. Remmert, *Non-Archimedean Analysis*, Springer-Verlag (1984).

[2] K. Buzzard, Eigenvarieties, in preparation.

[3] J. Cassels, A. Fröhlich, *Algebraic Number Theory*, Academic Press, Boston (1967).

[4] G. Chenevier, *Familles p-adiques de formes automorphes et applications aux conjectures de Bloch-Kato*, PhD, Univ. Paris VII, 2003.

[5] R. Coleman, *Classical and overconvergent modular forms*, Invent. Math. 124 (1996) 215–241.

[6] R. Coleman, *p-adic Banach spaces and families of modular forms.* Invent. Math. 127 (1997), 417–479.

[7] R. Coleman and B. Mazur, *The eigencurve.* In *Galois representations in arithmetic algebraic geometry* (Durham 1996), CUP 1998, 1–113.

[8] F. Diamond and R. Taylor, *Non-optimal levels for mod l modular representations of* $\mathrm{Gal}(\overline{\mathbf{Q}}/\mathbf{Q})$, Invent. Math. 115 (1994), 435–462.

[9] G. Stevens, *Overconvergent modular symbols*, preprint.

[10] F. Gouvêa, *Arithmetic of p-adic modular forms*, Springer LNM 1304 (1988).

[11] F. Gouvêa, B. Mazur, *Families of modular eigenforms.* Math. Comp. 58 (1992), 793–805.

[12] D. Jacobs, *Slopes of compact Hecke operators*, PhD, University of London, 2003.

[13] H. Jacquet and R. Langlands, *automorphic forms on* GL(2), Springer LNM 114 (1970).

[14] N. Katz, *p-adic properties of modular schemes and modular forms*, Modular functions of one variable, III (Proc. Internat. Summer. School, Univ. Antwerp, Antwerp, 1972) (Berlin), Springer, 1973, 69–190. LNM Vol. 350.

[15] T. Miyake, *Modular forms*, Springer Verlag, Berlin, 1989.

[16] J.-P. Serre, *Endomorphismes complètement continus des espaces de Banach p-adiques*, Publ. Math. IHES 12 (1962) 69–85.

Kevin Buzzard
Imperial College
660 Huxley Building
South Kensington Campus
London
e-mail: buzzard@imperial.ac.uk

Progress in Mathematics, Vol. 224, 45–52

$\mathbb{Q}$-curves and Abelian Varieties of GL_2-type from Dihedral Genus 2 Curves

Gabriel Cardona

Abstract. In this paper we review some of the results in [3] about curves of genus 2 with dihedral group of automorphisms and apply Ellenberg-Skinner's criterion (cf. [4]) to determine that one can find infinitely many of them with modular Jacobian.

Introduction

Let $E/\mathbb{Q}$ be an elliptic curve defined over $\mathbb{Q}$. We say that E is modular if, equivalently, E is parametrized over $\mathbb{Q}$ by a modular curve,

$$X_1(N) \overset{/\mathbb{Q}}{\twoheadrightarrow} E,$$

or E is a factor (up to $\mathbb{Q}$-isogeny) of a modular Jacobian,

$$J_1(N) \overset{\mathbb{Q}}{\sim} E \times A'.$$

The equivalence of the two conditions above could be expressed as: An elliptic curve is modular (as a curve) if, and only if, its Jacobian (which is equal to the curve itself) is modular (as an abelian variety).

If one tries to generalize the conditions above to curves of genus $g \geq 2$, one could ask if such a curve C is parametrized over $\mathbb{Q}$ by a modular curve,

$$X_1(N) \overset{/\mathbb{Q}}{\twoheadrightarrow} C,$$

or ask if its Jacobian, $J(C)$, is a factor (up to $\mathbb{Q}$-isogeny) of a modular Jacobian,

$$J_1(N) \overset{\mathbb{Q}}{\sim} J(C) \times A'.$$

In this case, this two conditions happen to be non-equivalent. In fact, the first condition implies the second one, but the converse is not true in general. In other words: If a curve of genus $g \geq 2$ is modular, then its Jacobian is modular, but the converse is not true.

If we look for modular curves of genus $g \geq 2$, there are results of finiteness by Baker-González-González-Poonen (cf. [1]). In this paper we prove that there are infinitely many curves of genus 2 defined over $\mathbb{Q}$ with modular Jacobian.

Supported by grants BFM2000-0794-C02-02, HPRN-CT-2000-00114 and 2002SGR-00148.

1. Dihedral curves of genus 2

We will only consider what we will call dihedral curves of genus 2, that is, curves of genus 2 with group of automorphisms isomorphic to D_8 or D_{12}. We will also restrict ourselves to curves defined over $\mathbb{Q}$.

1.1. Classification over $\overline{\mathbb{Q}}$

Let $\mathcal{M}_2$ be the moduli variety that classifies curves of genus 2 up to isomorphism. The moduli points corresponding to dihedral genus 2 curves form two 1-dimensional subvarieties of $\mathcal{M}_2$, one for the case D_8 and another one for the case D_{12}. In fact, both subvarieties can be parametrized by an absolute invariant and, moreover, given a rational point on these subvarieties one can find a curve of genus 2 defined over $\mathbb{Q}$ that represents the moduli point.

Proposition 1.1. *(cf. [3, Section 2]) The set of points in $\mathcal{M}_2$ corresponding to curves with group of automorphisms isomorphic to D_8 or D_{12} and defined over $\mathbb{Q}$ is parametrized by*

$$t \in \begin{cases} \mathbb{Q} \setminus \{0, 1/4, 9/100\}, & \text{in the } D_8 \text{ case,} \\ \mathbb{Q} \setminus \{0, 1/4, -1/50\}, & \text{in the } D_{12} \text{ case.} \end{cases}$$

Given a curve C, the corresponding parameter t can be obtained as the absolute invariant

$$t = t(C) = \begin{cases} \dfrac{8c_6(6c_4 - c_2^2) + 9c_{10}}{900c_{10}}, & \text{in the } D_8 \text{ case,} \\ \dfrac{3c_4c_6 - c_{10}}{50c_{10}}, & \text{in the } D_{12} \text{ case,} \end{cases}$$

where c_i are the Clebsch invariants. Given such a $t \in \mathbb{Q}$, the curve of genus 2 given by

$$y^2 = x^5 + x^3 + tx,$$

in the D_8 case, or

$$y^2 = x^6 + x^3 + t,$$

in the D_{12} case, is a representative, defined over $\mathbb{Q}$, of the moduli point corresponding to t.

1.2. Classification over $\mathbb{Q}$

Let C be a dihedral curve of genus 2 defined over $\mathbb{Q}$. The group of automorphisms of C is endowed with a natural action of the absolute Galois group of $\mathbb{Q}$, $G_{\mathbb{Q}} = \mathrm{Gal}(\overline{\mathbb{Q}}/\mathbb{Q})$. What we do is classify all possible actions of $G_{\mathbb{Q}}$ on a group A isomorphic to D_8 or D_{12} and determine which of them appear as the Galois action on the group of automorphisms of a genus 2 curve defined over $\mathbb{Q}$; for each of these we find the set of $\mathbb{Q}$-isomorphism classes of dihedral curves of genus 2 with Galois action on its group of automorphisms isomorphic to the given one.

Any action of $G_{\mathbb{Q}}$ on A is determined, up to equivalence, by a morphism $\rho : G_{\mathbb{Q}} \to \mathrm{Aut}(A)$, defined up to inner automorphisms. This action factors through a number field $K = \overline{\mathbb{Q}}^{\ker \rho}$ and also determines actions on certain subgroups and quotients of A. Namely, A has a single quotient isomorphic to V_4 and a single subgroup isomorphic to C_4 (in the case D_8) or C_3 (in the case D_{12}). The induced actions on these quotients and subgroups factor through certain fields $K_u = \mathbb{Q}(\sqrt{u})$ and $K_v = \mathbb{Q}(\sqrt{v})$.

Computing explicitly all the possibilities for ρ, up to inner automorphisms, we can find all possible Galois actions on A. In the following table we give all these possibilities and label each of them according to the structure of $\mathrm{Gal}(K/\mathbb{Q})$ and distinguishing those with isomorphic Galois group with a superindex.

$A \simeq D_8$

Type	K	K_u	K_v
I	$\mathbb{Q}$	$\mathbb{Q}$	$\mathbb{Q}$
$C_2^{(1)}$	K	$\mathbb{Q}$	$\mathbb{Q}$
$C_2^{(2)}$	K	$\mathbb{Q}$	K_v
$C_2^{(3)}$	K	K_u	K_v
C_4	K	K_u	$\mathbb{Q}$
$V_4^{(1)}$	K	$\mathbb{Q}$	K_v
$V_4^{(2)}$	K	K_u	K_v
D_8	K	K_u	K_v

$A \simeq D_{12}$

Type	K	K_u	K_v
I	$\mathbb{Q}$	$\mathbb{Q}$	$\mathbb{Q}$
$C_2^{(1)}$	K	K_u	$\mathbb{Q}$
$C_2^{(2)}$	K	$\mathbb{Q}$	K_v
$C_2^{(3)}$	K	K_u	K_v
C_3	K	$\mathbb{Q}$	$\mathbb{Q}$
C_6	K	K_u	$\mathbb{Q}$
V_4	K	K_u	K_v
$D_6^{(1)}$	K	$\mathbb{Q}$	K_v
$D_6^{(2)}$	K	K_u	K_v
D_{12}	K	K_u	K_v

From these tables it follows the action of $G_{\mathbb{Q}}$ on the group A is determined by the fields K, K_u, K_v.

In the case $A \simeq D_8$, due to our choices for K_u and K_v, the quaternion algebra $(u, -v)$ is trivial in $\mathrm{Br}_2(\mathbb{Q})$ and we can always write the fields in the following way:

$$K = \mathbb{Q}\left(\sqrt{v}, \sqrt{w\frac{1 \pm z\sqrt{u}}{2}}\right), \quad K_u = \mathbb{Q}(\sqrt{u}), \quad K_v = \mathbb{Q}(\sqrt{v}).$$

Then, A is isomorphic (as a $G_{\mathbb{Q}}$-group) to the group of automorphisms of a curve of genus 2 defined over $\mathbb{Q}$ if, and only if, the class of the quaternion algebra $(w, -v)$ in $\mathrm{Br}_2(\mathbb{Q})$ is trivial. Analogously, for the case $A \simeq D_{12}$ the condition is that the quaternion algebra $(u, -3v)$ is trivial in $\mathrm{Br}_2(\mathbb{Q})$ (cf. [3, Section 3]).

If the previous conditions hold, we can take $s, z \in \mathbb{Q}$ such that

$$1 - z^2 u = uvs^2$$

in the case $A \simeq D_8$, and

$$u^3 - z^2 = 3vs^2$$

in the case $A \simeq D_{12}$.

Theorem 1.2. *(cf. [3, Section 4]) Let $t \in \mathbb{Q} \smallsetminus \{0, 1/4, 9/100\}$ (resp. $t \in \mathbb{Q} \smallsetminus \{0, 1/4, -1/50\}$). The set of $\mathbb{Q}$-isomorphism classes of curves of genus 2 with $\mathrm{Aut}(C) \simeq D_8$ (resp. D_{12}) with absolute invariant t is in bijection with the set of tuples (A, e, d), where*

- *A is a $G_{\mathbb{Q}}$-group isomorphic to D_8 (resp. D_{12}) such that $(w,-v)=1\in \mathrm{Br}_2(\mathbb{Q})$ (resp. $(u,-3v)=1\in \mathrm{Br}_2(\mathbb{Q})$) and $K_u=\mathbb{Q}(\sqrt{t})$;*
- *$e\in\{\pm1\}$ if $A\simeq D_8$ and is of type $C_2^{(2)}$, $V_4^{(1)}$ or D_8; otherwise, $e=1$;*
- *$d\in\mathbb{Q}^*/H$, where*

$$H=\begin{cases}\mathbb{Q}^{*2}\cdot(\mathbb{Q}^*\cap K^{*2}), & \text{if } A\simeq D_8 \text{ and } K_u=\mathbb{Q},\\ \mathbb{Q}^{*2}\cdot(\mathbb{Q}^*\cap K_v^{*2}), & \text{if } A\simeq D_8 \text{ and } K_u\neq\mathbb{Q},\\ \mathbb{Q}^{*2}\cdot u\mathbb{Q}^{*2}, & \text{if } A\simeq D_{12} \text{ and } 3\nmid[K:\mathbb{Q}],\\ \mathbb{Q}^{*2}, & \text{if } A\simeq D_{12} \text{ and } 3\mid[K:\mathbb{Q}].\end{cases}$$

Given such a tuple, a curve of genus 2 representing the $\mathbb{Q}$-isomorphism class is given by:

$$C_{D_8}: dY^2=(1+2etz)X^6-8estvX^5+v(3-10etz)X^4 \\ +v^2(3-10etz)X^2+8estv^3X+v^3(1+2etz),$$

$$C_{D_{12}}: dY^2=27(t+2z)X^6-324svX^5+27v(t-10z)X^4 \\ +360sv^2X^3+9v^2(t+10z)X^2-36sv^3X+v^3(t-2z),$$

where the parameters appearing are chosen as in the previous paragraph.

2. $\mathbb{Q}$-curves from dihedral curves of genus 2

Let C be a genus 2 curve and W a subgroup of automorphisms of C. Then C/W is an elliptic curve if, and only if, W is generated by an involution of C different from the hyperelliptic involution. Moreover, two involutions provide isomorphic quotients if, and only if, they are $\mathrm{Aut}(C)$-conjugate. From these remarks we get that dihedral genus 2 curves have two non-isomorphic quotients. Since these isomorphism classes of elliptic quotients depend only on the isomorphism class of C, using the models given by Proposition 1.1, we can use the procedure in [2] to find their j-invariants as a function of t:

$$(j_1,j_2)=\begin{cases}\dfrac{2^6(3\pm10\sqrt{t})^3}{(1\pm2\sqrt{t})(1\mp2\sqrt{t})^2}, & \text{if } \mathrm{Aut}(C)\simeq D_8,\\[2ex] \dfrac{2^8 3^3(2\pm5\sqrt{t})^3(\mp\sqrt{t})}{(1\pm2\sqrt{t})(1\mp2\sqrt{t})^3}, & \text{if } \mathrm{Aut}(C)\simeq D_{12}.\end{cases}$$

By evaluating the second and third modular polynomial at these j-invariants we find that the two non-isomorphic elliptic quotients of a dihedral genus 2 curves are linked by an isogeny of degree 2 (in the D_8 case) or 3 (in the D_{12} case). Note also that the finite set of values of t giving quotients with complex multiplication can be found from the list of orders of quadratic fields with class number equal one or two.

Let $E = C/W$ be an elliptic quotient of a genus 2 curve C defined over $\mathbb{Q}$. Since ${}^{\sigma}E = {}^{\sigma}(C/W) = C/{}^{\sigma}W$, for every $\sigma \in G_{\mathbb{Q}}$, the Galois-conjugates of E are also elliptic quotients of C and, therefore, isogenous to E. In other words, the elliptic quotients of dihedral genus 2 curves defined over $\mathbb{Q}$ are elliptic $\mathbb{Q}$-curves of degree 2 (in the D_8 case) or 3 (in the D_{12} case).

Not only we get in this way "some" $\mathbb{Q}$-curves of degree 2 and 3, but also, in some sense, we get all of them. Recall that $\mathbb{Q}$-curves of prime degree p are parametrized by rational points on the modular curve $X^*(p)$. Using a procedure due to González-Lario (cf. [5]), we can find generators for $\mathbb{Q}(X^*(p))$ when this modular curve has genus zero or one. For the case $p = 2, 3$, we can take as generators:

$$h_0(\tau) = \begin{cases} \dfrac{\eta(\tau)^{24}}{\eta(2\tau)^{24}} + \dfrac{2^{12}\eta(2\tau)^{24}}{\eta(\tau)^{24}}, \\[2ex] \dfrac{\eta(\tau)^{12}}{\eta(3\tau)^{12}} + \dfrac{3^{6}\eta(3\tau)^{12}}{\eta(\tau)^{12}}, \end{cases} \qquad h(\tau) = \begin{cases} \dfrac{h_0(\tau)+104}{4(h_0(\tau)-152)}, & \text{if } p=2, \\[2ex] \dfrac{h_0(\tau)+42}{4(h_0(\tau)-66)}, & \text{if } p=3. \end{cases}$$

Note that $h_0(\tau)$ is the generator one obtains from González-Lario procedure, η being Dedekind's η-function and τ a variable taking values in the upper half complex plane, and $h(\tau)$ is chosen to make the following description easier. Expressing the Galois-conjugated j-invariants $j(\tau)$ and $j(p\tau)$ of two elliptic curves linked by an isogeny of degree p in terms of rational values for the function h, one gets the following expression:

$$(j(\tau), j(p\tau)) = \begin{cases} \dfrac{2^6(3 \pm 10\sqrt{h})^3}{(1 \pm 2\sqrt{h})(1 \mp 2\sqrt{h})^2}, & \text{if } p=2, \\[2ex] \dfrac{2^8 3^3(2 \pm 5\sqrt{h})^3(\mp\sqrt{h})}{(1 \pm 2\sqrt{h})(1 \mp 2\sqrt{h})^3}, & \text{if } p=3. \end{cases}$$

From the identity between the expressions obtained for (j_1, j_2), the j-invariants of the elliptic quotients of a dihedral genus 2 curve, and $(j(\tau), j(p\tau))$, we get the following result.

Theorem 2.1. *Let $E/\overline{\mathbb{Q}}$ be an elliptic curve with no complex multiplication. Then E is a $\mathbb{Q}$-curve of degree 2 (resp. 3) if, and only if, it is covered by some curve of genus 2 defined over $\mathbb{Q}$ with group of automorphisms isomorphic to D_8 (resp. D_{12}).*

3. Abelian varieties of GL_2-type from dihedral curves of genus 2

Let C be a dihedral curve of genus 2 defined over $\mathbb{Q}$. We have seen that C has elliptic quotients; therefore, its Jacobian $J = \mathrm{Jac}(C)$ splits up to isogeny as the product of two elliptic curves. Moreover, since $\mathrm{Aut}(C)$ is non-abelian, from the embedding $\mathrm{Aut}(C) \hookrightarrow \mathrm{End}(J)$ we get that $\mathrm{End}(J)$ is non-abelian and, therefore, the elliptic curves appearing as factors of J are isogenous and, assuming that the elliptic quotients of C have no complex multiplication, we get

$$\mathrm{End}(J) \simeq \mathbf{M}_2(\mathbb{Q}).$$

The image of $\mathrm{Aut}(C)$ in $\mathrm{End}(J)$ generates (as a $\mathbb{Q}$-vector space) the full algebra of endomorphisms; therefore, from the knowledge of the Galois action on $\mathrm{Aut}(C)$ we can compute the Galois action on $\mathrm{End}(J)$ and find the subalgebra of endomorphisms defined over $\mathbb{Q}$.

Theorem 3.1. *Let C be a dihedral curve of genus 2 defined over $\mathbb{Q}$ whose elliptic quotients have no complex multiplication, then the subalgebra $\mathrm{End}_{\mathbb{Q}}(J)$ of endomorphisms of the Jacobian of C defined over $\mathbb{Q}$ depends only on the Galois action on $\mathrm{Aut}(C)$ and is reduced to $\mathbb{Q}$ except in the following cases:*

$\mathrm{Aut}(C) \simeq D_8$			$\mathrm{Aut}(C) \simeq D_{12}$		
Type	$\mathrm{End}_{\mathbb{Q}}(J(C))$	$J(C) \overset{\mathbb{Q}}{\sim}$	*Type*	$\mathrm{End}_{\mathbb{Q}}(J(C))$	$J(C) \overset{\mathbb{Q}}{\sim}$
I	$\mathbf{M}_2(\mathbb{Q})$	E^2	I	$\mathbf{M}_2(\mathbb{Q})$	E^2
$C_2^{(2)}$	$\mathbb{Q}\times\mathbb{Q}$	$E_1 \times E_2$	$C_2^{(2)}$	$\mathbb{Q}\times\mathbb{Q}$	$E_1 \times E_2$
$C_2^{(1)}$	$\mathbb{Q}(\sqrt{-1})$	GL_2-*type*	C_3	$\mathbb{Q}(\sqrt{-3})$	GL_2-*type*
C_4	$\mathbb{Q}(\sqrt{-1})$	GL_2-*type*	$C_2^{(1)}$	$\mathbb{Q}(\sqrt{-3})$	GL_2-*type*
$C_2^{(3)}$	$\mathbb{Q}(\sqrt{2})$	GL_2-*type*	C_6	$\mathbb{Q}(\sqrt{-3})$	GL_2-*type*
			$C_2^{(3)}$	$\mathbb{Q}(\sqrt{3})$	GL_2-*type*

We recall that, given a $\overline{\mathbb{Q}}$-isomorphism class of dihedral curves of genus 2, the condition to exist a $\mathbb{Q}$-isomorphism class with a given Galois action on its group of automorphism is given in Section 1.2. For the cases appearing in the theorem above, the conditions are as follows.

$\mathrm{Aut}(C) \simeq D_8$		$\mathrm{Aut}(C) \simeq D_{12}$	
Type	Condition	Type	Condition
I	$t \in \mathbb{Q}^{*2}$	I	$t \in \mathbb{Q}^{*2}$
$C_2^{(2)}$	$t \in \mathbb{Q}^{*2}$	$C_2^{(2)}$	$t \in \mathbb{Q}^{*2}$
$C_2^{(1)}$	$t \in \mathbb{Q}^{*2}$	C_3	$t \in \mathbb{Q}^{*2}$
C_4	$(t,-1)=1$	$C_2^{(1)}$	$(t,-3)=1$
$C_2^{(3)}$	$(t,2)=1$	C_6	$(t,-3)=1$
		$C_2^{(3)}$	$(t,3)=1$

Since there are infinitely many values of t such that any of the conditions above hold, we get that there are infinitely many non-isomorphic curves of genus 2 with Jacobian of GL_2-type.

4. Results about modularity

Ellenberg and Skinner give in [4] sufficient conditions for a $\mathbb{Q}$-curve to be modular. Using their result it is possible to prove the modularity for an infinite set of quadratic $\mathbb{Q}$-curves of degree 2 or 3.

Proposition 4.1. *Let $t \in \mathbb{Q}^*$ and E_t a $\mathbb{Q}$-curve of degree $p = 2$ or $p = 3$ with j-invariant as in section 2. If*

$$\begin{aligned} \operatorname{ord}_3(t) &\leq 0, \\ \operatorname{ord}_3(t) &\equiv 0 \pmod 2, \\ t \cdot 3^{-\operatorname{ord}_3(t)} &\equiv 1 \pmod 3, \end{aligned}$$

in the case $p = 2$, or

$$\operatorname{ord}_3(t - 7) \geq 4,$$

in the case $p = 3$, then E_t is modular.

Proof. Using Ellenberg-Skinner's criterion (cf. [4, Theorem 5.1]), it is enough to prove that the given $\mathbb{Q}$-curves have potentially good reduction at primes over 3 and that $(b_{E_t})_3$ is trivial.

According to the conditions in the proposition, let

$$t = 3^{-2n}(1 + 3\,h_0/h_1),$$

if $p = 2$, or

$$t = 7 + 3^{n+4}\,h_0/h_1,$$

if $p = 3$, both with $n, h_0, h_1 \in \mathbb{Z}$, $n \geq 0$, $3 \nmid h_1$. Then, computing the norm of the j-invariants j_i, one gets

$$N_{\mathbb{Q}(\sqrt{t})/\mathbb{Q}}(j_i) = \begin{cases} \dfrac{2^{12}(300\,h_0 + 100\,h_1 - 3^{2n+2}\,h_1)^3}{(12\,h_0 + 4\,h_1 - 3^{2n}h_1)^3}, & \text{if } p = 2, \\ \dfrac{2^{16}(3^{n+4}\,h_0 + 7\,h_1)(25 \cdot 3^{n+2}\,h_0 + 19\,h_1)^3}{(4 \cdot 3^{n+1}\,h_0 + h_1)^4}, & \text{if } p = 3. \end{cases}$$

Since in both cases the denominator is not divisible by 3, the corresponding elliptic curves have potentially good reduction.

From the formulas in [6, Theorem 3.1] one gets that

$$b_{E_t} = (t, p)$$

in the two cases $p = 2$ and $p = 3$. Then, using the expressions for t as above, we get

$$(t, 2)_3 = (3, 2)_3^{-2n}(1, 2)_3 = 1,$$

in the case of degree 2, and

$$(t, 3)_3 = (7, 3)_3 = 1,$$

in the case of degree 3. □

As a consequence, we can prove an infiniteness result on the set of modular abelian surfaces.

Corollary 4.2. *There are infinitely many non-isomorphic curves of genus 2 defined over $\mathbb{Q}$ with modular Jacobian.*

Proof. Consider, for example, the infinite set of $t \in \mathbb{Q} \setminus (\{0, 1/4, -1/50\} \cup \mathbb{Q}^{*2})$ such that $(t, -3) = 1$ and $\mathrm{ord}_3(t-7) \geq 4$. For each choice of t, there exists C_t, a dihedral curve of genus 2 defined over $\mathbb{Q}$, and E_t, a quadratic $\mathbb{Q}$-curve completely defined over $\mathbb{Q}(\sqrt{t})$, together with a covering $C_t \to E_t$ also defined over $\mathbb{Q}(\sqrt{t})$. From this setting it follows that the Jacobian $J_t = J(C_t)$ is $\mathbb{Q}$-isogenous to the abelian surface obtained from E_t by Weil's restriction of scalars,

$$J_t \overset{\mathbb{Q}}{\sim} \mathrm{Res}_{\mathbb{Q}(\sqrt{t})/\mathbb{Q}} E_t.$$

Since E_t is modular, also is $\mathrm{Res}_{\mathbb{Q}(\sqrt{t})/\mathbb{Q}} E_t$ and, therefore, J_t is modular. □

References

[1] Matthew Baker, Enrique González, Josep González and Bjorn Poonen. Finiteness results for modular curves of genus at least 2. Preprint, November 2002.

[2] Gabriel Cardona, Josep González, Joan-Carles Lario and Anna Rio. On curves of genus 2 with jacobian of GL_2-type. *Manuscripta Math.*, 98(1):37–54, 1999.

[3] Gabriel Cardona and Jordi Quer. Curves of genus 2 with group of automorphisms isomorphic to D_8 or D_{12}. Submitted, February 2002.

[4] Jordan S. Ellenberg and Chris Skinner. On the modularity of $\mathbb{Q}$-curves. *Duke Math. J.*, 109(1):97–122, 2001.

[5] Josep González and Joan-Carles Lario. Rational and elliptic parametrizations of $\mathbb{Q}$-curves. *J. Number Theory*, 72(1):13–31, 1998.

[6] Jordi Quer. $\mathbb{Q}$-curves and abelian varieties of GL_2-type. *Proc. London Math. Soc.*, 81(3):285–317, 2000.

Gabriel Cardona
Departament de Ciències Matemàtiques i Informàtica
Universitat de les Illes Balears
Ed. Anselm Turmeda, Campus UIB
Ctra. de Valldemossa, km. 7.5
E-07122 – Palma de Mallorca, Spain
e-mail: gabriel.cardona@uib.es

Progress in Mathematics, Vol. 224, 53–74

The old Subvariety of $J_0(NM)$

János A. Csirik

Abstract. Let N and M be distinct prime numbers and let $J_0(NM)$ be the Jacobian of the modular curve $X_0(NM)$. The **old part of** $J_0(NM)$ is defined as the span of the images of certain natural maps from $J_0(N)$ and $J_0(M)$ to $J_0(NM)$. Ribet [16] and Ling [9] determined the old part of $J_0(NM)$ for most pairs (N, M). In this paper, we use the structure of the kernel of the Eisenstein ideal in $J_0(N)$ (and $J_0(M)$) to complete the task of determining the old subvariety of $J_0(NM)$. Our results were previously known in the special case where neither N nor M are congruent to 1 modulo 16.

1. Introduction

For any positive integer n, let $X_0(n)$ denote the modular curve classifying elliptic curves with a cyclic subgroup of order n, and let $J_0(n)$ denote its Jacobian (see [4]).

Let N and M be distinct prime numbers. The curve $X_0(NM)$ has two naturally defined degeneracy morphisms π_1 and π_M to $X_0(N)$ (to be described specifically in Section 3). We thus obtain the abelian subvariety

$$A = \operatorname{im}((\pi_1^*, \pi_M^*) \colon J_0(N)^2 \to J_0(NM))$$

of $J_0(NM)$, and the analogously defined (with M playing the role of N) $B \subseteq J_0(NM)$. The **old part of** $J_0(NM)$ is the abelian subvariety of $J_0(NM)$ generated by A and B. This subvariety is of wide use in the theory of modular curves. For example, its complete description achieved in this paper answers questions posed in [11], [14] and [16].

Each of A and B was determined by Ribet in [14]. Therefore, to complete the description of the old part of $J_0(NM)$, we need to determine $A \cap B$ by exhibiting the points in $A \cap B$ via a simple description unrelated to the method of construction for A and B. This project was carried out partially by Ribet in [16], where he determined the odd part of $A \cap B$. In [9], Ling went on to obtain results about the even part of $A \cap B$ when neither N nor M is congruent to 1 modulo 16.

In Section 3, we prove that the module $A \cap B$ is contained in the image (of the cartesian square) of a certain submodule, namely the **kernel of the Eisenstein**

I thank Ken Ribet for many helpful conversations and for suggesting this problem to me. Some preliminary calculations were carried out using the computer program PARI–GP [1].

ideal, of $J_0(N)$ (to be described specifically in Section 2). The kernel of the Eisenstein ideal was first identified and studied by Mazur in [10]. The Galois structure and specific construction of that module was not completely known before it was determined in [2]. In Section 3 we build on the results of that paper to obtain

Theorem 1.1. *Let N and M be distinct primes. Let $n = (N-1)/\gcd(N-1,12)$, let $m = (M-1)/\gcd(M-1,12)$, and let $D^{-,-} = P_1 - P_N - P_M + P_{NM} \in J_0(NM)$. Then $A \cap B$ is the unique subgroup of order $\gcd(n,m)$ of the cyclic group generated by $D^{-,-}$.*

The symbols P_1, P_N, P_M and P_{NM} are the conventional names for the cusps of $X_0(NM)$. Their meaning will be explained in Section 3.

Section 4 contains some modular unit calculations that are used in Section 3.

This paper is based on the author's 1999 UC Berkeley thesis. The author thanks Ken Ribet for suggesting this problem and many helpful conversations, and an anonymous referee for numerous helpful suggestions. The software package PARI/GP [1] was used for some exploratory calculations.

2. The kernel of the Eisenstein ideal

In this section, we will summarize briefly some properties of the modular curves we will be using (for more details, see [4]), and we will explain the setup and relevant results of [2], which will be used frequently in later sections.

Let N be a positive integer. We shall consider the usual modular curves $X_0(N)$, $X_1(N)$ and $X(N)$, and their Jacobians $J_0(N)$, $J_1(N)$ and $J(N)$. These correspond to the moduli problems of classifying an elliptic curve with a cyclic subgroup of order N, an elliptic curve with a point of order N, and an elliptic curve with an embedding of $\mu_N \times \mathbb{Z}/N\mathbb{Z}$ (with its natural pairing) that is compatible with the Weil pairing, respectively. These curves are all defined over $\mathbb{Q}$, as are the usual degeneracy maps (which are Galois coverings) α: $X(N) \to X_1(N)$, β: $X_1(N) \to X_0(N)$ and $\gamma = \beta \circ \alpha$.

The curves $X_0(N)_{\mathbb{C}}$, $X_1(N)_{\mathbb{C}}$ and $X(N)_{\mathbb{C}}$ can also be regarded as compactified quotients of the complex upper half-plane $\mathfrak{H}^*/\Gamma_0(N)$, $\mathfrak{H}^*/\Gamma_1(N)$ and $\mathfrak{H}^*/\Gamma(N)$, respectively, where

$$\begin{aligned}
\Gamma_0(N) &= \left\{ \begin{pmatrix} a & b \\ c & d \end{pmatrix} \in \mathrm{SL}_2\mathbb{Z} \colon c \equiv 0 \pmod{N} \right\}, \\
\Gamma_1(N) &= \left\{ \begin{pmatrix} a & b \\ c & d \end{pmatrix} \in \mathrm{SL}_2\mathbb{Z} \colon c \equiv 0, d \equiv 1 \pmod{N} \right\}, \\
\Gamma(N) &= \left\{ \begin{pmatrix} a & b \\ c & d \end{pmatrix} \in \mathrm{SL}_2\mathbb{Z} \colon a \equiv 1, b \equiv 0, c \equiv 0, d \equiv 1 \pmod{N} \right\},
\end{aligned}$$

and these subgroups of $\mathrm{SL}_2\mathbb{Z}$ act on the complex upper half-plane $\mathfrak{H}$ via fractional linear transformations. The points introduced during the compactification are called cusps.

Now let N be an odd prime number, and let

$$r = (N-1)/2.$$

The curve $X_0(N)$ has two cusps, denoted 0 and ∞. They are both defined over $\mathbb{Q}$ and are distinguished by the fact that under the natural map $X_0(N)_{\mathbb{C}} = \mathfrak{H}^*/\Gamma_0(N) \to X(1)_{\mathbb{C}} = \mathfrak{H}^*/\mathrm{SL}_2\mathbb{Z}$, the cusp 0 is ramified with index N and the cusp ∞ is unramified.

The curve $X_1(N)$ has $N-1$ cusps that come in two groups. We shall use Klimek's notation in [7] for them. The cusps $P_1, P_2, \dots, P_r$ are defined over $\mathbb{Q}$ and are mapped to 0 under β: $X_1(N) \to X_0(N)$. The cusps $Q_1, Q_2, \dots, Q_r$ are defined over $\mathbb{Q}(\mu_N)^+$ (the maximal totally real subfield of the Nth cyclotomic field) and are mapped to ∞ under β. All the cusps of $X_1(N)$ are unramified with respect to β. The curve $X(N)$ has $(N^2-1)/2$ cusps and we use Shimura's notation in [13] to regard them as pairs $\pm\binom{x}{y}$ with $x, y \in \mathbb{F}_N$, not both equal to 0. In this representation, $\mathrm{Gal}(X(N)_{\mathbb{C}}/X_0(N)_{\mathbb{C}}) \cong \mathrm{PSL}_2\mathbb{F}_N$ acts naturally from the left. For $1 \le i \le r$, the cusps $\binom{*}{i}$ are all defined over $\mathbb{Q}(\mu_N)$ and map unramifiedly to P_i under α: $X(N) \to X_1(N)$. For $1 \le i \le r$, the cusps $\binom{i}{0}$ are all defined over $\mathbb{Q}(\mu_N)^+$ and map to Q_i under α with ramification index N.

Shimura's notation can be used to label the cusps of any modular curve. We shall now provide the translations to Shimura's system of all the names we use. On the curve $X_0(N)$, the cusps 0 and ∞ (respectively) are called $\binom{0}{1}$ and $\binom{1}{0}$ (respectively). On the curve $X_1(N)$, for any $1 \le t \le r$, our notation P_t corresponds to $\binom{0}{t}$, while Q_t corresponds to $\binom{t}{0}$.

The variety $J_0(N)$ possesses certain naturally defined endomorphisms T_ℓ (for all primes $\ell \ne N$) and w. These endomorphisms together with $\mathbb{Z}$ (the multiplications by integers) generate the Hecke ring $\mathbb{T}_N$ of endomorphisms of $J_0(N)$. In his celebrated article **Modular Curves and the Eisenstein Ideal** [10], Mazur defined the Eisenstein ideal I in $\mathbb{T}_N$ as the ideal generated by $1+w$ and the $1+\ell-T_\ell$ and used it to identify the possible rational torsion subgroups of elliptic curves defined over the rational numbers. The Galois module $J_0(N)(\overline{\mathbb{Q}})[I]$ plays an important role in [10] and later studies of the arithmetic geometry of the curve $X_0(N)$.

Recall that

$$n = (N-1)/\gcd(N-1, 12).$$

Mazur proved that

$$J_0(N)[I] \cong \mathbb{Z}/n\mathbb{Z} \times \mathbb{Z}/n\mathbb{Z}$$

as groups. The group $J_0(N)[I]$ has two noteworthy Galois-invariant subgroups. The cuspidal subgroup C is generated by the divisor $c = 0 - \infty$ (the formal difference of the two cusps of $X_0(N)$). The group C is cyclic of order n and is pointwise fixed by $\mathrm{Gal}(\overline{\mathbb{Q}}/\mathbb{Q})$. The Shimura subgroup Σ is a finite flat subgroup scheme of $J_0(N)$ such that

$$\Sigma(\overline{\mathbb{Q}}) = \ker(\beta^*\colon J_0(N) \to J_1(N)),$$

where β^* is induced by the degeneracy map β: $X_1(N) \to X_0(N)$ defined above. The group Σ is also cyclic of order n, but is isomorphic to μ_n as a group scheme.

In [2], we find an explicit construction of $J_0(N)[I]$ which identifies the action of $\mathrm{Gal}(\overline{\mathbb{Q}}/\mathbb{Q})$ on $J_0(N)[I]$ as follows. For any positive integer k, let χ_k denote the kth cyclotomic character $\mathrm{Gal}(\overline{\mathbb{Q}}/\mathbb{Q}) \to (\mathbb{Z}/k\mathbb{Z})^\times$ obtained via the identification $\mathrm{Gal}(\mathbb{Q}(\mu_k)/\mathbb{Q})) \cong (\mathbb{Z}/k\mathbb{Z})^\times$.

Theorem 2.1 (Thm. 1.1 of [2]**).** *$J_0(N)[I]$ has a basis e_1, e_2 over $\mathbb{Z}/n\mathbb{Z}$ such that*

a) $c = e_1 + 2e_2$;

b) e_1 *generates* Σ;

c) $\sigma \in \mathrm{Gal}(\overline{\mathbb{Q}}/\mathbb{Q})$ *acts via left multiplication by*

$$\begin{pmatrix} \chi_n(\sigma) & (1-\chi_{2n}(\sigma))/2 \\ 0 & 1 \end{pmatrix}$$

with respect to the given basis $e_1 = \binom{1}{0}$, $e_2 = \binom{0}{1}$.

If n is odd (equivalently $N \not\equiv 1 \pmod 8$) then $C \cap \Sigma = 0$, so $J_0(N)[I] \cong C \oplus \Sigma$ and this theorem is easy.

We shall now outline the construction of $J_0(N)[I]$ for $N \equiv 1 \pmod 8$.

Definition 2.2. *Let C_N denote the group $(\mathbb{Z}/N\mathbb{Z})^\times / \pm 1$.*

Since β: $X_1(N)_{\mathbb{C}} \to X_0(N)_{\mathbb{C}}$ is a cyclic Galois covering of degree r, it has a unique intermediate covering of $X_0(N)_{\mathbb{C}}$ of any degree dividing r. Because of its uniqueness, any such curve is defined over $\mathbb{Q}$ (for a thorough treatment of these intermediate curves, see [3, IV, §3]). We know from [10, II, §2] that the intermediate curve $X_2(N)_{\mathbb{C}} \to X_0(N)_{\mathbb{C}}$ of degree n (the **Shimura covering**) is the largest étale covering of $X_0(N)_{\mathbb{C}}$ through which β factors. (As remarked before, the cusps of $X_0(N)$ are not branch points for β; it is the points with $j = 0$ and $j = 1728$ that ramify in β.)

Definition 2.3. *Write n as*

$$n = 2^k v$$

where 2^k is the largest power of 2 that divides n (and $v = n/2^k$ is an odd integer). Set $z = 3$ if $N \equiv 1 \pmod 3$ and $z = 1$ otherwise. For future use, we also set $q = 3/z$ here.

Let

$$\phi\colon X_0^\#(N) \to X_0(N)$$

be the unique covering of degree 2^k through which β factors.

Let

$$J_0^\#(N) = \mathrm{Jac}(X_0^\#(N)).$$

Observe that the definitions of k, v, z imply that

$$r = 2^{k+1} z v.$$

Since 2^k divides n, the Shimura covering factors through ϕ. This implies that ϕ is étale and that $\Sigma_0 = \ker(\phi^*\colon J_0(N) \to J_0^\#(N))$ is contained in $\Sigma = \ker(J_0(N) \to J_2(N))$.

The Galois group $X_1(N)_{\mathbb{C}}$ over $X_0(N)_{\mathbb{C}}$ is isomorphic to C_N, with $\left(\begin{smallmatrix} s & f \\ Nh & t \end{smallmatrix}\right)$ mapping to the image of t in C_N. Let ν be a generator of C_N. We will abuse notation to lighten it, and let the same ν denote the generator of

$$\mathrm{Gal}(X_1(N)_{\mathbb{C}}/X_0(N)_{\mathbb{C}}) \cong C_N$$

and the corresponding generator of the Galois group of the function field extension; so that $(\nu f)(\tau) = f(\nu\tau)$ for all functions f on $X_1(N)$. Let Ω denote the set of 2^kth powers in C_N. Then Ω is the Galois group of $X_1(N)$ over $X_0^{\#}(N)$ and

$$\#\Omega = 2zv.$$

By its uniqueness property, $X_0^{\#}(N)$ is Galois over $X_0(N)$.

The curve $X_0^{\#}(N)$ is the coarse moduli space for the problem of classifying elliptic curves with a point of order N, where (E,P) and (E',P') are to be considered equivalent if there is an isomorphism δ: $E \to E'$ such that $\delta(P) = \pm b \cdot P'$ for some $b \in \Omega$.

In [2], we find a careful analysis of **units** (rational functions whose divisors are concentrated at the cusps) of $X_1(N)$ and $X_0^{\#}(N)$ which results, in [2, Thm. 4.11], in the construction of a unit f of $X_0^{\#}(N)$ whose divisor $\mathrm{div}(f)$ is divisible by 2^k in $\mathrm{Div}^0(X_0^{\#}(N))$. It is then shown that if $\mathcal{D}$ is the group generated by $d = \mathrm{div}(f)/2^k$ in $J_0^{\#}(N)$ then $A = (\phi^*)^{-1}\mathcal{D}$ is exactly the odd part of $J_0(N)[I]$. The explicitness of the construction then allows the Galois action to be determined as in Theorem 2.1.

3. The old subvariety of $J_0(NM)$

Let N and M be distinct primes. The modular curve $X_0(NM)$ classifies elliptic curves equipped with a subgroup of order NM. (Recall from elementary group theory that all abelian groups of order NM are cyclic.) Forgetting all but the N-primary part of such a subgroup yields a subgroup of order N, thereby giving rise to a degeneracy map

$$\pi_1\colon X_0(NM) \to X_0(N),$$

which is defined over $\mathbb{Q}$.

The Atkin–Lehner involution

$$w_M\colon X_0(NM) \to X_0(NM)$$

is another morphism defined over $\mathbb{Q}$ that can be associated to a moduli-theoretic operation. The operation is as follows. Let (E,C) denote a pair of an elliptic curve E and a subgroup C of order NM. Let C_N (respectively, C_M) be the N-primary (respectively, M-primary) subgroup of C, so that $C = C_N \times C_M$. Let γ: $E \to E'$ be the degree M isogeny with kernel C_M. Then it is easy to see that the group $\gamma(E[M])$ is cyclic of order M, and that the group $\gamma(C_N)$ is cyclic of order N. Then the operation we are looking for sends the pair (E,C) to the pair $(E', \gamma(E[M]) \times \gamma(C_N))$.

On $X_0(NM)_{\mathbb{C}} = \mathfrak{H}^*/\Gamma_0(NM)$, the map w_M is induced by the action of the matrix $\begin{pmatrix} -aN & b \\ -NM & M \end{pmatrix}$ acting on $\mathfrak{H}^*$ via the corresponding Möbius transformation, where a and b are integers chosen so that $bN - aM = 1$.

The curve $X_0(NM)$ has four cusps, called P_1, P_N, P_M and P_{NM} in the notation of [16]. The morphism $\pi_1\colon X_0(NM) \to X_0(N)$ sends P_1 and P_M to the cusp 0 on $X_0(N)$, with ramification indices M and 1, respectively. The map π_1 also sends P_N and P_{NM} to the cusp ∞ of $X_0(N)$, with ramification indices M and 1, respectively. The morphism w_M swaps the points P_1 and P_M, and it also swaps the points P_N and P_{NM}.

Now we can define the degeneracy map

$$\pi_M\colon X_0(NM) \to X_0(N)$$

by setting $\pi_M = \pi_1 \circ w_M$. This allows us to define the morphisms

$$\alpha = (\pi_1^*, \pi_M^*) \quad : \quad J_0(N) \times J_0(N) \to J_0(NM)$$

and

$$\beta = (\pi_1^*, \pi_N^*) \quad : \quad J_0(M) \times J_0(M) \to J_0(NM).$$

Since w_M^2 is the identity morphism on $X_0(NM)$, we can see that $w_M^*\alpha(R_1, R_2) = \alpha(R_2, R_1)$. (Note that we shall use w_M to denote w_M^* where it is not likely to cause confusion.)

Definition 3.1. *Let* $A = \operatorname{im}(\alpha)$ *and* $B = \operatorname{im}(\beta)$. *The* **old part** *of* $J_0(NM)$ *is the abelian subvariety of* $J_0(NM)$ *that is generated by* A *and* B.

Since $J_0(M)$ has good reduction everywhere away from M and β is defined over $\mathbb{Q}$, we can conclude that B has good reduction away from M. Similarly, A has good reduction away from N. Therefore, $A \cap B$ has good reduction everywhere and is finite (see [16, Theorem 3] and [6]).

It was proved in [14] that

$$\ker(\alpha) = \{(P, -P) \in J_0(N)^2 \colon P \in \Sigma\}.$$

We shall paraphrase this fact by saying that "the kernel of α is Σ, embedded anti-diagonally". The analogous result of course also holds for β.

This completes the description of A and B. To complete the description of the old part of $J_0(NM)$, we need to find $A \cap B$. The odd part of $A \cap B$ was found in [16]. We shall complete that description by proving Theorem 1.1. We shall proceed through a sequence of lemmas.

Lemma 3.2. *The preimage of the group* $A \cap B$ *under the map* α *is contained in* $J_0(N)[I]^2$.

Proof. Let $x \in J_0(N)^2$ be such that $\alpha(x) \in A \cap B$. Then there is some $y \in J_0(M)^2$ such that $\beta(y) = \alpha(x)$. The point $\beta(y) = \alpha(x)$ has finite order, since it is contained in the finite group $A \cap B$. Since B has good reduction at N, the point $\beta(y)$ must be unramified at N unless its order is divisible by N. But that cannot happen, since, by [16, Corollary 1], the odd part of the order of $\beta(y)$ must divide the quantity[1]

$$\gcd(\mathrm{num}((N-1)/24), \mathrm{num}((M-1)/24)),$$

which is not divisible by N. Therefore, $\beta(y) = \alpha(x)$ is unramified at N. Now, since $\ker(\alpha) = \Sigma$ (embedded anti-diagonally), this implies, as in [10, II, Lemma (16.5)], that x is unramified at N. This in turn implies by [17, Prop. 3.3] that $x \in J_0(N)[I]^2$ as required. □

Lemma 3.3. *The image of the divisor $c = 0 - \infty$ on $X_0(N)$ under the map $\pi_1^* - \pi_M^*$ is the divisor $(M-1)D^{-,-} = (M-1)(P_1 - P_N - P_M + P_{NM})$.*

Proof. This is immediate using the action of w_M on $X_0(NM)$ and the ramification indices of the cusps of $X_0(NM)$ over the cusps of $X_0(N)$. Indeed,

$$\pi_1^*(c) = \pi_1^*(0-\infty) = MP_1 + P_M - MP_N - P_{NM}$$

and

$$\pi_M^*(c) = \overline{w}_M^*(\pi_1^*(c)) = MP_M + P_1 - MP_{NM} - P_N.$$

The difference of the two displayed lines yields the result. □

If $M < 5$ then $B = 0$ and Theorem 1.1 is true. Therefore, we may assume from now on that $M \geq 5$.

Definition 3.4. *Let P denote the point e_2 in Theorem 2.1. If $N \equiv 1 \pmod 8$, let z be the non-trivial 2-torsion element in Σ.*

If $N \equiv 1 \pmod 8$ then $\Sigma[2] = C[2]$. Thus, we can express z as

$$z = \frac{n}{2} \cdot c. \tag{1}$$

Now that we have established that the image of $J_0(N)[I]^2$ is the only thing we need to consider when determining $A \cap B$, we prove a lemma that allows us to focus on an even smaller set.

Lemma 3.5. *Let R be any element of $J_0(N)[I]^2$ that maps to $A \cap B$ under α.*

- *If $N \not\equiv 1 \pmod 8$ then R is an integral multiple of $\pi_1^*(P) - \pi_M^*(P)$.*
- *If $N \equiv 1 \pmod 8$ then R lies in the $\mathbb{Z}$-linear span of $\pi_1^*(P) - \pi_M^*(P)$ and $\pi_1^*(z)$.*

Proof. Note that w_M acts on $A \cap B$ as multiplication by -1. Indeed, it is well known that $\beta \circ w_M = w_M \circ \beta$ (see for example [16, "Formulaire"]), and by (the M-analog of) Lemma 3.2, we know that w_M acts as -1 on $\beta^{-1}(A \cap B)$. Therefore,

$$-\alpha(R_1, R_2) = w_M \alpha(R_1, R_2) = \alpha(R_2, R_1)$$

and hence

$$0 = \alpha(R_1 + R_2, R_1 + R_2).$$

[1]For any non-zero rational number x, let $\mathrm{num}(x)$ denote the numerator of x, that is, the smallest positive integer n such that n/x is an integer.

Once again using the fact that $\ker(\alpha) = \Sigma$ (embedded anti-diagonally), we conclude that $R_1 + R_2$ is an element of $\Sigma[2]$. We now complete the proof by examining each of the possibilities that this allows for $R_1 + R_2$.

In case $R_1 + R_2 = 0$, we can write $R_1 = aP + \sigma$ for some integer a and some $\sigma \in \Sigma$. Then

$$\alpha(R_1, R_2) = \alpha(aP + \sigma, -aP - \sigma) = a\alpha(P, -P) = a(\pi_1^*(P) - \pi_M^*(P)).$$

If $N \equiv 1 \pmod 8$ then $\Sigma[2] \neq 0$, and therefore it can also happen that $R_1 + R_2 = z$. Then we can write $R_1 = z + aP + \sigma$ for some integer a and some $\sigma \in \Sigma$. Then $R_2 = -aP - \sigma$ and

$$\alpha(R_1, R_2) = \alpha(z, 0) + a\alpha(P, -P) = \pi_1^*(z) + a(\pi_1^*(P) - \pi_M^*(P)).$$

In either case, the lemma is proved. □

Let us assume from now on that

$$\boxed{N \equiv 1 \pmod 8.}$$

In the rest of this section, we shall frequently use the following argument to prove that various points on $J_0(NM)$ are not equal. Let J be an abelian variety defined over $\mathbb{Q}$ (for example, $J_0(NM)$). Given a prime number p, we can consider J to be defined over $\mathbb{Q}_p$ so that we can look at its Néron model $\mathcal{J}/\mathbb{Z}_p$.

For any point Q in $J(\mathbb{Q})$, we can consider it an element of $J(\mathbb{Q}_p)$. By the Néron mapping property, Q even extends to an element of $\mathcal{J}(\mathbb{Z}_p)$. Reducing modulo p, we can then regard Q as an element of $\mathcal{J}_p(\mathbb{F}_p)$, where $\mathcal{J}_p$ denotes the special fiber of $\mathcal{J}$. The scheme $\mathcal{J}_p$ is a group scheme over $\mathbb{F}_p$, but it is not necessarily irreducible. Let $\Phi_p(J)$ denote its group of irreducible components. Given our point $Q \in \mathcal{J}_p(\mathbb{F}_p)$, we can check which element of $\Phi_p(J)$ it maps to. The argument we shall often use proceeds as follows. Consider the function

$$\varpi\colon J(\mathbb{Q}) \to \Phi_p(J)$$

described above. Let $P, Q \in J(\mathbb{Q})$. If $\varpi(P) \neq \varpi(Q)$ then we must also have $P \neq Q$. Usually we shall show that $\varpi(P) \neq \varpi(Q)$ by first showing that $\varpi(P) = 0$ (that is to say, P maps to the component of the identity under reduction modulo p), and then showing that $\varpi(Q) \neq 0$.

This method can work only if $\Phi_p(J)$ has more than one element, which in turn can happen only if J has bad reduction at p. Since $J_0(NM)$ has good reduction away from N and M, we will always choose $p = N$ or $p = M$.

Now we prepare the ground for this method by proving that various points are defined over $\mathbb{Q}$ and by studying the components they map to when reduced modulo N or modulo M.

Lemma 3.6. *The points $\pi_1^*(P) - \pi_M^*(P)$ and $\pi_1^*(z)$ are defined over $\mathbb{Q}$, and hence so is every point of $A \cap B$.*

Proof. By Lemma 3.5, (and the fact that $J_0(NM)$ is defined over $\mathbb{Q}$), $\pi_1^*(P) - \pi_M^*(P)$ and $\pi_1^*(z)$ being defined over $\mathbb{Q}$ does imply that every point of $A \cap B$ is defined over $\mathbb{Q}$.

The point c is defined over $\mathbb{Q}$, and hence so is $z = (n/2)c$. Since the degeneracy map π_1 is also defined over $\mathbb{Q}$, we conclude that $\pi_1^*(z)$ is defined over $\mathbb{Q}$.

By [14, Theorem 4.3], the map $\pi_1^* - \pi_M^* : J_0(N) \to J_0(NM)$ (which is defined over $\mathbb{Q}$) factors through $J_0(N)/\Sigma \to J_0(NM)$ (also defined over $\mathbb{Q}$). Since the image of P in $J_0(N)/\Sigma$ is also defined over $\mathbb{Q}$ by Theorem 2.1, it follows that $\pi_1^*(P) - \pi_M^*(P)$ is defined over $\mathbb{Q}$. □

Lemma 3.7. *Every point of $A \cap B$ reduces to the identity component of $J_0(NM)$ modulo N (and modulo M).*

Proof. By symmetry, it suffices to prove the claim for the reduction modulo N.

The variety B has good reduction modulo N, therefore any point in $A \cap B \subseteq B(\mathbb{Q})$ reduces to the identity component modulo N. □

Lemma 3.8. *The point $\pi_1^*(z)$ does not reduce to the identity component of $J_0(NM)$ modulo N.*

Proof. Recall from (1) that $z = (n/2)c$. This will enable us to use the results of [10, Appendix I] to determine exactly where in the component group of $J_0(N)$ modulo N the point $\pi_1^*(z)$ will map. We have

$$\begin{aligned}\pi_1^*(z) &= \pi_1^*(\frac{n}{2}(0-\infty)) = \frac{n}{2}(MP_1 + P_M - MP_N - P_{NM}) = \\ &= \frac{n}{2}(M+1)(\overline{0} - \overline{\infty}).\end{aligned}$$

Here $\overline{0}$ (respectively $\overline{\infty}$) means a cusp of $X_0(NM)$ that reduces to the same irreducible component as the cusp 0 (respectively ∞) modulo N.

Using the notation of [10, Appendix I], we have $u \in \{0,1\}$ and $v \in \{0,1\}$, with

$$\begin{aligned} u = 1 &\iff N \equiv 7 \text{ or } 11 \pmod{12} \text{ and } M \equiv 1 \pmod 4 \\ v = 1 &\iff N \equiv 5 \text{ or } 11 \pmod{12} \text{ and } M \equiv 1 \pmod 3. \end{aligned}$$

We have assumed that $N \equiv 1 \pmod 8$, which excludes the possibilities $N \equiv 7, 11 \pmod{12}$. Therefore we must have $u = 0$, and

$$v = 1 \iff N \equiv 2 \pmod 3 \text{ and } M \equiv 1 \pmod 3.$$

Now that we have a good grip on the pair (u, v), we can look up in the table of [10, Appendix I] the order of the divisor $\overline{0} - \overline{\infty}$ in the component group of $J_0(NM)$ modulo N. The table below summarizes the possibilities (for brevity, we use the notation $x = (N-1)(M+1)$).

$N \pmod 3$	$M \pmod 3$	n	$\mathrm{ord}(\overline{0} - \overline{\infty})$	$\pi_1^*(z)$
1	1, 2	$(N-1)/12$	$x/12$	$x(\overline{0} - \overline{\infty})/24$
2	1	$(N-1)/4$	$x/4$	$x(\overline{0} - \overline{\infty})/8$
2	2	$(N-1)/4$	$x/12$	$x(\overline{0} - \overline{\infty})/8$

In each case, we can see that $\pi_1^*(z)$ does not reduce to the identity. This completes the proof. □

Lemma 3.9. *The point $D^{-,-}$ reduces to the identity component of $J_0(NM)$ modulo N and modulo M.*

Proof. Considering $D^{-,-}$ in the group of components modulo N, we have

$$D^{-,-} = P_1 - P_N - P_M + P_{NM} = \overline{0} - \overline{\infty} - \overline{0} + \overline{\infty} = 0.$$

The same proof works modulo M. □

Lemma 3.10. *The point $(\pi_1^* - \pi_M^*)P$ reduces to the identity component of $J_0(NM)$ modulo N.*

Proof. This is a direct consequence of the fact that $\pi_1^* - \pi_M^*$ annihilates the component group of $J_0(N)$. (See [15, Theorem 2] and [5, proof of Théorème 1].) □

In view of Lemma 3.5, the following theorem shall bring us closer to our goal of proving Theorem 1.1.

Theorem 3.11. *The divisor class of $(\pi_1^* - \pi_M^*)(vP)$ on $X_0(NM)$ is equal to the divisor class of $\frac{v(M-1)}{2}D^{-,-}$.*

REMARK. Note that since $\pi_1^* - \pi_M^*$ annihilates $\Sigma \subset J_0(N)$ and P is "half of c modulo Σ", the Theorem 3.11 is consistent with the statement of the Lemma 3.3. However, the Theorem does not follow yet, since there are numerous ways in which we could take half of $v(M-1)D^{-,-}$, only one of which is consistent with the theorem.

Since the definition of P involved the cover $X_0^{\#}(N)$ of $X_0(N)$, we must pause here to define a cover of $X_0(NM)$ that will allow us to determine $(\pi_1^* - \pi_M^*)P$.

Let $X_{1,0}(N, M)$ denote the modular curve corresponding to the congruence group $\Gamma_1(N) \cap \Gamma_0(M)$. The curve $X_{1,0}(N, M)$ is defined over $\mathbb{Q}$ and corresponds to the moduli problem of classifying elliptic curves equipped with a point of order N and a subgroup of order M. The natural degeneracy map

$$X_{1,0}(N, M) \to X_0(NM)$$

will be denoted by β, by an abuse of notation that is intended to remind the reader to the similarity of this map to the previously defined β: $X_1(N) \to X_0(N)$. Specifically, both maps β have the moduli-theoretic interpretation of taking a point of order N and replacing it by the subgroup of order N that it generates.

There is also a degeneracy map

$$\pi_1 \colon X_{1,0}(N, M) \to X_1(N),$$

corresponding to the natural transformation "forget the level M structure" between the corresponding moduli functors. Then we have a commutative diagram of curves and maps defined over $\mathbb{Q}$ as follows:

$$\begin{array}{ccc} X_{1,0}(N,M) & \xrightarrow{\beta} & X_0(NM) \\ \downarrow{\scriptstyle \pi_1} & & \downarrow{\scriptstyle \pi_1} \\ X_1(N) & \xrightarrow{\beta} & X_0(N). \end{array}$$

The curve $X_{1,0}(N,M)$ has $2N-2$ cusps, which shall be denoted P_i^0, P_i^∞, Q_i^0, Q_i^∞, where the index i is allowed to range over $1, 2, \ldots, r$. The map π_1: $X_{1,0}(N,M) \to X_1(N)$ sends the points P_i^0 and P_i^∞ to the point P_i, with ramification indices M and 1, respectively. Similarly, the same map π_1 sends the points Q_i^0 and Q_i^∞ to the point Q_i, with ramification indices M and 1, respectively.

The map β: $X_{1,0}(N,M) \to X_0(NM)$ takes the cusps P_1^0, P_2^0, ..., P_r^0 to the cusp P_1 of $X_0(NM)$; P_1^∞, P_2^∞, ..., P_r^∞ are sent to P_M; Q_1^0, Q_2^0, ..., Q_r^0 go to the cusp P_N; Q_1^∞, Q_2^∞, ..., Q_r^∞ all map to P_{NM}. None of the four cusps of $X_0(NM)$ is a branch point of β.

Since β: $X_{1,0}(N,M) \to X_0(NM)$ is a cyclic Galois covering of degree r, it has a unique intermediate covering of $X_0(NM)$ of any degree dividing r (and this intermediate covering is also defined over $\mathbb{Q}$). As in the definition of $X_0^\#(N)$, we can use this fact to define the curve $X_0^\#(N,M)$.

Definition 3.12. *Let*

$$\phi\colon X_0^\#(N,M) \to X_0(NM)$$

be the unique covering of degree 2^k that factors through

$$\beta\colon X_{1,0}(N,M) \to X_0(NM).$$

Let $J_0^\#(N,M) = \mathrm{Jac}(X_0^\#(N,M))$.

The curve $X_0^\#(N,M)$ has a moduli interpretation similar to that of $X_0^\#(N)$, with an extra subgroup of order M thrown in.

We have again a degeneracy map π_1: $X_0^\#(N,M) \to X_0^\#(N)$ (corresponding to forgetting the level M structure) that makes the following diagram commute

$$\begin{array}{ccc} X_0^\#(N,M) & \xrightarrow{\phi} & X_0(NM) \\ \downarrow{\scriptstyle \pi_1} & & \downarrow{\scriptstyle \pi_1} \\ X_0^\#(N) & \xrightarrow{\phi} & X_0(N). \end{array} \tag{2}$$

The curve $X_0^\#(N,M)$ has 2^{k+2} cusps. Every cusp of $X_0(NM)$ has 2^k cusps of $X_0^\#(N,M)$ lying over it. To simplify the notation, we shall refer to a cusp of $X_0^\#(N,M)$ by the name of any cusp of $X_{1,0}(N,M)$ that lies over it.

We can now use $X_0^\#(N,M)$ to deal with π_1^*P, but in order to study $\pi_M^*P = w_M^*\pi_1^*P$, we shall need an analog of w_M on the curve $X_0^\#(N,M)$. Let

$$\lambda\colon X_0^\#(N,M) \to X_0^\#(N,M)$$

be the morphism induced by the action of the matrix $\begin{pmatrix} Mb & 1 \\ aNM & M \end{pmatrix}$ on the complex upper half-plane (here again, a and b are integers chosen so that $bM - aN = 1$). The morphism λ has the following moduli interpretation: if (E, P_N, C_M) is a triplet of an elliptic curve E, a point $P \in E$ of order N, and a cyclic subgroup $C_M \subset E[M]$ of order M corresponding to a point Q of $X_0^{\#}(N, M)$, then λQ corresponds to the triplet $(E/C_M, P_N/C_M, E[M]/C_M)$.

We now let $\overline{w}_M = \lambda^{-1}$. The morphism $\overline{w}_M$ covers w_M in the sense that the following diagram is commutative:

$$\begin{array}{ccc} X_0^{\#}(N,M) & \xrightarrow{\overline{w}_M} & X_0^{\#}(N,M) \\ \downarrow{\scriptstyle \pi_1} & & \downarrow{\scriptstyle \pi_1} \\ X_0(NM) & \xrightarrow{w_M} & X_0(NM). \end{array}$$

(One might notice that λ also covers w_M and wonder why we set $\overline{w}_M = \lambda^{-1}$ instead of $\overline{w}_M = \lambda$. The reason is that although $\overline{w}_M = \lambda$ would also work, the function we would have to consider to prove Theorem 3.14 would be much more complicated.)

The morphism $\overline{w}_M$ sends the cusps P_i^0, P_i^∞, Q_i^0, Q_i^∞ (respectively) to the cusps P_i^∞, $P_{i/M}^0$, Q_{Mi}^∞, Q_i^0 (respectively), where i/M is understood to mean division modulo N and where all indices are to be taken in C_N.

We now have the tools to deal with $(\pi_1^* - \pi_M^*)P$. Before we move on to the proof of Theorem 3.11, we shall need the following lemma.

Lemma 3.13. *The unique non-trivial 2-torsion element in the kernel of*

$$\phi^*: J_0(NM) \to J_0^{\#}(N, M)$$

is the point $\pi_1^*(z)$.

Proof. We can use the fact that z is a 2-torsion element of the kernel ϕ^*: $J_0(N) \to J_0^{\#}(N)$ and the commutativity of the diagram (2) to conclude that $\pi_1^*(z)$ is a 2-torsion element of the kernel of ϕ^*: $J_0(NM) \to J_0^{\#}(N, M)$.

The fact that $\pi_1^*(z)$ is not trivial follows from Lemma 3.8.

Since the map ϕ: $X_0^{\#}(N, M) \to X_0(NM)$ is Galois with cyclic Galois group, there are no other non-trivial 2-torsion points in its kernel. □

Let us now take for granted the following theorem, the proof of which shall be the subject of Section 4.

Theorem 3.14. *The divisors*

$$(1 - \overline{w}_M^*)\pi_1^*(d) \qquad \textit{and} \qquad \phi^*\left(\frac{v(M-1)}{2} D^{-,-}\right)$$

on $X_0^{\#}(N, M)$ are linearly equivalent.

This theorem allows us to conclude our proof as follows.

Proof of Theorem 3.11. Consider the following commutative diagram.

$$\begin{array}{ccc} J_0(NM) & \xrightarrow{\phi^*} & J_0^{\#}(N,M) \\ \uparrow{\scriptstyle 1-w_M^*} & & \uparrow{\scriptstyle 1-\overline{w}_M^*} \\ J_0(NM) & \xrightarrow{\phi^*} & J_0^{\#}(N,M) \\ \uparrow{\scriptstyle \pi_1^*} & & \uparrow{\scriptstyle \pi_1^*} \\ J_0(N) & \xrightarrow{\phi^*} & J_0^{\#}(N). \end{array}$$

By definition, the point vP maps to $d \in J_0^{\#}(N)$. The left-hand vertical map $(1-w_M^*)\circ\pi_1^* = \pi_1^*-\pi_M^*$ sends the point vP to $(\pi_1^*-\pi_M^*)(vP)$. By the commutativity of the diagram and Theorem 3.14, we may therefore deduce that $(\pi_1^*-\pi_M^*)(vP) - \left(\frac{v(M-1)}{2}D^{-,-}\right)$ lies in the kernel of the top map ϕ^*. Furthermore, by Lemma 3.3 and the remark after it, we can see that

$$2\left((\pi_1^*-\pi_M^*)(vP) - \left(\frac{v(M-1)}{2}D^{-,-}\right)\right) = 0.$$

By Lemma 3.13, $(\pi_1^*-\pi_M^*)(vP) - \left(\frac{v(M-1)}{2}D^{-,-}\right)$ is equal to either 0 or $\pi_1^*(z)$. However, reduction modulo N lands in the identity component of $J_0(NM)$ for $(\pi_1^*-\pi_M^*)(vP)$ (by Lemma 3.10), and for $D^{-,-}$ (by Lemma 3.9), and away from the identity component for $\pi_1^*(z)$ (by Lemma 3.8). Therefore

$$(\pi_1^*-\pi_M^*)(vP) = \left(\frac{v(M-1)}{2}D^{-,-}\right)$$

and we have now completed the proof of Theorem 3.11. □

Proof of Theorem 1.1. In this proof, allow N and M to be any pair of distinct primes. Observe that if $N < 11$ then $A = 0$ and Theorem 1.1 is true. Similarly, we may assume that $M \geq 11$.

If $N \not\equiv 1 \pmod 8$ (equivalently, n is odd), then by Lemmas 3.2 and 3.5, the group $A \cap B$ is spanned by a multiple of $(\pi_1^*-\pi_M^*)(P)$.

On the other hand, if $N \equiv 1 \pmod 8$ (equivalently, n is even), then by Lemmas 3.2 and 3.5, the group $A \cap B$ is spanned by a multiple of $(\pi_1^*-\pi_M^*)(P)$ and $\pi^*(z)$. However, by Lemmas 3.7 and 3.8, in fact the group $A \cap B$ is spanned by multiples of $(\pi_1^*-\pi_M^*)(P)$ alone.

The multiples of $(\pi_1^*-\pi_M^*)(P)$ are exactly the image of the group $J_0(N)[I]/\Sigma$ under the map $\pi_1^*-\pi_M^*$. Since $\pi_1^*-\pi_M^*$ is injective on $J_0(N)[I]/\Sigma$, we can see that the image X_N has order n. We will now show that this image lies entirely in the cyclic group spanned by $D^{-,-}$.

We shall consider separately the image of the 2-primary part and the odd part of $J_0(N)[I]/\Sigma$. The odd part is generated by $2^kP = 2^{k-1}c$, and its image does lie among the multiples of $D^{-,-}$ by Lemma 3.3. The 2-primary part of $J_0(N)[I]/\Sigma$ (which is non-trivial only if n is even) is generated by vP. An application of Theorem 3.14 completes our argument.

Running the same argument again, but exchanging the roles of N and M, we find that $A \cap B$ can also be found in the subgroup X_M of order m of the multiples of $D^{-,-}$. Therefore, $A \cap B$ must be the intersection of our groups X_N and X_M. This completes the proof. □

4. A unit calculation on $X_0^{\#}(N, M)$

This section is devoted to proving Theorem 3.14 and thereby completing the proof of Theorem 1.1. The proof involves the construction of a unit (e in Definition 4.2) in the style of [2], which is to say, using results from [8]. The divisor of this unit e is exactly the divisor asserted to be principal in Theorem 3.14.

Recall that a **unit** of a modular curve is a rational function on the curve that has its divisor concentrated at the cusps. (It is a unit of the ring of rational maps from the noncuspidal points of the curve to the affine line.) In [8], Kubert and Lang studied the units of $X(NM)$. We briefly recall their results here, using $\mathbb{Z}/NM\mathbb{Z} \times \mathbb{Z}/NM\mathbb{Z}$ as the indexing group instead of their $\frac{1}{NM}\mathbb{Z}/\mathbb{Z} \times \frac{1}{NM}\mathbb{Z}/\mathbb{Z}$. Let $e = (e_1, e_2)$ be a pair of integers such that not both of e_1 and e_2 are divisible by NM. One can use the classical Weierstrass σ and Dedekind η functions to define the Klein form $k_e(\tau)$ on $\mathfrak{H}$. This form enjoys the properties

$$\forall \alpha = \begin{pmatrix} a & b \\ c & d \end{pmatrix} \in \mathrm{SL}_2\mathbb{Z}, \qquad k_e(\alpha\tau) = (c\tau + d)^{-1} k_{e\alpha}(\tau) \tag{K1}$$

(where $e\alpha$ denotes usual matrix multiplication) and

$$\forall f = (f_1, f_2) \in NM\mathbb{Z} \times NM\mathbb{Z}, \qquad k_{e+f}(\tau) = \varepsilon(e, f) k_e(\tau), \tag{K2}$$

where

$$\varepsilon(e, f) = (-1)^{\frac{f_1 f_2}{N^2M^2} + \frac{f_1}{NM} + \frac{f_2}{NM}} \exp\left(\frac{\pi i}{N^2M^2}(e_1 f_2 - e_2 f_1)\right).$$

These Klein forms are then used to define for all $e = (e_1, e_2) \in (\mathbb{Z} \times \mathbb{Z}) \setminus (NM\mathbb{Z} \times NM\mathbb{Z})$ the **Siegel function**

$$g_e(\tau) = k_e(\tau)\eta^2(\tau).$$

Recall that

$$\forall \alpha = \begin{pmatrix} a & b \\ c & d \end{pmatrix} \in \mathrm{SL}_2\mathbb{Z}, \qquad \eta^2(\alpha\tau) = \psi(\alpha)(c\tau + d)\eta^2(\tau), \tag{N}$$

where ψ is defined by its values on the two standard generators of $\mathrm{SL}_2\mathbb{Z}$ as

$$\psi\left(\begin{pmatrix} 1 & 1 \\ 0 & 1 \end{pmatrix}\right) = \exp\left(\frac{\pi i}{6}\right), \qquad \psi\left(\begin{pmatrix} 0 & 1 \\ -1 & 0 \end{pmatrix}\right) = \exp\left(\frac{\pi i}{2}\right) = i.$$

Now [8, Chapter 4, Theorem 1.3] says

Theorem 4.1. *The functions of the following form are all units of $X(NM)_{\mathbb{C}}$ are exactly the functions of the form*

$$g = c\prod_{e\in E} g_e(\tau)^{m(e)},$$

for some constant c and some finite set $E \subseteq \mathbb{Z}\times\mathbb{Z}$, where the $m(e)$ satisfy the conditions

$$\sum_{e\in E} m(e) \equiv 0 \pmod{12}, \tag{U1}$$

$$\sum_{e=(e_1,e_2)\in E} e_1^2 m(e) \equiv 0 \pmod{NM}, \tag{U2}$$

$$\sum_{e=(e_1,e_2)\in E} e_1e_2 m(e) \equiv 0 \pmod{NM}, \tag{U3}$$

$$\sum_{e=(e_1,e_2)\in E} e_2^2 m(e) \equiv 0 \pmod{NM}. \tag{U4}$$

The order of such a g at a cusp $P = \binom{x}{y}$ of $X(NM)$ is as follows. Pick some $\alpha \in \mathrm{PSL}_2(\mathbb{Z}/NM\mathbb{Z})$ such that $\alpha\binom{1}{0} = P$, and let $\hat{\alpha}$ be some lift of α to $\mathrm{SL}_2\mathbb{Z}$. Let $(c_1(e), c_2(e)) = e\hat{\alpha}$ be the components of the usual matrix product of e and $\hat{\alpha}$; thus, we may take $c_1(e) = xe_1 + ye_2$. Then

$$\mathrm{ord}_P(g) = \sum_{e\in E} m(e)\frac{NM}{2}B_2\left(\frac{c_1(e) \bmod NM}{NM}\right), \tag{3}$$

where $B_2(X) = X^2 - X + 1/6$ is the second Bernoulli polynomial, and we used $x \bmod NM$ to denote the smallest non-negative residue of x modulo NM. For details and a derivation using the q-expansion of g, see [8, Chapter 2, §3].

Let $e \in E$ and $\alpha \in \mathrm{SL}_2\mathbb{Z}$. From (K1) and (N), we conclude that

$$g_e(\alpha\tau) = \psi(\alpha)g_{e\alpha}(\tau), \tag{4}$$

so using (U1) and the fact that $\psi(\alpha)^{12} = 1$ for any α, for $g(\tau) = c\prod_{e\in E} g_e(\tau)^{m(e)}$ we have

$$g(\alpha\tau) = c\prod_{e\in E} g_{e\alpha}(\tau)^{m(e)}. \tag{5}$$

By (K2), if $e \equiv e' \pmod{NM}$, then $g_e/g_{e'}$ is a root of unity. By (K1) with $\alpha = \begin{pmatrix} -1 & 0 \\ 0 & -1 \end{pmatrix}$, if $e + e' = (0,0)$, then $g_e/g_{e'} = -1$ (here we also used the fact that with this α, the ψ and the $c\tau + d$ factors in (N) are both -1, and so they multiply to 1). Hence all units of $X(NM)_{\mathbb{C}}$ mentioned in Theorem 4.1 can be put into the form $g = c' \prod_{e \in E'} g_e(\tau)^{m(e)}$, with

$$
\begin{array}{llllll}
E' = & \{ & (0,1), & (0,2), & \dots, & (0,r'), & \\
& (1,0), & (1,1), & (1,2), & \dots, & & (1,NM-1), \\
& (2,0), & (2,1), & (2,2), & \dots, & & (2,NM-1), \\
& \vdots & & & & & \\
& (r',0), & (r',1), & (r',2), & \dots, & & (r',NM-1) \ \},
\end{array}
$$

where we used $r' = (NM-1)/2$.

We now proceed to give a proof of Theorem 3.14.

Definition 4.2. *Recall from Section 2 that Ω denotes the set of 2^kth powers in C_N. For any integer y that is not divisible by N,*

- *let Ω_y denote the representatives in the interval $[1, r]$ of the elements of the coset $y\Omega$ of Ω;*
- *let J_y denote the set of integers x in the interval $[1, (NM-1)/2]$ such that*
  - *M does not divide x, and*
  - *x maps to Ω_y under the natural surjection $\mathbb{Z} \to C_N$;*
- *let*

$$e(x) = \prod_{j \in J_M} g^q_{(0,j)}.$$

We shall need a definition and some lemmas before we can show that div(e) is the divisor mentioned in Theorem 3.14.

Definition 4.3. *For any integer $x \in [1, r]$, let $\phi(x)$ be the element of $[1, r]$ that satisfies the congruence*

$$M\phi(x) \equiv \pm x \pmod{N}.$$

(In other words, $\phi(x)$ is the representative for x/M in C_N.)

Definition 4.4. *As in [13, §2], for an odd positive integer u, let*

$$\{\cdot\}_u : \mathbb{Z} \to \{0, 1, \dots, (u-1)/2\}$$

be the function defined by

$$\{a\}_u \equiv \pm a \pmod{u}.$$

For brevity, let $\{\cdot\}$ denote $\{\cdot\}_N$.

Convention 4.5. *For P a statement, let $[P]$ be 1 if P is true, 0 if P is false.*

The following two lemmas will be useful later. Their proofs are very easy and will not be given here.

Lemma 4.6. *For any integer y that is not divisible by N,*

(a) $$\sum_{j\in J_y} 1 = 2zv(M-1),$$

(b) $$\sum_{j\in J_y} j = \frac{1}{2}zvN(M-1)(M+1) + \sum_{b\in\Omega_y}(\widetilde{b} - M\phi(\widetilde{b})),$$

(c) $$\sum_{j\in J_y} j^2 = \frac{1}{6}zvN^2(M-1)(M+1)M + \sum_{b\in\Omega_y}(M\widetilde{b}^2 - M^2\phi(\widetilde{b})^2).$$

Lemma 4.7. *Let y be a real number. Then*

$$M\left(\sum_{i=0}^{M-1} B_2\left(\frac{y+iN}{NM}\right)\right) - B_2\left(\frac{y}{N}\right) = 0.$$

We shall need one last lemma.

Lemma 4.8. *Let c and d be relatively prime integers. Assume that c is divisible by NM and that d maps into Ω under the map $\mathbb{Z} \to C_N$. If y is any integer that is not divisible by N, we have*

$$\sum_{j\in J_y}\left\lfloor\frac{dj}{NM}\right\rfloor \equiv (1+d)\sum_{j\in J_y} j \pmod 2.$$

Proof. All the displayed congruences below are modulo 2 (the notation $a \bmod b$ for integers a, b carries its usual meaning).

$$\begin{aligned}\sum_{j\in J_y}\left\lfloor\frac{dj}{NM}\right\rfloor &\equiv -d\sum_{j\in J_y} j + d\sum_{j\in J_y} j - NM\sum_{j\in J_y}\left\lfloor\frac{dj}{NM}\right\rfloor \\ &= -d\sum_{j\in J_y} j + \sum_{j\in J_y}(dj \bmod NM)\end{aligned}$$

Now, $dj \bmod NM$ is either $\{dj\}_{NM}$ or $NM - \{dj\}_{NM}$, depending on whether $dj < NM/2$ or not. Hence, using Convention 4.5, we get

$$\begin{aligned}\sum_{j\in J_y}\left\lfloor\frac{dj}{NM}\right\rfloor &\equiv -d\sum_{j\in J_y} j + d\sum_{j\in J_y}\{dj\}_{NM} \\ &\quad + \sum_{j\in J_y}[dj \bmod NM > NM/2](-\{dj\}_{NM} + NM - \{dj\}_{NM}) \\ &\equiv m,\end{aligned}$$

where $m = \sum_{j\in J_y}[dj \bmod NM > NM/2]$, and we used that $\sum_{j\in J_y}\{dj\}_{NM} = \sum_{j\in J_y} j$, since d maps to Ω. Then

$$(-1)^m\prod_{j\in J_y} j \equiv \prod_{j\in J_y}\{dj\}_{NM} \equiv \prod_{j\in J_y}(dj) \pmod{NM},$$

which we can divide through by $\prod_{j\in J_y} j$, since no element of J_y is divisible by either N or M, to obtain

$$(-1)^m \equiv d^{\#J_y} \pmod{NM}.$$

Since $M-1$ divides $\#J_y$, we get that

$$d^{\#J_y} \equiv 1 \pmod{M}$$

by Fermat's Little Theorem. On the other hand, since $2(\#\Omega)$ divides $\#J_y$, we obtain

$$d^{\#J_y} \equiv 1 \pmod{N}.$$

(Already raising to the $(\#\Omega)$th power will send $d \in \Omega$ to $1 \in C_N$, which corresponds to $d \equiv \pm 1 \pmod{N}$.) Therefore, m must be even and the proof of the lemma is complete. □

Claim 4.9. *The function $e(\tau)$ is a unit on $X_0^{\#}(N, M)$.*

Proof. We just need to check conditions (U1–4) of Theorem 4.1, and that $e(\tau)$ remains invariant under the action of any matrix $\left(\begin{smallmatrix} a & b \\ c & d \end{smallmatrix}\right) \in \mathrm{SL}_2\mathbb{Z}$ with $c \equiv 0 \pmod{NM}$ and d mapping to Ω under the map $\mathbb{Z} \to C_N$.

By the definition of the function e and Lemma 4.6(a), (U1) is equivalent to

$$2zvq(M-1) \equiv 0 \pmod{12}.$$

This is always satisfied since $2zq = 6$ and $M-1$ is even.

It is clear from the definition of e that conditions (U2) and (U3) are satisfied, because 0 is divisible by NM.

Lemma 4.6(c) shows us that condition (U4) is equivalent to

$$\frac{1}{6} zvqN^2(M-1)(M+1)M + q \sum_{b\in\Omega_y} (M\widetilde{b}^2 - M^2\phi(\widetilde{b})^2) \equiv 0 \pmod{NM}.$$

The condition modulo M is obviously satisfied. For the condition modulo N, observe that the first term is clearly divisible by N, whereas the second term is divisible by N since for any $b \in \Omega_y$,

$$\widetilde{b}^2 \equiv M^2\phi(\widetilde{b})^2 \pmod{N},$$

and hence

$$\sum_{b\in\Omega_y} (M\widetilde{b}^2 - M^2\phi(\widetilde{b})^2) \equiv \sum_{b\in\Omega_y} (M\widetilde{b}^2 - \widetilde{b}^2) \equiv (M-1) \sum_{b\in\Omega_y} \widetilde{b}^2 \equiv 0 \pmod{N},$$

where the last congruence used [2, Lemma 4.7].

To check invariance under $\left(\begin{smallmatrix} a & b \\ c & d \end{smallmatrix}\right)$, we proceed as follows. First observe that $\left(\begin{smallmatrix} a & b \\ c & d \end{smallmatrix}\right)$ indeed permutes the indices of functions in $e(\tau)$ (after reducing the indices to E'). The only thing we need to show is that the transformation factor arising from reducing the indices to E' is 1. Using (K2), this reduces to showing that

$$c\sum_{j\in J_M} j - d\frac{c}{NM}\sum_{j\in J_M} j^2 + NM\sum_{j\in J_M}\left\lfloor\frac{dj}{NM}\right\rfloor \equiv 0 \pmod{2NM}.$$

Since NM divides each of c, $\sum j^2$ (see above in our proof that (U4) is satisfied) and NM (respectively), it must divide the first, second and third term (respectively) of the above expression. Therefore, we need only check that

$$c\sum_{j\in J_M} j - d\frac{c}{NM}\sum_{j\in J_M} j^2 + NM\sum_{j\in J_M}\left\lfloor\frac{dj}{NM}\right\rfloor \equiv 0 \pmod{2}.$$

We have, using Lemma 4.8 and Lemma 4.6 parts (b) and (c),

$$\begin{aligned}
&c\sum_{j\in J_M} j - d\frac{c}{NM}\sum_{j\in J_M} j^2 + NM\sum_{j\in J_M}\left\lfloor\frac{dj}{NM}\right\rfloor \\
\equiv\ & c\sum_{b\in\Omega_M}(\widetilde{b}-\phi(\widetilde{b})) - dc\sum_{b\in\Omega_M}(\widetilde{b}-\phi(\widetilde{b})) + (1+d)\sum_{b\in\Omega_M}(\widetilde{b}-\phi(\widetilde{b})) \\
\equiv\ & (d+1)(c+1)\sum_{b\in\Omega_M}(\widetilde{b}-\phi(\widetilde{b})) \pmod{2}.
\end{aligned}$$

Since c and d are relatively prime integers, at least one of them is odd, and therefore the number $(d+1)(c+1)$ must be even. This concludes our proof of the claim. □

Claim 4.10. *The divisor of $e(\tau)$ is*

$$(1-\overline{w}_M^*)\pi_1^*(d) - \phi^*\left(\frac{v(M-1)}{2}D^{-,-}\right).$$

Proof. As was mentioned before, we shall prove here that the divisor above is equal to the divisor of the function $e(\tau)$.

First, we identify the divisors $\pi_1^*(d) \in J_0^\#(N,M)$ and $\overline{w}_M^*\pi_1^*(d) \in J_0^\#(N,M)$. We do this by first recalling what the divisor $d \in J_0^\#(N)$ is. For any $1 \le t \le r$, let

$$A_t = \frac{q}{2^k}\left(\sum_{b\in\Omega}\frac{2^kN}{2}B_2\left(\frac{\{\widetilde{b}t\}}{N}\right) + \sum_{b\in C_N}\frac{-N}{2}B_2\left(\frac{\{\widetilde{b}t\}}{N}\right)\right).$$

Then, by the definition of d, for any $1 \le t \le r$ we have

$$\mathrm{ord}_{P_t}(d) = A_t, \qquad \mathrm{ord}_{Q_t}(d) = 0,$$

and therefore

$$\mathrm{ord}_{P_t^0}(\pi_1^*(d)) = MA_t, \qquad \mathrm{ord}_{Q_t^0}(\pi_1^*(d)) = 0$$
$$\mathrm{ord}_{P_t^\infty}(\pi_1^*(d)) = A_t, \qquad \mathrm{ord}_{Q_t^\infty}(\pi_1^*(d)) = 0.$$

By considering the action of $\overline{w}_M^*$ on the cusps, we also obtain that

$$\mathrm{ord}_{P_t^0}(\overline{w}_M^*\pi_1^*(d)) = A_{Mt}, \qquad \mathrm{ord}_{Q_t^0}(\overline{w}_M^*\pi_1^*(d)) = 0$$
$$\mathrm{ord}_{P_t^\infty}(\overline{w}_M^*\pi_1^*(d)) = MA_t, \qquad \mathrm{ord}_{Q_t^\infty}(\overline{w}_M^*\pi_1^*(d)) = 0.$$

Before we turn our attention to $\mathrm{div}(e)$, we note that

$$\mathrm{ord}_{P_t^0}(D^{-,-}) = 1, \qquad \mathrm{ord}_{Q_t^0}(D^{-,-}) = -1$$
$$\mathrm{ord}_{P_t^\infty}(D^{-,-}) = -1, \qquad \mathrm{ord}_{Q_t^\infty}(D^{-,-}) = 1.$$

We shall now show for each cusp of $J_0^\#(N,M)$ that the divisor mentioned in the statement of the claim and $\mathrm{div}(e)$ have the same order.

For a cusp labeled Q_t^∞ for some $1 \le t \le r$ (having ramification index 1 over the cusp of $X(1)$), we observe that it can be represented by $\binom{t}{NM}$ in Shimura's notation and therefore

$$\mathrm{ord}_{Q_t^\infty}(\mathrm{div}(e)) = q(\#J_M)\frac{1}{2}B_2(0) = \frac{2zvq(M-1)}{12} = \frac{v(M-1)}{2}.$$

Since $\pi_1^*(d)$ and $\overline{w}_M^*\pi_1^*(d)$ have order 0 at this cusp, this proves our claim for the cusps Q_t^∞.

For a cusp labeled Q_t^0 for some $1 \le t \le r$ (having ramification index M over the cusp of $X(1)$), we observe that it can be represented by $\binom{t}{N}$ in Shimura's notation and therefore

$$\begin{aligned}\mathrm{ord}_{Q_t^0}(\mathrm{div}(e)) &= q\sum_{j\in J_M}\frac{M}{2}B_2\left(\frac{Nj \bmod NM}{NM}\right) = q\sum_{j\in J_M}\frac{M}{2}B_2\left(\frac{j \bmod M}{M}\right)\\ &= qM(\#\Omega_M)\frac{1}{2}\sum_{i=1}^{M-1}B_2\left(\frac{i}{M}\right) = \frac{-v(M-1)}{2},\end{aligned}$$

where we used the fact that $\sum_{i=1}^{M-1}B_2(i/M)$ is equal to $(1-M)/(6M)$. Since $\pi_1^*(d)$ and $\overline{w}_M^*\pi_1^*(d)$ have order 0 at this cusp, this proves our claim for the cusps Q_t^0.

For a cusp labeled P_t^∞ for some $1 \le t \le r$ (having ramification index N over the cusp of $X(1)$), we observe that it can be represented by $\binom{1}{M\phi(t)}$ in Shimura's notation. Therefore, as above,

$$\mathrm{ord}_{P_t^\infty}(\mathrm{div}(e)) = q\sum_{j\in J_M}\frac{N}{2}B_2\left(\frac{\phi(t)j \bmod N}{N}\right).$$

Therefore,

$$\begin{aligned}&\mathrm{ord}_{P_t^\infty}(\mathrm{div}(e) - \pi_1^*(d) + \overline{w}_M^*\pi_1^*(d))\\ &= \frac{qN}{2}\sum_{j\in J_M}B_2\left(\frac{\phi(t)j \bmod N}{N}\right) + (1-M)\left(q\sum_{b\in\Omega}\frac{N}{2}B_2\left(\frac{\{\widetilde{bt}\}}{N}\right) + \frac{qr}{12\cdot 2^k}\right)\end{aligned}$$

$$= \frac{qN}{2}(1-M)\left(\sum_{b\in\Omega_{M\phi(t)}} B_2\left(\frac{\widetilde{b}}{N}\right) - \sum_{b\in\Omega_t} B_2\left(\frac{\widetilde{b}}{N}\right)\right) + \frac{qr(1-M)}{12\cdot 2^k}$$

$$= \frac{qN}{2}(1-M)\cdot 0 + \frac{-2qzv(M-1)}{12} = \frac{-v(M-1)}{2},$$

where the above argument used the fact that modulo N, the set J_M can be considered as $M-1$ copies of the set Ω_M. Then we used the fact that $M\phi(t)$ and t map to the same element in C_N to cancel out the last two nasty-looking sums. Thereby we completed the proof for the cusps P_t^∞.

It remains for us to consider the cusps labeled P_t^0 for some $1 \le t \le r$ (having ramification index NM over the cusp of $X(1)$). We observe that such a cusp can be represented by $\binom{0}{t}$ in Shimura's notation (where we add N to t if it is otherwise divisible by M). Then

$$\operatorname{ord}_{P_t^0}(\operatorname{div}(e)) = q\sum_{j\in J_M}\frac{NM}{2}B_2\left(\frac{tjM \bmod NM}{NM}\right).$$

Therefore

$$\operatorname{ord}_{P_t^0}(\operatorname{div}(e) - \pi_1^*(d) + \overline{w}_M^*\pi_1^*(d)) = \frac{qNM}{2}\sum_{j\in J_M} B_2\left(\frac{tjM \bmod NM}{NM}\right)$$

$$+\frac{qNM}{2}\sum_{b\in\Omega} B_2\left(\frac{\widetilde{\{bt\}}}{N}\right) - \frac{qNM}{2^{k+1}}\sum_{b\in C_N} B_2\left(\frac{\widetilde{b}}{N}\right)$$

$$-\frac{qN}{2}\sum_{b\in\Omega} B_2\left(\frac{\widetilde{\{btM\}}}{N}\right) + \frac{qN}{2^{k+1}}\sum_{b\in C_N} B_2\left(\frac{\widetilde{b}}{N}\right).$$

We shall separate the above sum of five terms into groups. First of all, note that the sum of the third and the fifth terms is equal to

$$\frac{qN}{2}(1-M)\frac{1}{2^k}\sum_{b\in C_N} B_2\left(\frac{\widetilde{b}}{N}\right) = \frac{qN}{2}(1-M)\frac{1}{2^k}\frac{-r}{6N} = \frac{v(M-1)}{2}.$$

Therefore, to complete the proof of our claim, it suffices to show that the sum of the first, second and fourth terms is zero. This boils down (after division by $qN/2$) to showing that

$$M\sum_{j\in J_{Mt}} B_2\left(\frac{j}{NM}\right) + M\sum_{b\in\Omega_t} B_2\left(\frac{\widetilde{b}}{N}\right) - \sum_{b\in\Omega_{Mt}} B_2\left(\frac{\widetilde{b}}{N}\right) = 0.$$

However, the above equation is just the sum of the conclusions of Lemma 4.7 when we allow the y of the Lemma to run over all elements of Ω_M. □

Proof of Theorem 3.14. This theorem follows from Claims 4.9 and 4.10. □

References

[1] C. Batut, D. Bernardi, H. Cohen, M. Olivier, *PARI–GP*, computer software, 1987–2003

[2] J.A. Csirik, *The Kernel of the Eisenstein Ideal*, J. of Number Theory **92** 2 (2002), 348–375

[3] P. Deligne and M. Rapoport, *Les schémas de modules de courbes elliptiques*, pages 143–316 in P. Deligne and W. Kuyk (eds.), *Modular Functions of One Variable II*, Lecture Notes in Mathematics, 349, Springer, 1973

[4] F. Diamond and J. Im, *Modular forms and modular curves*, pages 39–133 in *Seminar on Fermat's Last Theorem (Toronto, ON 1993–1994)*, Canadian Mathematical Society Conference Proceedings, Vol. 17, 1995

[5] B. Edixhoven, *L'action de l'algèbre de Hecke sur les groupes de composantes des jacobiennes des courbes modulaires est "Eisenstein"*, Astérisque, No. 196–197 (1992), 159–170

[6] J.-M. Fontaine, *Il n'y a pas de variété abélienne sur* $\mathbb{Z}$, Invent. math. **81** (1985), 515–538

[7] P.E. Klimek, *Modular functions for* $\Gamma_1(N)$, Ph.D. dissertation, Berkeley, 1975

[8] D.S. Kubert and S. Lang, *Modular Units*, Grundlehren der mathematischen Wissenschaften 244, Springer, 1981

[9] S. Ling, *The old subvariety of* $J_0(pq)$ *and the Eisenstein kernel in Jacobians*, Israel Journal of Mathematics, **84** (1993), 365–384

[10] B. Mazur, *Modular curves and the Eisenstein ideal*, Inst. Hautes Études Sci. Publ. Math. **47**, 1977, 33–186. See also the Errata on pages 187–188 of [12] for an important correction for line 18 of page 105. Also note that some minor errors in Appendix I are corrected in [5, Section 4.4.1]

[11] B. Mazur, *Rational isogenies of prime degree*, Invent. math. **44** (1978), 129–162

[12] B. Mazur and A. Wiles, *Class fields of abelian extensions of* $\mathbb{Q}$, Invent. math. **76** (1984), 179–330

[13] A.P. Ogg, *Rational points on certain elliptic modular curves*, pages 221–231 in *Analytic number theory*, Proc. Sympos. Pure Math., Vol. XXIV, AMS, 1973

[14] K.A. Ribet, *Congruence relations between modular forms*, in *Proceedings of the ICM 1983*, 503–514

[15] K.A. Ribet, *On the component groups and the Shimura subgroup of* $J_0(N)$, Exp. No. 6, 10, in *Séminaire de Théorie des Nombres, 1987–1988 (Talence, 1987–1988)*, Univ. Bordeaux I, Talence.

[16] K.A. Ribet, *The old subvariety of* $J_0(pq)$, in G. van der Geer, F. Oort, J. Steenbrink (eds.), *Arithmetic Algebraic Geometry*, Progress in Mathematics **89** (1990), 293–307

[17] K.A. Ribet, *Torsion points on* $J_0(N)$ *and Galois representations*, pages 145–166 in *Arithmetic theory of elliptic curves (Cetraro, 1997)*, Lecture Notes in Mathematics, 1716, Springer, 1999

János A. Csirik
AT&T Labs–Research
180 Park Ave.
Florham Park, NJ 07932-0971, USA
e-mail: janos@pobox.com

Progress in Mathematics, Vol. 224, 75–80

Irreducibility of Galois Actions on Level 1 Siegel Cusp Forms

Luis V. Dieulefait

Abstract. In this article we consider the Galois representations associated by Taylor and Weissauer to a genus two Siegel cusp form f, in the case of level 1 and multiplicity one. After extending (by elementary arguments) to a set of density 1 set of primes a result ensuring maximal image obtained in a previous article, we derive two irreducibility results for these modular Galois representations. In particular we show that under some technical, effective conditions, the representations are defined over the field of coefficients of f at least for $\ell \geq 4k - 5$ (k = weight of f).

1. Review of previous results

Let f be a cuspidal weight k Siegel modular form for $\mathrm{Sp}(4, \mathbb{Z})$ which is a Hecke eigenform and denote a_n its eigenvalues.
Let

$$Z_f(s) := \zeta(2s - 2k + 4) \sum_{n=1}^{\infty} \frac{a_n}{n^s}$$

be the spinor zeta function. Then Z_f has an Euler product of the form

$$Z_f(s) = \prod Q_p(p^{-s})^{-1},$$

where Q_p is the polynomial:

$$Q_p(x) = 1 - a_p x + (a_p^2 - a_{p^2} - p^{2k-4})x^2 - a_p p^{2k-3} x^3 + p^{4k-6} x^4.$$

Then we have the following result of Taylor [6], completed by Weissauer [9]:

Theorem 1.1. *Let f be a Siegel modular form of weight $k > 2$ on the full Siegel modular group* $\mathrm{Sp}(4, \mathbb{Z})$ *which is a cusp form and a simultaneous eigenform for all Hecke operators $T(n)$. Furthermore, assume that the automorphic representation corresponding to f has multiplicity one. Let E be the number field generated by*

Supported by a MECD postdoctoral fellowship from the Ministerio de Educación y Cultura at the Centre de Recerca Matemática, by the Arithmetic Geometry European Research Network at the University Paris 13 and by the GTEM European Research Network at the Institut de Mathématiques de Jussieu.

the eigenvalues a_n. For any prime number ℓ and any extension λ of ℓ to E there exists a continuous Galois representation

$$\rho_{f,\lambda} : G_{\mathbb{Q}} \to \mathrm{GSp}(4, \overline{E}_\lambda)$$

such that the following holds: the representation $\rho_{f,\lambda}$ is unramified outside ℓ and

$$\det(Id - x\rho_{f,\lambda}(\mathrm{Frob}\, p)) = Q_p(x)$$

for every $p \neq \ell$. If $\rho_{f,\lambda}$ is absolutely irreducible, then it is defined over E_λ.

For such an f and under the hypothesis of absolute irreducibility, the maximal possible image for $\rho_{f,\lambda}$ is the group

$$A_\lambda^k = \{g \in \mathrm{GSp}(4, \mathcal{O}_{E_\lambda}) : \det(g) \in (\mathbb{Q}_\ell^*)^{4k-6}\},$$

where $\mathcal{O}_{E_\lambda}$ is the ring of integers of E_λ.
If we consider the residual mod λ Galois representation $\bar{\rho}_{f,\lambda}$, then its image is contained in

$$\{g \in \mathrm{GSp}(4, \mathbb{F}_\lambda) : \det(g) \in (\mathbb{F}_\ell^*)^{4k-6}\}.$$

The representations $\rho_{f,\lambda}$ are realized in the tale cohomology of the Siegel variety X_1 of level 1. From results of Faltings and Chai-Faltings, it holds (cf. [9]):

Proposition 1.2. *Let f be a Siegel cusp form as in Theorem 1.1. Then $\rho_{f,\lambda}$ is crystalline and has Hodge-Tate weights $\{2k-3, k-1, k-2, 0\}$.*

Weissauer proved (see [8]) that any genus 2 Siegel cusp form that is not in the image of the Saito-Kurokawa lift (and more generally any irreducible cuspidal automorphic representations π of $\mathrm{GSp}(4, \mathbb{A})$ such that π is not Cuspidal Associated to Parabolic and π_∞ is a discrete series representation), verifies the generalized Ramanujan conjecture. Thus, $Pol_p(x) := x^4 Q_p(1/x)$, the characteristic polynomial of $\rho_{f,\lambda}(\mathrm{Frob}\, p)$, has all its roots of the same absolute value $\sqrt{p^{2k-3}}$.
Using these results and the classification of maximal proper subgroups of $\mathrm{PSp}(4, F)$ with F a finite field, in [2] we found that under certain conditions on a Siegel cusp form f the images of the family of Galois representations attached to f as in Theorem 1.1 are generically large. In order to state this result, let us introduce two technical conditions:

Definition 1.3. We say that a genus two Siegel cusp form f is untwisted if there exists a prime p such that $a_p \neq 0$ and $\mathbb{Q}(b_p) = E$, where $b_p = a_p^2 - a_{p^2} - p^{2k-4}$ is the quadratic coefficient of $Pol_p(x)$.

The representations $\rho_{f,\lambda}$ being symplectic, we know that for every prime p, the roots of $Pol_p(x)$ come in reciprocal pairs $\alpha, p^{2k-3}/\alpha$, $\beta, p^{2k-3}/\beta$. Let d_p be the invariant of $Pol_p(x)$ appearing in the standard factorisation:

$$Pol_p(x) = (x^2 - (a_p/2 + \sqrt{d_p})x + p^{2k-3})(x^2 - (a_p/2 - \sqrt{d_p})x + p^{2k-3}).$$

The second condition is given in term of this invariant: we require that there exists a prime p such that:

$$\mathbb{Q}(d_p) = E \quad \text{and} \quad \sqrt{d_p} \notin E. \tag{1.1}$$

Assuming (1.1), consider the (infinite) set of primes λ in E such that

$$\ell \nmid \operatorname{disc}(d_p) \quad \text{and} \quad d_p \notin (\mathbb{F}_\lambda)^2 . \tag{1.2}$$

The following is the main result of [2]:

Theorem 1.4. *Let f be a Siegel cusp form (and Hecke eigenform) of weight $k > 3$ for the full Siegel modular group* $\mathrm{Sp}(4, \mathbb{Z})$ *verifying the conditions:*

- *Multiplicity one.*
- *Not of Saito-Kurokawa type.*
- *Untwisted* (*see Definition* 1.3).

Assume Serre's conjecture (3.2.4?) *(cf.* [4]*). Then the images of the Galois representations $\rho_{f,\lambda}$ are A_λ^k, for almost every λ.*
Without assuming Serre's conjecture, take a prime p such that (1.1) *holds. Then for all but finitely many primes verifying* (1.2) *the image of $\rho_{f,\lambda}$ is A_λ^k.*

Remark: Thanks to a lemma of Serre, to prove maximal image it suffices to prove the maximality of the image of the corresponding residual representation. In the analysis done in [2], the only case of non-maximal image that was not dealt in a completely satisfactory way (except if one is willing to accept Serre's conjecture) is the case where the residual representation is reducible with two two-dimensional irreducible components, both having the same determinant.

2. Improvement of Theorem 1.4

Assume that a Siegel cusp form f verifies all the conditions of Theorem 1.4, including Condition (1.1) for some prime p. This implies that (1.2) holds for infinitely many primes λ of E, and that for all λ in a cofinite set Λ of this infinite set of primes the image of $\rho_{f,\lambda}$ is A_λ^k.
Let us use this fact, together with Cebotarev density theorem and the compatibility of the family $\{\rho_{f,\lambda}\}$ to obtain more information on the images of the λ-adic representations for those $\lambda \notin \Lambda$.
Let λ be any prime of E in Λ and let M_λ be the corresponding maximal ideal of $\mathcal{O}_E$. The image of $\rho_{f,\lambda}$ is "as large as possible", so there exists an element $g \in G_\mathbb{Q}$ such that the characteristic polynomial of $\rho_{f,\lambda}(g)$ is:

$$x^4 + (2 - \ell)x^2 + 1 = (x^2 + \sqrt{\ell}x + 1)(x^2 - \sqrt{\ell}x + 1) .$$

Using the fact that the representation $\rho_{f,\lambda}$ is continuous and Cebotarev density theorem, we can find a prime q such that the image of Frob q is as close as we want to the image of g, in particular from the definition of d_q if Frob q is close enough to g we have

$$d_q - \ell \in (M_\lambda)^2 .$$

Therefore, as divisors in E, λ divides d_q but λ^2 does not. Letting λ run through the infinite set of primes Λ we obtain this way infinitely many d_q and it is clear (simply because each of them has only finitely many prime divisors) that the extension of E generated by their square roots is infinite, and a fortiori:

$$[E(\{\sqrt{d_p}\}) : E] = \infty . \tag{2.1}$$

We have generated infinitely many substantially different d_p verifying condition (1.1), except for the fact that many of them may not generate E. Nevertheless, by restricting to primes λ in E of residue class degree 1 we see that the set of primes λ verifying (1.2) for at least one of these p has density one. Applying Theorem 1.4 we obtain:

Theorem 2.1. *Let f be a Siegel cusp form verifying all conditions of Theorem* 1.4, *including Condition* (1.1). *Then, the image of the attached Galois representations $\rho_{f,\lambda}$ are maximal, namely A_λ^k, for all primes λ in E except at most for a set of Dirichlet density* 0.

3. Irreducibility as a family

For the case of a cusp form of Saito-Kurokawa type f, it is well known that the representations $\rho_{f,\lambda}$ are reducible, moreover, it holds:

$$\rho_{f,\lambda} = \rho_{g,\lambda} \oplus \chi^{k-1} \oplus \chi^{k-2}$$

where $\rho_{g,\lambda}$ are the Galois representations associated by Deligne to a classical cusp form g of weight $2k-2$, and level 1.

On the other hand, we have already mentioned that in the complementary case, the Galois representations attached to f are pure (Ramanujan conjecture holds). From this fact, because $4k-6$ is not multiple of 4, it is easy to see that, for any λ, $\rho_{f,\lambda}$ cannot be reducible with a one-dimensional component. Moreover, if reducible, $\rho_{f,\lambda}$ should have two two-dimensional irreducible components both with the same determinant: χ^{2k-3} (because the representations are pure and have conductor 1, see [2] for more details).

It is conjectured that, for arbitrary level, the representations $\rho_{f,\lambda}$ will have this reducibility type only in the case of a cusp form f that is a weak endoscopic lift. This case does not occur in level 1, and this conjecture for level 1 cusp forms can be stated simply by saying that the family $\{\rho_{f,\lambda}\}$ can be reducible only in the Saito-Kurokawa case. In what follows, we will prove this conjecture (only in the level 1 case).

So let f be a Siegel cusp form of level 1 and weight k as in Theorem 1.1, not of Saito-Kurokawa type, and assume that the family of Galois representations $\{\rho_{f,\lambda}\}$ is reducible (i.e., that all members of the family are reducible). Thus, for every prime λ in E, we have:

$$\rho_{f,\lambda}^{s.s.} = \pi_{1,\lambda} \oplus \pi_{2,\lambda}$$

where the representations $\pi_{1,\lambda}$ and $\pi_{2,\lambda}$ are both irreducible and have determinant χ^{2k-3}. Applying Proposition 1.2 we see that both representations are crystalline

and have the following Hodge-Tate numbers: $\{0, 2k-3\}$ and $\{k-2, k-1\}$, respectively. Moreover, having both the same determinant, their traces are precisely those appearing in the factorization of $Pol_p(x)$ used to define d_p, namely, $a_p/2 \pm \sqrt{d_p}$. Let us consider the twisted family of representations: $\pi'_{2,\lambda} := \pi_{2,\lambda} \otimes \chi^{-k+2}$. Let u be a prime in E dividing 3. The residual representation $\bar{\pi}'_{2,u}$ is odd, unramified outside 3 and takes values in $\mathrm{GL}(2, F)$ with F a finite field of characteristic 3. Then, by a well-known result of Serre (cf. [3], p. 710) the only possibility is $\bar{\pi}'_{2,u} = 1 \oplus \chi$. In particular, the action of the inertia subgroup I_3 is given by $1 \oplus \chi$, and the case of fundamental characters of level 2 is excluded.
Thus, because the representation $\pi'_{2,u}$ is crystalline with Hodge-Tate numbers $\{0, 1\}$ and 3 is sufficiently large with respect to these weights ($3 >$ max. weight $+1 = 2$) we can apply the results of Breuil (cf. [1], Chapter 9) on crystalline representations to conclude that locally $\pi'_{2,u}|_{G_3}$ is reducible. Now, an application of the result of Skinner and Wiles (see [5]) on deformations of residually reducible Galois representations proves that $\pi'_{2,u}$, and therefore the family $\{\pi'_{2,\lambda}\}$, is modular. These representations being crystalline with Hodge-Tate numbers $\{0, 1\}$ and conductor 1, one immediately gets for the values of Serre's weight and level: 2 and 1, respectively. Thus, we get a contradiction since the space of classical cusp forms $S_2(1)$ does not contain any non-trivial element.

Theorem 3.1. *Let f be a level* 1 *Siegel cusp form as in Theorem* 1.1. *Suppose that the family of Galois representations $\{\rho_{f,\lambda}\}$ is reducible. Then, f is of Saito-Kurokawa type.*

Remark: In particular, we conclude that the reducibility of $\{\rho_{f,\lambda}\}$ is equivalent to the failure of the Ramanujan conjecture for f of level 1.

4. A stronger irreducibility result under condition (1.1)

Theorem 4.1. *Let f be a Siegel cusp form of level* 1 *and weight k verifying the conditions of Theorem* 1.4, *including Condition* (1.1). *Then, for any prime $\ell \geq 4k-5$ and λ a prime in E above ℓ, the representation $\rho_{f,\lambda}$ is (absolutely) irreducible.*

Proof. Suppose that for $\ell_0 \geq 4k-5$, $\lambda_0 \mid \ell_0$, ρ_{f,λ_0} is reducible. As we explained in Section 3, f being not of Saito-Kurokawa type, it must hold: $\rho^{s.s}_{f,\lambda_0} = \pi_{1,\lambda_0} \oplus \pi_{2,\lambda_0}$, these two two-dimensional irreducible components having the same determinant. π_{1,λ_0} is crystalline and has Hodge-Tate numbers $\{0, 2k-3\}$. Then, because $\ell_0 \geq 4k-5$, we can apply the results of Taylor (cf. [7]) asserting that this representation is potentially modular and that there exists a compatible family of continuous Galois representations $\{\Pi_{1,\lambda}\}$ containing π_{1,λ_0}. Moreover, if we call $c_p = \mathrm{trace}(\pi_{1,\lambda_0}(\mathrm{Frob}\ p))$, for $p \neq \ell_0$, we have $c_p = \mathrm{trace}(\Pi_{1,\lambda}(\mathrm{Frob}\ p))$ for every $\lambda \nmid p$, and it follows from the construction done in [7] (these representations extend the representations attached by Blasius and Rogawski to a Hilbert modular form) that these traces $\{c_p\}$ generate a finite extension of $\mathbb{Q}$.

On the other hand, we know how to express the values c_p in terms of the characteristic polynomial of $\rho_{f,\lambda}(\mathrm{Frob}\, p)$: $c_p = a_p/2 \pm \sqrt{d_p}$, and as we have seen in Section 2, Condition (1.1) (together with the rest of conditions in Theorem 1.4) implies that $[E\{\sqrt{d_p}\} : E] = \infty$ (Formula (2.1)), and as a consequence: $[E\{c_p\} : E] = \infty$. Thus, we obtain a contradiction and the result follows.

Corollary 4.2. *Let f be a Siegel cuspform of level* 1 *and weight k verifying the conditions of Theorem* 1.4, *including Condition* (1.1), *or a form of Saito-Kurokawa type. Then, for every $\ell \geq 4k - 5$, and $\lambda \mid \ell$ a prime in E, the representation $\rho_{f,\lambda}$ is defined over E_λ:*

$$\rho_{f,\lambda} : G_{\mathbb{Q}} \rightarrow \mathrm{GSp}(4, E_\lambda)\,.$$

Remark: In the "opposite" case, i.e., when $E(\{\sqrt{d_p}\}) = E$, one expects the representations $\rho_{f,\lambda}$ to be reducible (as usual, assuming that we are not in the Saito-Kurokawa case, with two two-dimensional irreducible components of the same determinant). Nevertheless, the traces of the two-dimensional components at p being $a_p/2 \pm \sqrt{d_p} \in E$, and being both odd representations, it is well known that they have a model defined over E_λ.

References

[1] C. Breuil, *p-adic Hodge theory, deformations and local Langlands,* notes of a course held at the CRM, Bellaterra, Spain, July 2001.

[2] L. Dieulefait, *On the images of the Galois representations attached to genus* 2 *Siegel modular forms*, J. Reine Angew. Math., **553** (2002), 183–200.

[3] J-P. Serre, *Oeuvres,* vol. III, (1986) Springer-Verlag.

[4] J-P. Serre, *Sur les représentations modulaires de degré* 2 *de* $\mathrm{Gal}(\bar{\mathbb{Q}}/\mathbb{Q})$, Duke Math. J., **54** (1987), 179–230.

[5] C. Skinner and A. Wiles, *Residually reducible representations and modular forms,* IHES Publ. Math., **89** (2000).

[6] R. Taylor, *On the ℓ-adic cohomology of Siegel threefolds*, Invent. Math., **114** (1993), 289–310.

[7] R. Taylor, *On the meromorphic continuation of degree two L-functions*, preprint, (2001).

[8] R. Weissauer, *The Ramanujan Conjecture for genus two Siegel modular forms (An application of the Trace Formula),* preprint, (1994).

[9] R. Weissauer, Four-dimensional Galois representations, preprint, (2000).

Luis V. Dieulefait
Centre de Recerca Matemàtica
Apartat 50
E-08193, Bellaterra (Barcelona), Spain
e-mail: `Ldieulefait@crm.es`

Progress in Mathematics, Vol. 224, 81–91

On Elliptic K-curves

Noam D. Elkies

Abstract. For any field K, a "K-curve" is an elliptic curve E defined over some finite separable extension of K such that every Galois conjugate of E is isogenous with E over a separable closure of K. A K-curve must either have complex multiplication (CM) or be isogenous to a collection of K-curves parametrized by a K-rational point on some modular curve $\mathrm{X}^*(N)$. We give the first proof of this result, obtained in 1993 but not heretofore published. We also indicate some of the reasons for interest in K-curves, and in particular in $\mathbf{Q}$-curves, and give some computational examples and open questions.

1. Introduction

Motivation and definition of K-curves. Fix a field K, and let M be a separable closure of K, with Galois group $G = \mathrm{Gal}(M/K)$. An elliptic curve E/K is, among other things, a source of "Galois representations": homomorphisms from G to $\mathrm{GL}_2(\mathbf{Z}_l)$ for all but finitely many primes l (in this case, all l not equal to the characteristic of K). From this point of view, it is natural to consider the following generalization: E need not be defined over K, only over M, but every G-conjugate of E must be isogenous with E over M. Such an elliptic curve E is said to be a *K-curve*. For example, a $\mathbf{Q}$-curve is an elliptic curve over $\bar{\mathbf{Q}}$ that is $\bar{\mathbf{Q}}$-isogenous with all its $\mathrm{Gal}(\bar{\mathbf{Q}}/\mathbf{Q})$-conjugates. Gross introduced the term "$\mathbf{Q}$-curve" in [Gro, p. 32] with a much more restrictive definition: E had to have complex multiplication (CM), to be defined over the Hilbert class field of its CM field, and to be isogenous *over that class field* with all its conjugates. The definition we gave reflects current practice and interest in such curves, as we shall soon see.

While we formulate the definition of K-curves using an algebraic closure, in fact all the conjugate curves and the isogenies between them must be defined over a finite Galois extension K'. For each $g \in \mathrm{Gal}(K'/K)$, choose a nonzero isogeny from E to $g(E)$, and let $d(g)$ be the degree of this isogeny. Suppose that E is a K-curve without complex multiplication (CM), and let l be a prime that is neither equal to the characteristic of K nor a factor of any $d(g)$. Then the l^n-torsion points of E and its Galois conjugates $g(E)$, together with nonzero isogenies from $g(E)$ to E, yield a homomorphism from G to $\mathrm{PGL}_2(\mathbf{Z}_l)$.

Supported in part by the Packard Foundation and the National Science Foundation.

In general we cannot obtain a homomorphism to $\mathrm{GL}_2(\mathbf{Z}_l)$ due to the ambiguity in the choice of isogenies. (Since we excluded CM curves, this ambiguity is limited to a scalar multiple, which is why we still have a well-defined map to PGL_2.) Nevertheless, many of the constructions, conjectures and results known for elliptic curves over K extend nicely to K-curves. For example, Ribet showed [Rib1] that every elliptic curve dominated by some modular curve $\mathrm{X}_1(N)$ is a $\mathbf{Q}$-curve, and conjectured that conversely every $\mathbf{Q}$-curve is dominated by some $\mathrm{X}_1(N)$; this would generalize the proof by Wiles, Taylor, et al. that all elliptic curves defined over $\mathbf{Q}$ are dominated by some $\mathrm{X}_0(N)$. This conjecture is now proved in some cases [ES], and has arithmetic applications that parallel the proof of Fermat's Last Theorem using the modularity of Frey curves. In another direction, the methods of [Elk1, Elk2] have been adapted to prove that, when K is a real number field, some large classes of K-curves have infinitely many primes of supersingular reduction [Jao].

Parametrizations of K-curves. It is well known that the K-rational points of the modular curve $\mathrm{Y}(1) = \mathrm{X}(1) - \{\text{cusps}\}$ parametrize elliptic curves over K up to M-isomorphism. In 1993 I obtained a similar description of non-CM K-curves. For each positive integer N, let $\mathrm{X}^*(N)$ be the quotient of the modular curve $\mathrm{X}_0(N)$ by its group of Atkin-Lehner involutions, a group isomorphic with $(\mathbf{Z}/2\mathbf{Z})^r$ where r is the number of distinct prime factors of N. Then the K-rational points of $\mathrm{X}^*(N)$, other than the cusps, parametrize collections of K-curves. Conversely, every non-CM K-curve E is isogenous to the curves in such a collection. Moreover, one such N is squarefree and is a factor of all integers N' such that E is isogenous to the K-curves arising from some K-rational point of $\mathrm{X}^*(N')$.

The first proof of this result, together with some computational data and open questions, is contained in a 1993 preprint [Elk4] that was never published. Meanwhile Ribet [Rib2] obtained a cohomological proof that generalizes to some abelian varieties of dimension higher than 1. One of the conjectures made in [Elk4], and repeated in [Elk5, p. 61], has since been disproved computationally by Galbraith [Gal1]. But some of the contents of [Elk4], including the original proof, may still be of interest. I thank the editors of the present volume for soliciting this paper, which includes the parts of [Elk4] that have not yet been superseded, corrects and refines some details, and puts these results in the context of relevant developments during the past decade. I thank the editors also for their patience while I took much longer than expected to complete the paper; I thank Jordan Ellenberg for providing a number of important references, without which the process might have taken even longer; and I thank J. González and J.-C. Lario for reading and correcting the paper prior to publication.

Plan of the paper. We begin by describing the modular curves $\mathrm{X}^*(N)$ and explaining how they parametrize elliptic K-curves. We then state and prove the theorem that these parametrizations account for all non-CM K-curves up to isogeny. We then consider the question of which squarefree N can yield non-CM K-curves for a number field K, in particular for $K = \mathbf{Q}$, and give some numerical examples.

2. A theorem on K-curves

The modular curve $X^*(N)$ and elliptic K-curves. Let N be a positive integer. Recall that $Y_0(N)$ is the modular curve parametrizing pairs of elliptic curves with a cyclic N-isogeny, a curve which can be completed to a projective curve $X_0(N)$ by the addition of one or more cusps. Denote by r the number of distinct prime factors $l|N$. We define $X^*(N)$ to be the quotient of the modular curve $X_0(N)$ by the group[1] $W(N) \cong (\mathbf{Z}/2\mathbf{Z})^r$ generated by its r Atkin-Lehner involutions w_l. Likewise the affine curve $Y^*(N)$ is the quotient of $Y_0(N) = X_0(N) - \{\text{cusps}\}$ by the same group. As a Riemann surface, $Y^*(N)$ is the quotient of the upper half-plane by a group we shall call $\Gamma^*(N)$, a congruence subgroup of the group $PGL_2^+(\mathbf{Q})$ of projective linear transformations with $\mathbf{Q}$-coefficients and positive determinant. We are mainly interested in the case that N is squarefree. In this case $\Gamma^*(N)$ is a maximal congruence subgroup of $PGL_2^+(\mathbf{Q})$ (and indeed these are all the maximal congruence subgroups of $PGL_2^+(\mathbf{Q})$ up to conjugation); also $X^*(N) - Y^*(N)$ consists of a unique cusp. The curves $X^*(N)$ and $Y^*(N)$ have been investigated also for some N that are not squarefree, such as $N = l^2$ for some prime l (as in [Mom1]), when $X_0(N)$ is isomorphic with the quotient of $X(l)$ by the split Cartan subgroup of $PSL_2(\mathbf{Z}/l\mathbf{Z})$, while $X^*(N)$ is isomorphic with the quotient of $X(l)$ by the normalizer of that subgroup.

We review the description of $W(N)$ in terms of isogenies. Let k be any field whose characteristic is not a factor of N, and P a k-rational point on $Y_0(N)$. Such points parametrize cyclic N-isogenies $\phi : E \to E'$ defined over k, up to $\overline{k}$-isomorphism. A "cyclic N-isogeny" is one whose kernel is a cyclic group of order N. It follows that for each factorization $N = N_1 N_2$ the kernel of ϕ has a unique subgroup of size N_1. Hence ϕ factors uniquely as $\phi_2 \circ \phi_1$ where the ϕ_i are cyclic N_i-isogenies. Being unique, the ϕ_i are again defined over k. Likewise ϕ factors uniquely as $\varphi_1 \circ \varphi_2$ for some cyclic N_i-isogenies φ_i defined over k. Suppose now that $\gcd(N_1, N_2) = 1$. Denote by $\tilde{\phi}_1$ the dual isogeny of ϕ_1, which is a cyclic N_1-isogeny to E defined over k. Then $\varphi_2 \circ \tilde{\phi}_1$ is a cyclic N-isogeny defined over k. It thus corresponds to a k-rational point on $Y_0(N)$, which we provisionally call $P(N_1)$. This is well defined, even though ϕ was defined only up to twist, because twisting ϕ (that is, replacing ϕ by a cyclic N-isogeny isomorphic with ϕ over $\overline{k}$ but not over k) twists the isogeny $\varphi_2 \circ \tilde{\phi}_1$ but does not change the corresponding point on $Y_0(N)$. For example, if $N_1 = 1$ then $\varphi_2 \circ \tilde{\phi}_1 = \phi$, so $P(1) = P$. For $N_1 \neq 1$ we readily see that the map $P \mapsto P(N_1)$ is an involution of $Y_0(N)$. These involutions all commute; together with the identity map $P \mapsto P(1)$, they constitute the group $W(N)$. In particular, for each prime $l|N$ the Atkin-Lehner

[1] This is a subgroup of the group $B(N)$ of all automorphisms of $X_0(N)$ that come from the normalizer of $\Gamma_0(N)$ in $PGL_2^+(\mathbf{Q})$. It is known that $W(N) = B(N)$ unless $4|N$ or $9|N$. In particular, $W(N) = B(N)$ when N is squarefree. Thus the notation $B(N)$ could be used in [GL], where N is assumed squarefree throughout; but we need a different notation for the group generated by the Atkin-Lehner involutions, because we shall also consider a few N for which this group is properly contained in $B(N)$.

involution w_l is obtained by taking for N_1 the highest power of l that divides N. If $\gcd(N_1, N_1') = 1$ then $P(N_1)(N_1') = P(N_1N_1')$. In general $P(N_1)(N_1') = P(N_1'')$ for the unique factor $N_1''|N$ such that $\gcd(N_1'', N/N_1'') = 1$ and $N_1N_1'N_1''$ is a square. In the special case that N is squarefree, all factorizations $N = N_1N_2$ satisfy the condition $\gcd(N_1, N_2) = 1$. In this case we may identify $W(N)$ with U/U^2, where $U \subset \mathbf{Q}^*$ is the subgroup generated by the prime factors of N, and then the map $P \mapsto P(N_1)$ is identified with the class of N_1 mod U^2.

Now take $k = K$, and again assume that $N \neq 0$ in K. Let P^* be a K-rational point on $\mathrm{Y}^*(N)$, and $\mathcal{P}$ its preimage under the quotient map $\mathrm{Y}_0(N) \to \mathrm{Y}^*(N)$. This $\mathcal{P}$ is then a set of M-rational points that is stable under G. The elliptic curves that these points parametrize are all related by isogenies defined over M whose degrees are factors of N. In particular, they are K-curves. More specifically, let K' be the minimal field of definition of all the points in $\mathcal{P}$; then K'/K is a $(\mathbf{Z}/2)^\rho$ extension for some $\rho \leq r$ (i.e., K' is the compositum of at most r separable quadratic extensions of K). Indeed, $\mathrm{Gal}(K'/K)$ injects into $W(N)$, unless P^* is a ramification point of the cover $\mathrm{Y}^*(N)/\mathrm{Y}_0(N)$, in which case $\mathrm{Gal}(K'/K)$ injects into $W(N)/\{1, w\}$ where $\{1, w\}$ is the stabilizer of any point in $\mathcal{P}$. In this ramified case, the points in $\mathcal{P}$ all parametrize N-isogenies between CM curves; if K has characteristic zero, all these curves have the same endomorphism ring, whose discriminant is a factor of $4N$.

A description of K-curves. We have just seen that K-rational points on $\mathrm{X}^*(N)$ yield K-curves. In addition, any elliptic curve over M isogenous to a K-curve is again a K-curve, and every CM elliptic curve over M is a K-curve. If K has positive characteristic, we include the supersingular curves among the CM ones, so a curve has CM if and only if its j-invariant is algebraic over the prime field.

Our main result is that these constructions exhaust the possible K-curves. More precisely:

Theorem. *Let K be any field and let E be a K-curve without complex multiplication. Then there exists a squarefree N, depending only on the M-isogeny class of E, such that: E is isogenous over M with a K-curve arising from a K-rational point on $\mathrm{Y}^*(N)$, and if for some N' there is a K-rational point on $\mathrm{Y}^*(N')$ that parametrizes K-curves isogenous with E then $N|N'$.*

Proof. Let E/M be a non-CM K-curve. Let $[E]$ be the M-isogeny class of E, and for every rational prime l let $[E]_l$ be the subclass consisting of curves l^n-isogenous to E for some $n \geq 0$. We identify M-isomorphic curves. Then $[E]_l$ may be regarded as an infinite connected graph whose vertices and edges are respectively the elliptic curves in $[E]_l$ and the l-isogenies between them. In the absence of CM, the graph $[E]_l$ is a tree. If K has positive characteristic p, and $l = p$, then $[E]_l$ degenerates to the infinite directed path

$$\cdots \to E^{p^{-1}} \to E \to E^p \to E^{p^2} \to \cdots$$

(which may terminate on the left because M contains only separable extensions of K); otherwise $[E]_l$ is an infinite undirected tree of degree $l+1$. We have for each l a projection $\pi_l : [E] \to [E]_l$ such that $\pi_l(F)$ is the unique curve in $[E]_l$ admitting an isogeny to F of degree prime to l. The product over all primes l of these projections identifies $[E]$ with the restricted product of the $[E]_l$ relative to $\{E\}$; i.e., the possible collections $\{\pi_l(F) : l \text{ prime}\}$ are precisely those for which all but finitely many images are E itself, and each such collection determines a unique $F \in [E]$.

Let $\langle E \rangle$ be the finite set of elliptic curves G-conjugate to E; by assumption $\langle E \rangle \subset [E]$. Since π_l commutes with G, the graph $[E]_l$ inherits an action of G, which permutes its finite subset $\pi_l(\langle E \rangle)$. Note that this subset is the singleton $\{E\}$ for all but finitely many l.

For each l we construct a vertex or an edge of the tree $[E]_l$ fixed by G, which we shall call the *center* of $\pi_l(\langle E \rangle)$. This will be the central vertex or edge of the longest path joining two points in $\pi_l(\langle E \rangle)$. There may be more than one longest path, but for a finite subset of a tree one readily shows that all longest paths have the same central vertex or edge, so the center is well defined. For those l such that $\pi_l(\langle E \rangle) = \{E\}$ the center is of course E itself; if K has characteristic $p > 0$ then this is necessarily the case at $l = p$. Of the remaining finitely many primes l, let r be the number of those for which the center is an edge of $[E]_l$, and let N be the product of these r primes.

We first prove the claim that N is an isogeny invariant; equivalently, that for each $F \in [E]$ the center of $\pi_l(\langle F \rangle)$ is an edge if and only if the center of $\pi_l(\langle E \rangle)$ is an edge. In fact, we show that in this case the centers of $\pi_l(\langle E \rangle)$ and $\pi_l(\langle F \rangle)$ coincide. Let this edge be the l-isogeny between the elliptic curves $E_1, E_2 \in [E]_l$. There exists a Galois transformation $g \in G$ that switches E_1 with E_2, because if there were no such g then all of $\pi_l(\langle E \rangle)$ would lie on one side of the edge $\{E_1, E_2\}$. Since this edge is fixed by G, necessarily $g^2(E_1) = E_1$ and $g^2(E_2) = E_2$. But then there cannot be any vertex of $[E]_l$ fixed by G, nor a fixed edge other than $\{E_1, E_2\}$: a fixed edge or vertex must be at the same distance from E_1 and $E_2 = g(E_1)$, which happens only for $\{E_1, E_2\}$ itself. Hence $\{E_1, E_2\}$ is the center of $\pi_l(\langle F \rangle)$ for all $F \in [E]$, as claimed.

There are 2^r ways to choose a vertex of each of these edges. Every such choice determines a unique curve F in $[E]$ whose image under each π_l is either the central vertex or on the central edge of $\pi_l(\langle E \rangle)$. Let S be the set of these 2^r curves, which is stable under G. For each $F \in S$ there is a unique $F' \in S$ such that $\pi_l(F') \neq \pi_l(F)$ for each $l|N$. The cyclic N-isogeny from F to F' yields an M-rational point $P \in \mathrm{Y}_0(N)$. As F varies over S, the resulting points P vary over the orbit of P under $W(N)$. Hence the image of P in $\mathrm{Y}^*(N)$ is invariant under G, and is therefore defined over K.

We have thus obtained a K-rational point on $\mathrm{Y}^*(N)$ parametrizing K-curves isogenous with E. It remains to show that, for any K-rational point P^* on some $\mathrm{Y}^*(N')$ that parametrizes K-curves isogenous with E, the level N' is a multiple of N. If this were not the case, there would be a prime $l|N$ that did not divide N'.

But then the $2^{r'}$ K-curves parametrized by P^* (where r' is the number of distinct prime factors of N') would all have the same image under π_l, so the tree $[E]_l$ would have a vertex fixed under G – and we have already seen that this does not happen. This contradiction completes the proof. □

Note that S, and thus the K-rational point on $\mathrm{Y}^*(N)$, may depend not only on the isogeny class $[E]$ but also on the choice of E. This can already occur for E defined over K: then $N = 1$, $S = \{E\}$, and our point on $\mathrm{Y}^*(1) = \mathrm{Y}(1)$ is just the j-invariant $j(E)$; but if there is some other curve E'/K isogenous with E then we could start with E' and get $j(E')$ instead. More generally, let N' be a multiple of N (other than N itself) with $\gcd(N, N'/N) = 1$, and consider a K-rational point P^* on $\mathrm{Y}_0(N')/W(N)$ whose stabilizer in $W(N')$ is trivial. Then the $W(N')$ orbit of P^* yields $[W(N') : W(N)]$ K-rational points on $\mathrm{Y}^*(N)$ that parametrize different but isogenous collections of K-curves. Such P^* already exist when K is a number field, as long as N, N' are chosen so that $\mathrm{X}_0(N')/W(N) \cong \mathbf{P}^1$.

As a corollary of our theorem, E is isogenous to a curve F defined over some field K' such that K'/K is a normal extension with Galois group $(\mathbf{Z}/2\mathbf{Z})^\rho$ for some $\rho \leq r$. Let $G' \lhd G$ be the corresponding normal subgroup $\mathrm{Gal}(M/K')$ of $\mathrm{Gal}(M/K)$. Then G' is the kernel of a homomorphism $d : G \to (\mathbf{Q}^*/\mathbf{Q}^{*2})$, defined as follows: for any $E_1 \in \langle E \rangle$ and any $g \in G$, take $d(g)$ to be the degree of a isogeny between E_1 and $g(E_1)$, modulo $\mathbf{Q}^{*2}$. We readily check directly (that is, independently of our arguments involving $[E]_l$ etc.) that $d(g)$ is well defined, and remains so even if we allow $E_1 \in [E]$. The invariant N can then be described as the product of the primes that occur in the image $d(G)$. The homomorphism d was the starting point of Ribet's approach in [Rib2] (which applies to any K, though [Rib2] deals only with $K = \mathbf{Q}$): he constructs a $\mathbf{Q}^*$-valued 2-cocycle whose class in $H^2(G, \mathbf{Q}^*)$ is an obstruction to the existence of an isogenous curve defined over K, and uses properties of the 2-cocycle to obtain an isogenous curve F defined over a $(\mathbf{Z}/2\mathbf{Z})^\rho$ extension of K.

3. The sets $S^*(D)$

A boundedness conjecture. For any integer $D > 0$ let $S^*(D)$ be the set of positive integers N such that $\mathrm{Y}^*(N)$ has a non-CM rational point over some number field K with $[K : \mathbf{Q}] \leq D$. It is tempting to formulate a "boundedness conjecture": $S^*(D)$ is finite for all D. As noted in [Elk5, p. 61, footnote 25], this is part of the natural extension to discrete arithmetic subgroups of $\mathrm{PGL}_2^+(\mathbf{Q})$ of Serre's conjecture for arithmetic subgroups of $\mathrm{SL}_2(\mathbf{Z})$ ([Ser]; see 6.6 on page 395 of Serre's Collected Works, Vol. III, and the correction to 395.6 on p. 712). For example, when $D = 1$ this conjecture asserts that there are only finitely many curves $\mathrm{Y}^*(N)$ with non-CM points defined over $\mathbf{Q}$.

Even this first case of the conjecture will require new ideas. On the surface, it may seem comparable to the problem of describing the $\mathbf{Q}$-rational points on all curves $X_0(N)$ (that is, to describing all cyclic isogenies defined over $\mathbf{Q}$), a problem which has been solved completely, see for instance the "remarks on isogenies" preceding the Antwerp tables [BK]. But this was possible only because the Jacobian of $X_0(N)$ has a quotient of positive dimension and rank zero for all but finitely many N. This is not expected to be true for the Jacobian of $X^*(N)$ for any squarefree N: the L-series attached to each $\Gamma^*(N)$ eigenform of weight 2, whether new or old, has a functional equation with sign -1. [When N is not squarefree, the Jacobian of $X^*(N)$ may have quotients of positive dimension and rank zero; Momose ([Mom1, Mom2], see also [Mom3]) has used such factors to obtain several infinite families of integers N such that the only $\mathbf{Q}$-rational points $Y^*(N)$ are CM.] We thus cannot mimic the known arguments using Eisenstein quotients of $X_0(p)$; nor can we expect to directly use Chabauty's method [Cha], because the analytic rank of the Jacobian of $X^*(N)$ must equal or exceed its genus.

The set $S^*(1)$ certainly contains all N such that $X^*(N)$ has genus zero: such a curve is isomorphic with $\mathbf{P}^1$ already over $\mathbf{Q}$, because $X^*(N)$ always has at least one $\mathbf{Q}$-rational point, namely the image of the cusp at infinity of $X_0(N)$.[2] Likewise $N \in S^*(1)$ if $X^*(N)$ has genus one (that is, if $X^*(N)$ is one of the "involutory" curves of [M-SD]) and positive rank. If $X^*(N)$ has genus one and $\mathbf{Q}$-rank zero then N is still in $S^*(1)$ unless the cusps and CM points of $X^*(N)$ entirely account for its rational torsion, and this can be effectively checked in each individual case. For any other given N there are only finitely many rational points on $X^*(N)$, but these may (and in a few instances do – see below) include points that are neither cusps nor CM, and it seems very hard to exclude this possibility even for specific squarefree N. For general D we can adapt the arguments of [Abr, AH] to show that, for all but finitely many N (which can be effectively bounded in terms of D), the D-th symmetric power of $X^*(N)$ satisfies the criteria of Faltings' theorem [Fal] on subvarieties of abelian varieties and thus has finitely many rational points, and therefore that $X^*(N)$ has only finitely many points of degree D. But this does not allow us to decide whether $N \in S^*(D)$, because there is at present no way to provably list all such points.

In [Elk4] we gave an example, reproduced in [Elk5, p. 62–63], of one way to have $N \in S^*(1)$ even when $X^*(N)$ has genus greater than 1. If the curve $X^*(N)$ has genus 2 then it has an involution, which may take a cusp or CM point to an unexpected rational point on $X^*(N)$. The explicit example in [Elk4, Elk5] shows that this happens for $N = 191$, the largest prime for which $X^*(N)$ has genus 2. In [Elk4] I made the rash conjecture, again repeated in [Elk5, p. 61], that this might be the only source of rational points on $X^*(N)$ other than cusps and CM points, once the genus of $X^*(N)$ exceeds 1. But this guess was disproved by S. Galbraith [Gal1] who computed explicit models for many curves $X^*(N)$ and found rational non-CM points on the curves $Y^*(N)$ for $N = 137$ and $N = 311$ (with $X^*(N)$ of genus 4 in both cases), thus showing that $S^*(1)$ contains the primes 137 and 311.

[2] Many of these genus-zero curves $X^*(N)$ arise in "monstrous moonshine", with N the order of an element of the Monster.

See also [Gal2], which contains similar results for some composite N; for instance, $125 \in S^*(1)$.

Which $N \leq 67$ are in $S^*(1)$? We close with some concrete examples of computations that determine whether a given small integer N is in $S^*(1)$. We stop at 67, which is the smallest N for which $\mathrm{X}^*(N)$ has genus 2, and also the smallest for which we do not know whether $N \in S^*(1)$.

For most positive $N < 67$, we can prove that $N \in S^*(1)$ using only the genus and rank information in [BK]. Of the curves $\mathrm{X}^*(N)$ with $N < 67$, all but 12 have genus zero, and thus yield $N \in S^*(1)$. Of the 12 values of N that yield a curve of genus one, 7 are squarefree, namely 37, 43, 53, 57, 58, 61, 65. For each of these, $\mathrm{X}^*(N)$ has rank 1, so again $N \in S^*(1)$. [See [GL] for much more on the curves $\mathrm{X}^*(N)$ of genus 0 and 1, and the corresponding $\mathbf{Q}$-curves, in the case of N squarefree.] The remaining 5 curves $\mathrm{X}^*(N)$, with $N \in \{40, 48, 52, 63, 64\}$, are elliptic curves of rank zero. These are not covered by the results of [Mom1, Mom2], so we must handle them individually. In each case, we begin by identifying the elliptic curve $\mathrm{X}^*(N)$ and its finite group of rational points. We may then test every rational point to see if it is a cusp or CM point. Alternatively, we could carefully count the rational cusps and CM points, and compare the total, call it $c(N)$, with the number of rational points on $\mathrm{X}^*(N)$. This is in principle easier than computing with explicit equations for $\mathrm{X}^*(N)$, but in practice it can be too easy to err in calculating $c(N)$. Moreover, if it turns out that $N \in S^*(1)$ then explicit computation is still needed if we wish to recover the resulting isogeny class of non-CM $\mathbf{Q}$-curves.

We find that of these five values of N only one, namely 63, belongs to $S^*(1)$, with two non-CM rational points on $\mathrm{Y}^*(63)$. It turns out that in each case the value of $c(N)$ suffices to decide whether $N \in S^*(1)$, because all the elliptic curves whose conductor is a proper factor of N and have at least $c(N)$ rational points have the same number of rational points.[3] [The conductor of the elliptic curve $\mathrm{X}^*(N)$ must be a proper factor of N, because the unique cuspform of weight 2 for $\Gamma^*(N)$ is an oldform. If it were a newform, $\mathrm{X}^*(N)$ would have an odd functional equation, as noted earlier.] We next give some more computational details for $N = 40$, the smallest integer not in $S^*(1)$, and $N = 63$, the smallest integer that is contained in $S^*(1)$ even though $\mathrm{X}^*(N)$ is an elliptic curve of rank zero. We conclude by giving the translation of the question "is $67 \in S^*(1)$?" into a question on the rational points of an explicit genus-2 curve over $\mathbf{Q}$.

The case $N = 40$. An invariant differential on the elliptic curve $\mathrm{X}^*(40)$ is $\phi_{40}\, dq/q$, where

$$\phi_{40} = (\eta_2\eta_{10})^2 - 2(\eta_4\eta_{20})^2$$

and $\eta_m = \eta(m\tau)$. This ϕ_{40} is an oldform corresponding to the level-20 form $(\eta_2\eta_{10})^2$, so $\mathrm{X}^*(40)$ is isogenous to the elliptic curve $\mathrm{X}_0(20)$ of rank zero. We

[3] Specifically, $c(40) = c(63) = 6$, $c(48) = 8$, $c(52) = 7$, and $c(64) = 4$. These values, together with the tables of elliptic curves of low conductor, even suffice to determine $\mathrm{X}^*(N)$, except for $N = 64$, where we find that $\mathrm{X}^*(64)$ is isomorphic with the curve 32-A2(A): $Y^2 = X^3 - X$.

find that $X^*(40)$ is isomorphic to $X_0(20)$, with its function field generated by

$$X = \left(\frac{\eta_2\eta_{10}}{\eta_4\eta_{20}}\right)^2 + \left(\frac{\eta_4\eta_{20}}{\eta_2\eta_{10}}\right)^2$$

and $Y = (-q\,dX/dq)/\phi_{40}$, satisfying

$$Y^2 = X(X+4)(X^2+4).$$

This curve has 6 rational torsion points: the two points at infinity, which are the cusps of $X^*(40)$; the two points with $Y = 0$, which are CM points of discriminants -16 and -40 for $X = 0$, $X = -4$ respectively; and the two points $(X, Y) = (1, \pm 5)$, which are CM points of discriminant -15 and -60. Having thus accounted for all the rational points, we conclude that $40 \notin S^*(1)$ as claimed.

The case $N = 63$. [For the curve $X_0(63)$ itself, see [Elk3].] Let ϕ_{21} be the unique normalized cuspform of weight 2 for $\Gamma_0(21)$, and set

$$X = \left(\frac{\eta_7\eta_9}{\eta_1\eta_{63}}\right) + \left(\frac{\eta_1\eta_{63}}{\eta_7\eta_9}\right), \qquad Y = -\frac{1}{2}\left(X + \frac{q\,dX/dq}{\phi_{63}}\right),$$

where $\phi_{63}(\tau) = \phi_{21}(\tau) - 3\phi_{21}(3\tau)$. Then X, Y generate the function field of $X^*(63)$ and satisfy the equation $Y^2 + XY = X^3 - 4X - 1$, showing that this curve is isomorphic with $X_0(21)$. There are 8 rational points. Of these, the point at infinity and the 4-torsion point $(-1, 2)$ are cusps, the 2- and 4-torsion points $(2, -1)$ and $(-1, -1)$ are CM of discriminant -27, while the 2- and 4-torsion points $(-2, 1)$ and $(5, -13)$ are CM of discriminant -35. This leaves the 2-torsion point $(-1/4, -1/8)$ and the 4-torsion point $(5, 8)$, which yield non-CM $\mathbf{Q}$-curves.

Let E be one of these $\mathbf{Q}$-curves. Then the Galois orbit $\langle E \rangle$ has size 4, but since $3^2 \| 63$ there is a curve E_0 that is 3-isogenous to E and for which $\langle E_0 \rangle$ is a conjugate pair of curves related by 7-isogeny. [In terms of the notations we introduced to prove our theorem, the center of $\pi_3(\langle E \rangle)$ is a vertex of $[E]_3$, which is $\pi_3(E_0)$. Geometrically, $X^*(63)$ is a normal cover of $X^*(7)$, and thus parametrizes conjugate pairs of elliptic curves related by 7-isogeny with special 3-torsion structure; algebraically, $\Gamma^*(7)$ has a normal congruence subgroup conjugate in $\mathrm{PSL}_2^+(\mathbf{Q})$ to $\Gamma^*(63)$.] Using the Hauptmoduln $h_7 = (\eta_1/\eta_7)^4$, $h_7^* = (h_7 + 7)^2/h_7$ for $X_0(7)$ and $X^*(7)$, we calculate that E_0 corresponds to $h_7^* = 17/8$ when $(X, Y) = (-1/4, -1/8)$ and to $h_7^* = 35/8$ when $(X, Y) = (5, 8)$. Solving for h_7, we also find that E_0 can be defined over $\mathbf{Q}(\sqrt{-391})$ in the former case and over $\mathbf{Q}(\sqrt{-15})$ in the latter. In each case, the j-invariant of E_0 is not an algebraic integer (its denominator is a large power of 2), so E_0 and E cannot be CM curves.

The case $N = 67$. Using the methods of [Elk5] (see in particular pp. 61–62, where we obtained an equation for the genus-2 curve $X^*(191)$), we find the hyperelliptic model

$$u^2 - (t^3 + t^2 + 1)u + t^3 + t = 0$$

for $X^*(67)$. The two points at infinity are the cusp and the CM point of discriminant -11; the remaining eight rational CM points are the four pairs with $t =$

$-2, -1, 0, 1$. There are no further rational points with $t = m/n$ and $|m|, |n| < 10^4$. If indeed the above ten points exhaust $(X^*(67))(\mathbf{Q})$ then 67 is the smallest squarefree number not contained in $S^*(1)$. This problem may be within reach of current methods for provably listing all rational points on a genus-2 curve [Fly].

References

[Abr] Abramovic, D.: *Subvarieties of Abelian Varieties and of Jacobians of Curves.* Ph. D. thesis, Harvard 1991.

[AH] Abramovic, D., Harris, J.: Abelian varieties and curves in $W_d(C)$. *Compositio Math.* **78** (1991) #2, 227–238.

[BK] Birch, B.J., Kuyk, W., ed.: *Modular Functions of One Variable IV.* Berlin: Springer 1975 (LNM **476**).

[Cha] Chabauty, C.: Sur les points rationnels des variétés algébriques de genre supérieur à l'unité. *C. R. Acad. Sci. Paris* **212** (1941), 882–885.

[Elk1] Elkies, N.D.: The existence of infinitely many supersingular primes for every elliptic curve over **Q**, *Invent. Math.* **89** (1987) #3, 561–568.

[Elk2] Elkies, N.D.: Supersingular primes for elliptic curves over real number fields, *Compositio Math.* **72** (1989) #2, 165–172.

[Elk3] Elkies, N.D.: The automorphism group of the modular curve $X_0(63)$, *Compositio Math.* **74** (1990) #2, 203–208.

[Elk4] Elkies, N.D.: Remarks on elliptic K-curves. Preprint, 1993.

[Elk5] Elkies, N.D.: Elliptic and modular curves over finite fields and related computational issues. Pages 21–76 in *Computational Perspectives on Number Theory: Proceedings of a Conference in Honor of A.O.L. Atkin* (D.A. Buell and J.T. Teitelbaum, eds.; Providence: American Mathematical Society, 1998).

[Ell] Ellenberg, J.S.: Galois representations attached to **Q**-curves and the generalized Fermat equation $A^4 + B^2 = C^p$. *Amer. J. Math.*, to appear.

[ES] Ellenberg, J.S., Skinner, C.: On the modularity of **Q**-curves. *Duke Math. J.* **109** (2001) #1, 97–122.

[Fal] Faltings, G.: Diophantine approximation on abelian varieties. *Ann. of Math.* (2) **133** (1991) #3, 549–576.

[Fly] Flynn, V.E.: Coverings of curves of genus 2. Pages 65–84 in *Algorithmic Number Theory (Leiden, 2000)* (Proceedings of ANTS-IV, W. Bosma, ed.), Berlin: Springer 2000 *Lect. Notes in Computer Sci.* **1838**.

[Gal1] Galbraith, S.G.: Rational points on $X_0^+(p)$. *Experimental Math.* **8** (1999) #4, 311–318.

[Gal2] Galbraith, S.G.: Rational points on $X_0^+(N)$ and quadratic **Q**-curves. *J. Th. des Nombres de Bordeaux* **14** (2002) #1, 205–219.

[GL] González, J., Lario, J.-C.: Rational and Elliptic Parametrizations of **Q**-curves. *J. Number Theory* **72** (1998), 13–31.

[Gro] Gross, B.H.: *Arithmetic on elliptic curves with complex multiplication.* Berlin: Springer 1980 (LNM **776**).

[Jao] Jao, D.: *Supersingular primes for rational points on modular curves.* Ph. D. thesis, Harvard 2003.

[M-SD] Mazur, B., Swinnerton-Dyer, H.P.F.: Arithmetic of Weil curves, *Invent. Math.* **25** (1974), 1–61.

[Mom1] Momose, F.: Rational points on the modular curves $X_{\text{split}}(p)$. *Compositio Math.* **52** (1984) #1, 115–137.

[Mom2] Momose, F.: Rational points on the modular curves $X_0^+(p^r)$. *J. Fac. Sci. Univ. Tokyo, Sect. IA Math.* **33** (1986) #3, 441–466.

[Mom3] Momose, F.: Rational points on the modular curves $X_0^+(N)$. *J. Math. Soc. Japan* **39** (1987) #2, 269–286.

[Rib1] Ribet, K.A.: Abelian varieties over **Q** and modular forms. CPAM preprint, Berkeley 9/92. Also pages 53–79 in 1992 Proceedings of KAIST Mathematics Workshop (Taejon: Korea Advanced Institute of Science and Technology).

[Rib2] Ribet, K.A.: Fields of definition of abelian varieties with real multiplication. Pages 107–118 in *Arithmetic geometry* (Tempe, AZ, 1993; N. Childress and J.W. Jones, eds. = *Contemp. Math.* **174**, Amer. Math. Soc., Providence, RI, 1994).

[Ser] Serre, J.P.: Répresentations l-adiques. Pages 177–193 in *Algebraic number theory: papers contributed for the Kyoto International Symposium* (S. Iyanaga, ed.), Tokyo: Japan Society for the Promotion of Science, 1977. [= #112 (pages 384–400) in *Œuvres III* (Springer: Berlin 1986).]

Noam D. Elkies
Department of Mathematics
Harvard University
Cambridge, MA, 02138 USA
e-mail: elkies@math.harvard.edu

[illegible]

[M?] [illegible] (1978) [illegible]

[Sh] Shimura, G.: Introduction to the Arithmetic Theory of Automorphic Functions. Publ. Math. Soc. Japan 11. Tokyo: Iwanami Shoten [illegible] Princeton Univ. Press 1971.

Progress in Mathematics, Vol. 224, 93–104

$\mathbb{Q}$-curves and Galois Representations

Jordan S. Ellenberg

Introduction and definitions

Let K be a number field, Galois over $\mathbb{Q}$. A $\mathbb{Q}$-*curve* over K is an elliptic curve over K which is isogenous to all its Galois conjugates. The current interest in $\mathbb{Q}$-curves, it is fair to say, began with Ribet's observation [27] that an elliptic curve over $\bar{\mathbb{Q}}$ admitting a dominant morphism from $X_1(N)$ must be a $\mathbb{Q}$-curve. It is then natural to conjecture that, in fact, all $\mathbb{Q}$-curves are covered by modular curves. More generally, one might ask: from our rich storehouse of theorems about elliptic curves over $\mathbb{Q}$, which ones generalize to $\mathbb{Q}$-curves?

In this paper, we discuss recent progress towards several problems of this type, and some Diophantine applications. We will also state several open problems which seem both interesting and accessible to existing methods.

Remark 1. Elliptic curves with complex multiplication supply a natural population of $\mathbb{Q}$-curves. Indeed, the original use of the term "$\mathbb{Q}$-curve", by Gross [13], referred to CM curves exclusively. The arithmetic of CM curves is much more fully understood than that of curves without extra endomorphisms. For that reason, we will assume hereafter that our $\mathbb{Q}$-curves are *not* CM.

One might think of the class of $\mathbb{Q}$-curves as the "mildest possible generalization" of the class of elliptic curves over $\mathbb{Q}$. Many structures on elliptic curves over $\mathbb{Q}$ are invariant under isogeny. And since a $\mathbb{Q}$-curve E/K has an isogeny class which is fixed by $\mathrm{Gal}(\bar{\mathbb{Q}}/\mathbb{Q})$, we should expect that any isogeny-invariant structure of elliptic curves over $\mathbb{Q}$ can be defined for $\mathbb{Q}$-curves as well.

The structure we have chiefly in mind is the ℓ-adic Galois representation

$$\rho_{C,\ell} : \mathrm{Gal}(\bar{\mathbb{Q}}/\mathbb{Q}) \to \mathrm{End}(T_\ell C)$$

attached to an elliptic curve C over $\mathbb{Q}$. The isomorphism class of $\rho_{E,\ell}$ depends only on the isogeny class of C, and so one expects to find ℓ-adic representations of $\mathrm{Gal}(\bar{\mathbb{Q}}/\mathbb{Q})$ attached to $\mathbb{Q}$-curves as well. Indeed, given a $\mathbb{Q}$-curve E/K, there exists a representation

$$\rho_{E,\ell} : \mathrm{Gal}(\bar{\mathbb{Q}}/\mathbb{Q}) \to \bar{\mathbb{Q}}_\ell^* \,\mathrm{GL}_2(\mathbb{Q}_\ell)$$

Partially supported by NSA Young Investigator Grant MDA905-02-1-0097.

such that the restriction $\rho_{E,\ell}|K$ agrees, up to multiplication by scalars, with the action of $\mathrm{Gal}(\bar{K}/K)$ on $T_\ell E$. We define $\rho_{E,\ell}$ as follows. For each $\sigma \in \mathrm{Gal}(\bar{\mathbb{Q}}/\mathbb{Q})$, let $\mu_\sigma : E^\sigma \to E$ be an isogeny. Then we can define

$$\tilde{\rho}_{E,\ell}(\sigma) \in \mathrm{GL}(T_\ell E)$$

by the rule

$$\tilde{\rho}_{E,\ell}(\sigma)(P) = \mu_\sigma(P^\sigma).$$

Now $\tilde{\rho}_{E,\ell}$ is not a homomorphism from $\mathrm{Gal}(\bar{\mathbb{Q}}/\mathbb{Q})$ to $\mathrm{GL}(T_\ell E)$. However, it is easily checked that

$$\tilde{\rho}_{E,\ell}(\sigma)\tilde{\rho}_{E,\ell}(\tau)\tilde{\rho}_{E,\ell}(\sigma\tau)^{-1}$$

acts on $T_\ell E$ in the same way as does

$$c_E(\sigma,\tau) = (1/\deg \mu_{\sigma\tau})\mu_\sigma \mu_\tau^\sigma \hat{\mu}_{\sigma\tau} \in (\mathrm{Hom}(E,E) \otimes_{\mathbb{Z}} \mathbb{Q})^* = \mathbb{Q}^*.$$

In particular, the projectivizations of the $\tilde{\rho}_{E,\ell}(\sigma)$ fit together into a projective Galois representation

$$\mathbb{P}\rho_{E,\ell} : \mathrm{Gal}(\bar{\mathbb{Q}}/\mathbb{Q}) \to \mathrm{PGL}_2(\mathbb{Q}_\ell).$$

It is easy to check that c_E is a cocycle in $H^2(\mathrm{Gal}(\bar{\mathbb{Q}}/\mathbb{Q}), \mathbb{Q}^*)$. A theorem of Tate shows that c_E is trivial when the coefficients are extended from $\mathbb{Q}$ to $\bar{\mathbb{Q}}$, and this gives the existence of $\rho_{E,\ell}$. One point of view is that $\rho_{E,\ell}$ is the ℓ-adic Galois representation attached to an abelian variety of GL_2-type, which admits E as a factor over $\bar{\mathbb{Q}}$ [27, §6].

Note that the description above defines $\rho_{E,\ell}$ only up to twisting by scalars. In some cases there may be a "best" choice among the twists of $\rho_{E,\ell}$, but in this paper we will always take the projectivization $\mathbb{P}\rho_{E,\ell}$ as our real object of interest. This projective Galois representation is an invariant of the *geometric* isogeny class of E.

What kinds of $\mathbb{Q}$-curves are there?

A priori, a $\mathbb{Q}$-curve can be defined over an arbitrarily complicated number field. Of course, a $\mathbb{Q}$-curve defined over a large number field L might be geometrically isogenous to a curve over a smaller number field K. In fact, if we are considering $\mathbb{Q}$-curves only up to isogeny, we may assume that K is a compositum of quadratic extensions of $\mathbb{Q}$. (Of course, this would be false if we allowed our $\mathbb{Q}$-curves to have CM!)

Theorem 2 (Elkies,[6]). *Let E be a $\mathbb{Q}$-curve (without complex multiplication) over a number field L. Then E is geometrically isogenous to a $\mathbb{Q}$-curve E'/K, where K is a Galois extension of $\mathbb{Q}$ with Galois group $(\mathbb{Z}/2\mathbb{Z})^r$.*

See [28] for a slightly more general version of this theorem with a different proof.

One naturally wishes to place $\mathbb{Q}$-curves in a moduli space. The presence of discrete invariants (for instance, the minimal degree of an isogeny between

E and its Galois conjugate) ensures that no single connected moduli space will parametrize all $\mathbb{Q}$-curves. However, if all discrete invariants are held constant, the category of $\mathbb{Q}$-curves is indeed parametrized by a modular curve.

To fix ideas, consider the class of $\mathbb{Q}$-curves E which are defined over quadratic fields K, and which admit cyclic isogenies of degree d to their Galois conjugates. We call such an E a *$\mathbb{Q}$-curve of degree d.* Let E^σ/K be the Galois conjugate of E. Then the isogeny $\phi : E \to E^\sigma$ is represented by a point of $X_0(d)(K)$. Moreover, the Galois conjugate P^σ represents an isogeny $\phi^\sigma : E^\sigma \to E$. Since E does not have CM, ϕ^σ can only be $\pm\hat{\phi}$. In other words, P^σ is the image of P under the Atkin-Lehner involution w_d. It follows that the image of P in the quotient curve $X_0(d)/w_d$ is defined over $\mathbb{Q}$.

Let N be a squarefree positive integer, and let $X^*(N)$ be the quotient of $X_0(N)$ by the group generated by all Atkin-Lehner involutions $w_p, p|N$. If x is a point of $X^*(N)(\mathbb{Q})$ lying under a point y of $X_0(N)(K)$ for some number field K, and if E/K is an elliptic curve parametrized by y; then the Galois conjugates of E lie in the orbit of E under the Atkin-Lehner involutions. In particular, the Galois conjugates of E are all isogenous to E, and E is a $\mathbb{Q}$-curve.

What Elkies in fact proves in [6] is that every non-CM $\mathbb{Q}$-curve is isogenous to one parametrized, as above, by a rational point of $X^*(N)$. So the title question of this section can be rephrased as follows: what are the rational points on $X^*(N)$? More precisely: what are the rational points on $X^*(N)$ which are neither cusps nor CM points?

We first observe that, as with the modular curves $X_0(N)$, the curves $X^*(N)$ have genus greater than 2 when N is sufficiently large. González and Lario [12] give a complete list of the N for which $X^*(N)$ has genus 0 or 1; there are 43 values of N for which $X^*(N)$ is rational and 38 values such that $X^*(N)$ has genus 1. In many of the cases where $X^*(N)$ is rational, explicit equations for the corresponding family of $\mathbb{Q}$-curves have been worked out: see [12], [14], [16], [17], [24].

Since $X^*(N)$ has finitely many points for N sufficiently large, it is natural to make the following conjecture, which was first formulated (in somewhat greater generality) by Elkies. We call a point on $X^*(N)$ which is neither cuspidal nor CM an *exceptional point.*

Conjecture 3. There are only finitely many positive squarefree integers N such that $X^*(N)(\mathbb{Q})$ contains an exceptional point.

Several examples of exceptional points are known: [6] exhibits an exceptional point on $X^*(191)$, and Galbraith ([9],[10]) gives several more examples of exceptional points on $X^*(N)$, with N as large as 311.

We should emphasize that the method used by Mazur in [21] to control rational points on $X_0(N)$ is not directly applicable to Conjecture 3. Mazur's argument bounds the rational points on a curve X by means of a morphism

$$X \to A$$

where A is an abelian variety with $|A(\mathbb{Q})| < \infty$. But each simple factor A of the Jacobian of $X^*(N)$ is associated to a newform of weight 2 and level N whose functional equation has sign -1. Under the Birch-Swinnerton-Dyer conjecture, every such A has infinite Mordell-Weil group, so Mazur's argument does not apply. Note that this is the same difficulty that arises in the older problem of showing that the modular curve $X_{\text{non-split}}(p)$ does not have rational points for p large enough.

On the other hand, Mazur's method may apply if we specify the field of definition K of the $\mathbb{Q}$-curve. For example, Mazur shows in [21, §8] that, if K is a quadratic imaginary field, and N is a sufficiently large prime which is inert in K, then $X_0(N)(K)$ is empty. In particular, there are no $\mathbb{Q}$-curves over K of degree N. More generally, let K be a quadratic field, χ the corresponding quadratic Dirichlet character, and σ a generator for $\mathrm{Gal}(K/\mathbb{Q})$. Let N be a prime. We denote by $X_0^\chi(N)/\mathbb{Q}$ the modular curve admitting an isomorphism

$$\phi : X_0^\chi(N) \times_{\mathbb{Q}} K \to X_0(N) \times_{\mathbb{Q}} K$$

satisfying $\phi^\sigma = w_N \circ \phi$. Note that ϕ induces a bijection between $X_0^\chi(N)(\mathbb{Q})$ and the points P of $X_0(N)(K)$ satisfying $P^\sigma = w_N P$. In particular, every $\mathbb{Q}$-curve of degree N over K is parametrized by a point of $X_0^\chi(N)(\mathbb{Q})$.

The cuspidal points of $X_0^\chi(N)$ are no longer defined over $\mathbb{Q}$, so it is not immediately clear that $X_0^\chi(N)$ has any rational points at all, even locally.

Problem A: For which χ and N does $X_0^\chi(N)$ have rational points over every completion of $\mathbb{Q}$?

Noam Elkies pointed out to me that the answer to Problem A is certainly not "all χ and N"; for instance, the genus-0 curve $X_0^\chi(5)$ is a Brauer-Severi curve which is rational if and only if 5 is a norm from K.

Even if $X_0^\chi(N)$ has local points, one should be able to use Mazur's argument to obtain reduction information about the corresponding $\mathbb{Q}$-curves.

Problem B: Suppose K is a real quadratic field, and let N be a prime which splits in K. Suppose E is a $\mathbb{Q}$-curve over K of degree N. Show that E has potentially good reduction at all primes greater than 3.

The conditions on N and K imply that the Jacobian of $X_0^\chi(N)(\mathbb{Q})$ has quotients of the form A_f, where f is a cusp form whose functional equation has sign $+1$. By the results of Kolyvagin and Logachev, in order to show A_f has rank 0, it suffices to show that the special value of the L-function $L(f, s)$ at $s = 1$ is nonzero. This should be achievable by an averaging argument as in [8]. One then applies Corollary 4.3 of [21] to obtain the desired result. Can the arguments of [21, §5, 6, 7] be modified to show that under some circumstances there are no rational points on $X_0^\chi(N)$?

We have seen already that $\mathbb{Q}$-curves yield rational points on $X^*(N)$; the problem of describing a $\mathbb{Q}$-curve attached to a point of $X^*(N)(\mathbb{Q})$ is more subtle, due to the fact that $X^*(N)$ is not a fine moduli space. In particular, it is a delicate matter to determine a number field K over which all Galois conjugates of E and the isogenies between them can be defined. We will not treat this problem here, simply referring the reader to [24], [25], [11], and the paper of González, Lario, and Quer in this volume.

Modularity

We say a $\mathbb{Q}$-curve E is *modular* if $\rho_{E,\ell}$ is the ℓ-adic Galois representation attached to a cusp form on $\Gamma_1(N)$ for some N. Equivalently, E is a quotient of $J_1(N)_{\bar{\mathbb{Q}}}$. Note that this definition depends only on $\mathbb{P}\rho_{E,\ell}$.

Ribet showed in [27] that every elliptic curve over $\bar{\mathbb{Q}}$ which admitted a non-constant morphism from $J_1(N)$ for any N is a $\mathbb{Q}$-curve. He also conjectured the converse:

Conjecture 4 (Ribet). All $\mathbb{Q}$-curves are modular.

Naturally, progress towards this conjecture followed very closely upon the work of Wiles on modular deformations. Since this technique involves passing between ℓ-adic and mod-ℓ representations, we now introduce the mod ℓ representation

$$\bar{\rho}_{E,\ell} : \mathrm{Gal}(\bar{\mathbb{Q}}/\mathbb{Q}) \to \bar{\mathbb{F}}_\ell^* \,\mathrm{GL}_2(\mathbb{F}_\ell)$$

associated to a $\mathbb{Q}$-curve E. Defining this representation is subtle only when ℓ divides the degree μ_σ for some σ [7, Def. 2.14].

The first advance towards Conjecture 4 was made by Hasegawa, Hashimoto, and Momose [15]. Their idea was to consider a prime $\ell \geq 5$ such that one of the isogenies

$$\mu_\sigma : E^\sigma \to E$$

has squarefree degree which is a multiple of ℓ. Then the curve E/K admits a rational ℓ-isogeny, which means the image of G_K under $\bar{\rho}_{E,\ell}$ lies in a Borel subgroup. In fact, under certain circumstances the authors show that the image of $\mathrm{Gal}(\bar{\mathbb{Q}}/\mathbb{Q})$ under $\bar{\rho}_{E,\ell}$ is a dihedral group. But dihedral Galois representations are known classically to be modular, so one can use the theorems of Wiles, Taylor-Wiles, and Diamond to conclude that the deformation $\rho_{E,\ell}$ of $\bar{\rho}_{E,\ell}$ is modular as well. Hida, working independently, proved a similar result using a similar method in [18], as did Momose and Shimura [23]. A typical result is the following:

Theorem 5 (Hida, [18]). *Let E be a quadratic $\mathbb{Q}$-curve of degree $p \geq 5$ over a quadratic field unramified at p. Then E is modular.*

This approach generalizes quite nicely to Hilbert-Blumenthal abelian varieties over $\bar{\mathbb{Q}}$ which are isogenous to all their Galois conjugates, or $\mathbb{Q}$-HBAV's: see [19, Cor. 3.3].

The results of [15], [18], and [23] yield modularity for many classes of $\mathbb{Q}$-curves. However, these theorems do not apply to $\mathbb{Q}$-curves of degree 2, a case which has particularly nice Diophantine applications described below. The author and Chris Skinner took a different approach to proving modularity of $\mathbb{Q}$-curves in [7]. Instead of exploiting the isogenies between conjugates, [7] uses the mod 3 representation attached to E, which has solvable projective image and is therefore modular by the Langlands-Tunnell theorem. The lifting theorems of Wiles, as refined by Conrad, Diamond, and Taylor, then yield a modularity result for $\mathbb{Q}$-curves, under certain local conditions at 3. For instance, one has the following theorem.

Theorem 6 (Ellenberg-Skinner, [7]). *Let E/K be a $\mathbb{Q}$-curve with semistable reduction at all primes of K dividing 3, and suppose that $K/\mathbb{Q}$ is unramified at 3. Then E is modular.*

There are infinitely many $\mathbb{Q}$-curves whose modularity has not yet been proven. One example is the curve

$$\begin{aligned} E : y^2 \;=\; & x^3 + (-994708512\sqrt{5257}\sqrt{73} - 414461880\sqrt{5257} \\ & -4973542560\sqrt{73} - 1089620282520)x \\ & +36601957546560\sqrt{5257}\sqrt{73} + 5349307626327168\sqrt{5257} \\ & +55021459817878848\sqrt{73} + 32065347994985088. \end{aligned}$$

This curve is parametrized by a point of $X^*(6)$; we obtained the equation by specializing the explicit description of the universal $\mathbb{Q}$-curve over $X^*(6)$ given by Quer in [24, §6] to the point $a = 2^2 \cdot 3^2 \cdot 73$.

The final proof of the Shimura-Taniyama conjecture for elliptic curves rests crucially on the use of the "$3-5$ switch." The idea is as follows: suppose we wish to show an elliptic curve $C/\mathbb{Q}$ is modular, but the local behavior of C at 3 is so unpleasant that the 3-adic Tate module of C cannot be proved modular. Then one uses the rationality of the curve $X(5)$ to find a curve C' with an isomorphism of Galois modules

$$C[5] \cong C'[5]$$

and which has a "better" mod 3 representation. One then uses the 3-adic representation to show modularity of C', and the coincidence of mod 5 representations suffices to show that modularity of C' implies modularity of C.

This approach is unlikely to work for $\mathbb{Q}$-curves. For instance, let E be a $\mathbb{Q}$-curve of degree 2. Then the set of pairs (E', ϕ), where E' is a quadratic $\mathbb{Q}$-curve of degree 2 and ϕ is an isomorphism $\bar{\rho}_{E,5} \cong \bar{\rho}_{E',5}$, is parametrized by the rational points of a modular curve X, which is geometrically isomorphic to $X_0(2) \times_{X(1)} X(5)$. In order to execute the desired switch, one wants X to have many rational points. But in fact, X has genus 4. So one expects (though it is not clear how to prove!) that, for most E, there will be no other $\mathbb{Q}$-curve with the same mod 5 representation. In other words, no "switch" is available, and we are stuck with the 3-adic behavior of E, no matter how disagreeable. The proof of Ribet's conjecture in full, therefore, will require results on modular deformations on Galois representations which go beyond those currently available.

Problem C. New results on Artin's conjecture for icosahedral Galois representations have recently appeared ([1], [29].) Note especially that the result of [1] does not stipulate local conditions at 3. Can these theorems be used in place of the Langlands-Tunnell theorem to show modularity for some $\mathbb{Q}$-curves whose behavior over $\mathbb{Q}_3$ is very bad, but which, say, have good ordinary reduction at primes over 5?

Surjectivity of mod-p representations

The resolution of the Fermat problem rests on two crucial theorems concerning the Galois representations attached to elliptic curves over $\mathbb{Q}$. One, the modularity theorem of Wiles and Taylor-Wiles, we have already discussed. The other is the theorem of Mazur:

Theorem 7 (Mazur, [21]). *Let $C/\mathbb{Q}$ be an elliptic curve, and $p > 163$ a prime. Then $\bar{\rho}_{C,p}$ is irreducible.*

To what extent are theorems of this kind true for $\mathbb{Q}$-curves? The nice fact is that, in some sense, the situation for $\mathbb{Q}$-curves is *better* than that for elliptic curves over $\mathbb{Q}$.

Theorem 8 ([8]). *Let K be an imaginary quadratic field, and d a squarefree integer. There exists a constant $M_{K,d}$ such that, for all quadratic $\mathbb{Q}$-curves E/K of degree d, and all primes $p > M_{K,d}$, either*

- *$\mathbb{P}\bar{\rho}_{E,p}$ is surjective;*
- *E has potentially good reduction at all primes greater than 3.*

The natural conjecture here is that, for p large enough, $\bar{\rho}_{E,p}$ is surjective whenever E is not CM. In this theorem one should think of "has potentially good reduction everywhere" as a proxy for "is CM". (Compare with [22].)

The idea of the proof is as follows. Let G be a maximal proper subgroup of $\mathrm{PGL}_2(\mathbb{F}_p)$. As in Mazur's original argument, we are trying to control points on a certain modular curve X, whose rational points parametrize $\mathbb{Q}$-curves E over K such that $\bar{\rho}_{E,p}(\mathrm{Gal}(\bar{\mathbb{Q}}/\mathbb{Q})) \subset G$. In [21], G is a Borel subgroup, $K = \mathbb{Q}$, and $X = X_0(p)$; in [23], G is the normalizer of a split Cartan subgroup.

Now the main arithmetic step is to show the existence of a quotient

$$J(X) \to A$$

with $|A(\mathbb{Q})| < \infty$.

The proper maximal subgroups of $\mathrm{PGL}_2(\mathbb{F}_p)$ are the Borel subgroups, the normalizers of Cartan subgroups, and the exceptional subgroups isomorphic to A_4, S_4, or A_5. If G is anything other than the normalizer of a non-split Cartan subgroup, it is not hard to show the existence of an A as above by imitating existing results. So from now on let X be the modular curve parametrizing quadratic $\mathbb{Q}$-curves over K of degree d, whose mod p Galois representation takes image in the normalizer of a non-split Cartan subgroup of $\mathrm{PGL}_2(\mathbb{F}_p)$. If $J(X)$ has a quotient with Mordell-Weil rank 0, Theorem 8 follows after applying a formal immersion theorem of Darmon and Merel [3].

A result of Chen and Edixhoven ([2],[4]) shows that $J(X)$ is a quadratic twist of $J_0^{p-new}(dp^2)/w_p$. In particular, if f is a weight 2 cusp form of level dp^2 which does not come from level dp, and which satisfies $w_p f = f$, then $J(X)$ has a quotient which is isomorphic over K to the modular abelian variety A_f. Thanks to work of Kolyvagin and Logachev, in order to show that a quotient A has Mordell-Weil rank 0, we need only show non-vanishing of the special value of the L-function attached to A at the center of the functional equation. Precisely: for us, it will suffice to prove the following lemma.

Lemma 9. *Let χ be the quadratic Dirichlet character associated to $K/\mathbb{Q}$. For all sufficiently large p, there exists a weight 2 newform f of level p^2 such that $w_p f = f$ and $L(f \otimes \chi, 1) \neq 0$.*

In fact, more is true: as Duke observed in [5], one can use the Petersson formula to approximate the *average* value of $L(f \otimes \chi, 1)$ as f ranges over newforms satisfying the criteria in Lemma 9. In order to prove that $L(f \otimes \chi, 1)$ is nonzero for *some* such f, it suffices to prove that the average is nonzero for p large enough, which is immediate from the analytic estimate.

Problem D: Optimize the dependence of $M_{K,d}$ on K and d in Theorem 8.

Problem E: Extend Theorem 8 to $\mathbb{Q}$-curves over polyquadratic fields.

Diophantine problems

A solution to the Fermat equation $A^p + B^p + C^p = 0$ gives rise, via the Frey-Hellegouarch construction, to an elliptic curve over $\mathbb{Q}$ whose associated mod p Galois representation is unramified away from a small set of primes defined independently of A, B, C. Analogously, certain "twisted" versions of the Fermat equation give rise to $\mathbb{Q}$-curves whose mod p Galois representations satisfy similar non-ramification conditions.

Proposition 10. *Let d be a positive integer and let $\alpha \in \mathbb{Q}(\sqrt{d})$ be an algebraic integer of norm ± 2. Let A, B be integers. Then*

$$E = E_{A,B} : y^2 = x^3 + 4\alpha\sqrt{d}Ax^2 + 2\alpha^2(dA^2 - d^{-1/2}B)x$$

is a $\mathbb{Q}$-curve over $\mathbb{Q}(\sqrt{d})$ of degree 2. The isogeny $\mu : E \to E^\sigma$ has equation

$$(x, y) \mapsto (\alpha^{-2}(y^2/x^2), \alpha^{-3}yx^{-2}(2\alpha^2(dA^2 - d^{-1/2}B) - x^2))$$

and the automorphism $\mu^\sigma \mu$ of E is multiplication by the norm of α. The invariants of the given Weierstrass model of E are

$$E_4 = 32\alpha^2(5dA^2 + 3d^{-1/2}B)$$

and

$$\Delta = 512\alpha^6(dA^2 + d^{-1/2}B)(dA^2 - d^{-1/2}B)^2.$$

Note that the odd primes dividing Δ are precisely those dividing $(dA)^4 - dB^2$. Suppose (A, B, C) is an integral solution in coprime integers to

$$(dA)^4 - dB^2 = 2^n C^p.$$

Then it follows from the theory of the Tate curve that the mod p representation $E_{A,B}[p]$ is unramified away from 2 and primes dividing d. Moreover, if $3 \nmid d$, the representation $E_{A,B}[p]$ is modular, by [7]. Finally, in many cases we can show that $E_{A,B}[p]$ is surjective, by the argument of the previous section. By combining all these facts we can obtain Diophantine theorems. For instance, taking $d = -1$ and $\alpha = 1 + i$, we find ([8]) that the generalized Fermat equation $A^4 + B^2 = C^p$ has no solutions in coprime integers for $p \geq 211$.

Another interesting case arises from the elliptic curve

$$y^2 + 1 = 2x^4 \tag{1}$$

whose integral points are related to the problem of expressing π as a sum of rational arctangents [26, §A.12]. In particular, the point $(13, 239)$ corresponds to Machin's formula

$$\pi/4 = 4\arctan(1/5) - \arctan(1/239).$$

The problem of determining all integral solutions to (1) was first solved by Ljunggren [20]. Using the material in this section, one can reprove Ljunggren's result with a bonus: we can, in a sense, explain *why* there is an integral point on (1) with large height.

Setting $A = x, B = 2y$, we have

$$(2A)^4 - 2B^2 = 8;$$

setting $d = 2$ and $\alpha = \sqrt{2}$, we have a $\mathbb{Q}$-curve $E_{A,B}$ over $\mathbb{Q}(\sqrt{2})$ which has good reduction away from 2. By Tate's algorithm one finds that $E_{A,B}$ has conductor at most 2^{10}, and by [7] we know $E_{A,B}$ is modular. To find the possibilities for $E_{A,B}$, it thus suffices to find the $\mathbb{Q}$-curves among the modular abelian varieties of conductor dividing 2^{10}. The only difficulty is that there might be multiple $\mathbb{Q}$-curves $E_{A,B}$ lying in the same isogeny class. In fact, this is precisely what happens. There is an weight 2 newform f of level 1024 with $a_3(f) = \sqrt{2}, a_5(f) = -\sqrt{2}, a_7(f) = -2\ldots$ It turns out that the mod 5 Galois representation associated to this newform is reducible; that is, there are two $\mathbb{Q}$-curves, connected by a 5-isogeny, attached to the same modular form. One of these curves is $E_{1,2}$ and the other is $E_{13,-478}$. In other words, the existence of a large-height solution to Ljunggren's equation is "explained" by the presence of a rational 5-isogeny on one of the modular abelian varieties of conductor 2^{10}.

In general, the method outlined here should give information about integral solutions to equations of the form

$$A^4 - mB^2 \in \mathbb{Z}[1/N]^* \tag{2}$$

$$\text{and} \qquad A^4 - mB^2 = C^p \tag{3}$$

where m is a squarefree integer prime to 3.

If (A, B) is a solution to (2), the $\mathbb{Q}$-curve $E_{A,B}$ has good reduction away from primes $2mN$, and its conductor can thus be bounded by some constant M: if M is not too large, one can then look up in tables the weight 2 cusp forms of conductor dividing M whose associated abelian varieties are $\mathbb{Q}$-curves. Finally, one bounds the degree of an isogeny between $\mathbb{Q}$-curves of conductor dividing M as in [21, §5,6,7], thus classifying all $\mathbb{Q}$-curves associated to the modular abelian varieties under consideration.

Finding the solutions to (3) is more subtle; one still has that $E_{A,B}$ is modular, but all one can conclude is that $E_{A,B}[p]$ is a modular Galois representation with conductor less than some fixed M. If there exist surjective mod p modular representations V of that conductor, it may be quite difficult to control the set of $\mathbb{Q}$-curves E such that $E[p] \cong V$. On the other hand, one may be lucky enough to find, as in [8], that the computation of the conductor forces Galois to act non-surjectively on $E_{A,B}[p]$; in this case, one can hope to use Theorem 8 to get a Diophantine result.

Problem F. What Diophantine problems are associated to quadratic $\mathbb{Q}$-curves of small degree higher than 2? A natural approach is to work out a set of generators for the ring of modular forms on $X^*(d)$ for small d; if this ring has the form

$$\mathbb{C}[A, B, C]/f(A, B, C)$$

and C generates the ideal of cuspforms, then one should be able to classify triples of coprime integers (A, B, C) such that C is a pth power and $f(A, B, C) = 0$. The prototypical example is the case $d = 1$, in which case the ring of modular forms is given by

$$\mathbb{C}[E_4, E_6, \Delta]/(E_4^3 - E_6^2 = 1728\Delta)$$

and the Diophantine equation of interest is

$$x^3 - y^2 = z^p.$$

References

[1] K. Buzzard, M. Dickinson, N. Shepherd-Barron, and R. Taylor. On icosahedral Artin representations. *Duke Math. J.* 109 (2001), no. 2, 283–318.

[2] I. Chen. On relations between Jacobians of certain modular curves. *J. Algebra* 231 (2000), no. 1, 414–448.

[3] H. Darmon and L. Merel. Winding quotients and some variants of Fermat's last theorem. *J. Reine Angew. Math.* 490 (1997), 81–100.

[4] B. de Smit and B. Edixhoven. Sur un résultat d'Imin Chen. *Math. Res. Lett.* 7 (2000), no. 2–3, 147–153.

[5] W. Duke. The critical order of vanishing of automorphic L-functions with large level. *Invent. Math.* 119 (1995), no. 1, 165–174.

[6] N. Elkies. On elliptic K-curves. Preprint, 1993. Revised version published in this volume.

[7] J. Ellenberg and C. Skinner. On the modularity of $\mathbb{Q}$-curves. *Duke Math. J.* 109 (2001), no. 1, 97–122.

[8] J. Ellenberg. Galois representations attached to $\mathbb{Q}$-curves and the generalized Fermat equation $A^4 + B^2 = C^p$. To appear in *Amer. J. Math.*
[9] S.D. Galbraith. Rational points on $X_0^+(p)$. *Experiment. Math.* 8 (1999), no. 4, 311–318.
[10] S.D. Galbraith. Rational points on $X_0^+(N)$ and quadratic $\mathbb{Q}$-curves. *J. Théor. Nombres Bordeaux* 14 (2002), no. 1, 205–219.
[11] J. González. Isogenies of polyquadratic $\mathbb{Q}$-curves to their Galois conjugates. *Arch. Math. (Basel)* 77 (2001), no. 5, 383–390.
[12] J. González and J.-C. Lario. Rational and elliptic parametrizations of $\mathbb{Q}$-curves. *J. Number Theory* 72 (1998), no. 1, 13–31.
[13] B. Gross. *Arithmetic of elliptic curves with complex multiplication.* Lecture Notes in Mathematics 776, Springer, Berlin, 1980.
[14] Y. Hasegawa. $\mathbb{Q}$-curves over quadratic fields. *Manuscripta Math.* 94 (1997), no. 3, 347–364.
[15] Y. Hasegawa, K.-i. Hashimoto, and F. Momose. Modular conjecture for $\mathbb{Q}$-curves and QM-curves. *Internat. J. Math.* 10 (1999), no. 8, 1011–1036.
[16] T. Hibino and A. Umegaki. Families of elliptic $\mathbb{Q}$-curves defined over number fields with large degrees. *Proc. Japan Acad. Ser. A Math. Sci.* 74 (1998), no. 1, 20–24.
[17] T. Hibino and A. Umegaki. A family of elliptic $\mathbb{Q}$-curves defined over biquadratic fields and their modularity. *Acta Arith.* 88 (1999), no. 2.
[18] H. Hida. Modular Galois representations of "Neben" type. Preprint, 2000.
[19] H. Hida. Modularity problems of $\mathbb{Q}$-motives and base-change. Preprint, 2000.
[20] W. Ljunggren. Zur Theorie der Gleichung $x^2 + 1 = Dy^4$. *Avh. Norske Vid. Akad. Oslo. I.* 1942 (1942), no. 5, 27 pp.
[21] B. Mazur. Rational isogenies of prime degree. *Invent. Math.* 44 (1978), no. 2, 129–162.
[22] F. Momose. Rational points on the modular curves $X_{\text{split}}(p)$. *Compositio Math.* 52 (1984), no. 1, 115–137.
[23] F. Momose and M. Shimura. Moduli and modularity of (Q, F)-abelian varieties of GL_2-type. Algebraic number theory and related topics (Japanese) (Kyoto, 1998). *Surikaisekikenkyusho Kokyuroku* No. 1097 (1999), 151–160.
[24] J. Quer. $\mathbb{Q}$-curves and abelian varieties of GL_2-type. *Proc. London Math. Soc. (3)* 81 (2000), no. 2, 285–317.
[25] J. Quer. Fields of definition of $\mathbb{Q}$-curves. 21st Journées Arithmétiques (Rome, 2001). *J. Théor. Nombres Bordeaux* 13 (2001), no. 1, 275–285.
[26] P. Ribenboim. *Catalan's conjecture.* Academic Press, Boston, 1994.
[27] K. Ribet. Abelian varieties over $\mathbb{Q}$ and modular forms. In *Algebra and Topology 1992 (Taejŏn)*, 53–79. Korea Adv. Inst. Sci. Tech., (Taejŏn), 1992. Reprinted in this volume.
[28] K. Ribet. Fields of definition of abelian varieties with real multiplication. In *Arithmetic geometry (Tempe, AZ, 1993)*, 107–118, Contemp. Math., 174, Amer. Math. Soc., Providence, RI, 1994.
[29] R. Taylor. On icosahedral Artin representations II. To appear, *Amer. J. Math.*

Jordan S. Ellenberg
808 Fine Hall, Princeton University
Princeton, USA
e-mail: `ellenber@math.princeton.edu`

Progress in Mathematics, Vol. 224, 105–124

On the Local Behavior of Ordinary Modular Galois Representations

Eknath Ghate

Abstract. We show that if the restriction to the decomposition group at p of the p-adic Galois representation attached to one member of a Hida family of elliptic modular cusp forms is non-split then it is non-split for all but finitely many members of this family. We explain the relevance of this result to a question of Greenberg on the local splitting behavior of ordinary modular Galois representations.

1. Greenberg's question

In this paper we investigate a question of R. Greenberg concerning the local splitting behavior of the Galois representation attached to an ordinary primitive elliptic modular cusp form.

Let us state Greenberg's question. Let $f = \sum a(n, f)\, q^n$ denote a primitive elliptic modular cusp form of weight $k \geq 2$, level $N \geq 1$ and nebentypus $\chi : (\mathbb{Z}/N)^\times \to \mathbb{C}^\times$. Here primitive means that f is a normalized newform that is a common eigenform of all the Hecke operators. Let $\bar{\mathbb{Q}}$ denote an algebraic closure of $\mathbb{Q}$ and fix an embedding ι_∞ of $\bar{\mathbb{Q}}$ into $\mathbb{C}$. Let K_f denote the number field generated by the Fourier coefficients $a(n, f)$ of f via the embedding ι_∞. It is well known that K_f is either a totally real field or a CM field [16, Page 185].

Let p be a prime. Fix an embedding ι_p of $\bar{\mathbb{Q}}$ into $\bar{\mathbb{Q}}_p$ and let $\wp$ be the prime of $\bar{\mathbb{Q}}$ induced by this embedding. We write again $\wp$ for the restriction of $\wp$ to any subfield of $\bar{\mathbb{Q}}$. In particular $\wp$ is a prime of K_f. Let $K_{f,\wp}$ denote the completion of K_f at $\wp$. Eichler, Shimura and Deligne attach a Galois representation to f

$$\rho_f : \mathrm{Gal}(\bar{\mathbb{Q}}/\mathbb{Q}) \to \mathrm{GL}_2(K_{f,\wp})$$

which has the property that for all primes ℓ outside Np

$$\mathrm{tr}\, \rho_f(\mathrm{Frob}_\ell) = a(\ell, f) \quad \text{and} \quad \det \rho_f(\mathrm{Frob}_\ell) = \chi(\ell)\ell^{k-1}.$$

We write $\rho_{f,\wp}$ instead of ρ_f when we wish to emphasize that ρ_f takes values in $K_{f,\wp}$.

Let G_p denote the decomposition group at p determined by the embedding ι_p. We say that f is ordinary at $\wp$ (with respect to the embedding ι_∞) if $a(p, f)$ is a $\wp$-adic unit. If f is ordinary at $\wp$ then Deligne, Mazur-Wiles [12] and Wiles

[20, Theorem 2] have shown that the restriction of ρ_f to G_p is 'upper triangular'. More precisely let V be a model for $\rho_f|_{G_p}$; this is a two-dimensional vector space over $K_{f,\wp}$. Let $\bar{K}_{f,\wp} = \bar{\mathbb{Q}}_p$ be the algebraic closure of $K_{f,\wp}$. Then there exists a basis of $\bar{V} := V \otimes \bar{K}_{f,\wp}$ with respect to which $\rho_f|_{G_p}$ has the following shape:

$$\rho_f|_{G_p} \sim \begin{pmatrix} \delta & u \\ 0 & \epsilon \end{pmatrix} \tag{1.1}$$

where $\epsilon, \delta : G_p \to \bar{K}_{f,\wp}^{\times}$ are characters with ϵ unramified, and $u : G_p \to \bar{K}_{f,\wp}$ is a continuous function that is not necessarily a homomorphism. Moreover $\epsilon(\mathrm{Frob}_p)$ is equal to the unique $\wp$-adic unit root of the polynomial $x^2 - a(p,f)x + p^{k-1}\chi(p)$, where we take $\chi(p) = 0$ if $p \mid N$.

The map u has a cohomological interpretation. Let $c : G_p \to \bar{K}_{f,\wp}$ be the map defined by

$$c(g) = \epsilon^{-1}(g) \cdot u(g)$$

for all $g \in G_p$. Write $\bar{K}_{f,\wp}(\delta\epsilon^{-1})$ for $\bar{K}_{f,\wp}$ thought of as a G_p-module via the character $\delta\epsilon^{-1}$. Then it is easy to check that $c \in \mathrm{Z}^1(G_p, \bar{K}_{f,\wp}(\delta\epsilon^{-1}))$ is a 1-cocycle. Thus c represents a cohomology class

$$[c] \in \mathrm{H}^1(G_p, \bar{K}_{f,\wp}(\delta\epsilon^{-1})).$$

Let V_1 be a G_p-stable line in $\bar{V}$. A line $V_2 \subset \bar{V}$ is said to be *complementary* to V_1 if $V_1 \oplus V_2 = \bar{V}$. Let us say that $\rho_f|_{G_p}$ *splits*, or that ρ_f is *split at p*, if $\bar{V}$ has a G_p-stable line and a G_p-stable complementary line to it. The following easily proved lemma characterizes when $\rho_f|_{G_p}$ splits in terms of the vanishing of the cohomology class $[c]$ defined above.

Lemma 1. *$\rho_f|_{G_p}$ splits if and only if $[c] = 0$ in $\mathrm{H}^1(G_p, \bar{K}_{f,\wp}(\delta\epsilon^{-1}))$.*

Greenberg has asked the following question concerning the splitting behavior of the restriction of ρ_f to G_p.

Question 1. *Let f be a primitive cusp form that is ordinary at $\wp$. When is the representation $\rho_f|_{G_p}$ split? That is, when is the cocycle c that is defined in terms of the upper-shoulder map u in* (1.1) *a coboundary?*

The purpose of this paper is to provide some answers to Question 1. It is not difficult to see that if f has complex multiplication, or CM for short, then ρ_f splits at p (see Propositions 3 and 5 below). A natural guess for an answer to Question 1 is therefore:

$$(??) \quad \rho_f|_{G_p} \text{ splits if and only if } f \text{ has CM.} \tag{1.2}$$

As far as we are aware almost all evidence in support of the 'only if' direction in guess (1.2) is when f has weight 2 (more on this below). For this reason the following theorem concerning cusp forms of higher weight is of interest. To state it we need some notation. Assume for simplicity that p is prime to N. Let f_l for $l \geq 1$ denote the members of the Hida family containing f (see [9], [10] and Section 3 below). The form f_l has weight l, level dividing Np, and character $\chi\omega_p^{k-l}$ where $\omega_p : (\mathbb{Z}/p)^\times \to \mu_{p-1}$ is the Teichmüller character. Moreover, each f_l is ordinary at $\wp$, and, if $l = k$ then $\rho_{f_k} \sim \rho_f$. Now, following a suggestion of R. Greenberg, we prove (see Theorem 3 below):

Theorem 1. *If $\rho_{f_{l_0}}|G_p$ is not split for one weight $l_0 \geq 2$ then $\rho_{f_l}|G_p$ splits only for finitely many weights $l \geq 2$.*

It is a consequence of Hida's theory that the forms f_l for $l \geq 2$ either all have CM, or no f_l for $l \geq 2$ has CM (see, for instance, [5]). In particular if f does not have CM then f_2 does not have CM. In view of the weight 2 results to be described shortly one may now take $l_0 = 2$ and specify conditions under which the hypothesis of Theorem 1 holds. Then the theorem shows that under these conditions the 'only if' direction of (1.2) holds for all but finitely many of the f_l for $l \geq 2$.

While Theorem 1 does not answer Greenberg's question for the cusp form f we started with, it is nevertheless possible in some cases to specify the finite set of exceptional weights l for which $\rho_{f_l}|G_p$ is split: these are the weights l for which the quantity $T = (1+p)^l - 1$ is a zero of a certain power series $g(T)$. In Section 5 we illustrate this circle of ideas for $f = \Delta$, the unique primitive cusp form of level 1 and weight 12. Under a suitable hypothesis of total ordinariness on Δ_2, the weight 2 member of the family containing Δ, we obtain a criterion for the splitting behavior of $\rho_\Delta|_{G_p}$ in terms of the vanishing of a power series (see Theorem 4 below).

Let us now describe what can be said about the 'only if' direction in guess (1.2) when f has weight $k = 2$. Some of the results we describe below are explicitly in the literature and others are simple extensions of these results and so may be known to the expert. Assume first that p does not divide N. If $K_f = \mathbb{Q}$, so that f corresponds to an elliptic curve with good ordinary reduction at p, then Serre and Tate have proved that (1.2) holds (see [14]). It is possible to generalize this result as follows. We shall say that f is *totally ordinary* at p if f is ordinary at $\wp$ for each prime $\wp|p$ of K_f. In addition we say that ρ_f is *totally split* at p if $\rho_{f,\wp}$ splits at p for all such $\wp$. Then we prove (see Proposition 4 below):

Proposition 1. *Suppose f has weight 2 and that $p \nmid N$. Suppose that f is totally ordinary at p and that ρ_f is totally split at p. Then f has* CM.

The proof of Proposition 1 is based on a study of the endomorphism algebra of the abelian variety A_f associated to f by Shimura and uses the theory of p-divisible groups. It builds on the above-mentioned work of Serre and Tate [14]. Proposition 1 was proved independently by Emerton [4] and, when p splits completely in K_f, by Ribet [13].

Let π denote the automorphic representation corresponding to f and let π_p denote the local component of π at p. If $p \nmid N$ then π_p is in the unramified principal series. A statement similar to the proposition above also holds when $p|N$ and the power of p dividing N is the same as the power of p dividing the conductor of χ

in which case π_p is in the ramified principal series. One only needs to modify the notions 'totally ordinary' and 'totally split' appropriately. Since in this case

$$a(p,f)\overline{a(p,f)} = p^{k-1} = p,$$

the sum of the $\wp$-adic valuation and the $\overline{\wp}$-adic valuation of $a(p,f)$ is non-zero for any prime $\wp|p$ of K_f. We shall say that f is *totally ordinary* at p if for each such prime $\wp$ either $a(p,f)$ is a $\wp$-adic unit or it is a $\overline{\wp}$-adic unit. We shall further say that ρ_f is *totally split* at p if in addition either $\rho_{f,\wp}|_{G_p}$ splits, when $a(p,f)$ is a $\wp$-adic unit, or $\rho_{f,\overline{\wp}}|_{G_p}$ splits, when $a(p,f)$ is a $\overline{\wp}$-adic unit. We then have the following result (see Proposition 6 below):

Proposition 2. *Suppose that f has weight 2 and that $p|N$. Assume that the power of p dividing N is the same as the power of p dividing the conductor of χ. Suppose that f is totally ordinary at p and ρ_f is totally split at p. Then f has* CM.

If $p||N$ and p does not divide the conductor of χ (in this case π_p is a twist of the Steinberg representation) then ρ_f is ordinary at a prime $\wp$ of $\bar{\mathbb{Q}}$ lying over p exactly when $k=2$. Moreover in this case f does not have CM and it is a result of Ribet [13] that $\rho_{f,\wp}$ is not split at p for each prime $\wp$ of K_f lying over p. Finally if the exponent of the power of p dividing N is larger than one and unequal to the exponent of the power of p dividing the conductor of χ (this includes the cases where π_p is supercuspidal) and $k \geq 2$ then $a_p = 0$. So f is not ordinary at $\wp$ for any prime $\wp$ of $\bar{\mathbb{Q}}$ and Greenberg's question does not arise in this case.

The above discussion shows that much is known about Question 1 in the weight 2 case. However we stress that although all the above-mentioned results support (1.2) it is remarkable that the 'only if' direction in (1.2) appears still to be open even in weight 2. For instance say f is a non-CM form of weight 2 such that A_f is an abelian *surface*, i.e., K_f is a quadratic field. Say that $p \nmid N$ and $p = \wp\wp'$ splits in K_f. Suppose f is ordinary at $\wp$ but not ordinary at $\wp'$. Then after (1.2) one might expect that $\rho_{f,\wp}$ is not split at p, but this does not appear to be known.

In forthcoming joint work with V. Vatsal [5] we study the weight 1 specializations of Hida families. As a consequence we show that in a non-CM family, all but finitely many of the weight $l \geq 2$ specializations have non-split local Galois representation. This is a stronger result than Theorem 1 since we make no non-splitting assumption on any one member of the family. However, we do assume that p is odd, and work under some conditions on the mod p representation attached to the family. See [5] for further details.

We end this paper by giving an equivalent formulation of Question 1 in the case that N is prime to p in the setting of filtered modules. This was posed by Breuil during his lectures [1] at the CRM in Barcelona in the summer of 2001. As a final remark let us mention that much work has been done on the analog of Greenberg's question in the mod p setting. Here we refer the reader to the papers of Gross [8] and Coleman-Voloch [2].

2. Abelian varieties and weight 2

The main objective of this section is to investigate Greenberg's question when the cusp form f has weight $k = 2$. However since some of the results we present are valid for $k \geq 2$ we do not make any blanket assumption on the weight k.

Some of the results we present below occur explicitly in the literature and others may be well known to the experts. We collect together here these results with proofs for our as well as for the reader's convenience.

It is natural to break the problem up into two cases: $p \nmid N$ and $p|N$.

2.1. Case $p \nmid N$

Consider the first case. We have the following proposition.

Proposition 3. *Suppose that f is a* CM *form of arbitrary weight $k \geq 2$. Suppose that $p \nmid N$ and that f is ordinary at $\wp$. Then $\rho_{f,\wp}$ splits at p.*

Proof. Since f has CM, there is a quadratic Dirichlet character ψ such that $a(q, f) = a(q, f)\psi(q)$ for all but finitely many primes q. This means that

$$\rho_f \quad \sim \quad \rho_f \otimes \psi \tag{2.1}$$

where ψ is thought of as a character of $\mathrm{Gal}(\bar{\mathbb{Q}}/\mathbb{Q})$. Let F denote the quadratic extension of $\mathbb{Q}$ cut out by ψ and let $H = \mathrm{Gal}(\bar{\mathbb{Q}}/F)$. Then (2.1) implies that

$$\rho_f \sim \mathrm{Ind}_F^{\mathbb{Q}}(\phi)$$

for some Hecke character $\phi : H \to \bar{\mathbb{Q}}_p^{\times}$. Let σ denote the non-trivial element of $\mathrm{Gal}(F/\mathbb{Q})$ and let ϕ^σ denote the conjugate character of H. Then

$$\rho_f|_H \quad \sim \quad \begin{pmatrix} \phi & 0 \\ 0 & \phi^\sigma \end{pmatrix}. \tag{2.2}$$

Comparing determinants on H we see that

$$\phi\phi^\sigma = \chi\nu_p^{k-1} \tag{2.3}$$

where $\nu_p : \mathrm{Gal}(\bar{\mathbb{Q}}/\mathbb{Q}) \to \mathbb{Z}_p^\times$ is the p-adic cyclotomic character.

Let $\mathfrak{p}$ be the prime of F determined by the embedding ι_p. Let $H_\mathfrak{p}$ denote a decomposition subgroup of H at $\mathfrak{p}$. After (1.1) at most one of ϕ or ϕ^σ is ramified at $\mathfrak{p}$. On the other hand, since $p \nmid N$, χ is unramified at $\mathfrak{p}$. Since ν_p^{k-1} is ramified at $\mathfrak{p}$ at least one of ϕ or ϕ^σ is ramified at $\mathfrak{p}$. Thus exactly one of ϕ or ϕ^σ, say ϕ, is ramified at $\mathfrak{p}$. But this means that p splits in F. Indeed if it were inert, then ϕ would be unramified at $\mathfrak{p}^\sigma = \mathfrak{p}$ which is a contradiction. Likewise p cannot ramify in F. Thus $p = \mathfrak{p}\mathfrak{p}^\sigma$ splits in F and we have $G_p = H_\mathfrak{p} \subset H$ so that comparing (1.1) and (2.2) we see that $\rho_{f,\wp}$ splits at p.

The following simpler argument was suggested by the referee. Since we are working over a field of characteristic 0, to show $\rho_f|_{G_p}$ is semisimple it suffices to show that ρ_f is semisimple on an open subgroup of G_p. Now since f has CM, ρ_f is semisimple and abelian on H. But an abelian semisimple representation remains semisimple after arbitrary restriction. The proposition follows. □

What about a converse to the above proposition? We offer the following result when f has weight 2. The proof was inspired by an argument of Serre contained at the back of his book [14, Chapter IV, A.2.3–4]. The result below was proved independently by Emerton [4, Theorem 1.3] and a slightly weaker result (requiring p to split completely in K_f) may be found in Ribet [13, page 785].

Proposition 4. *Let f have weight 2. Suppose that $p \nmid N$. If f is ordinary at each prime $\wp$ of $\bar{\mathbb{Q}}$ lying over p and $\rho_{f,\wp}$ is split at p for each such $\wp$, then f is a* CM *form.*

Proof. Write $A = A_f$ for the abelian variety attached to f by Shimura. Recall that A is defined over $\mathbb{Q}$, is simple over $\mathbb{Q}$, and has multiplication by an order in the ring of integers of K_f (making it an abelian variety of GL_2-type). Since $p \nmid N$, A has good reduction at p, and the representation of $\mathrm{Gal}(\bar{\mathbb{Q}}/\mathbb{Q})$ on the ℓ-adic Tate module of A, where ℓ is a prime distinct from p, is unramified at p (cf. the criterion of Néron-Ogg-Shafarevich [15, Theorem 1]). Let $\tilde{A}$ denote the reduction of A at p. It is an abelian variety over $\mathbb{F}_p$. Let $\mathrm{End}(A)$ denote the ring of all the endomorphisms of A thought of as an abelian variety over $\bar{\mathbb{Q}}$, and likewise let $\mathrm{End}(\tilde{A})$ denote the ring of all the endomorphisms of $\tilde{A}$ thought of as an abelian variety over $\bar{\mathbb{F}}_p$. There is a natural map

$$\mathrm{End}(A) \to \mathrm{End}(\tilde{A}) \tag{2.4}$$

which is well known to be an injection. We claim that, under the hypotheses of the proposition, the map (2.4) is also a surjection after possibly replacing A by an isogenous abelian variety.

Let $A(p)$ denote the p-divisible group attached to A and likewise let $\tilde{A}(p)$ denote the p-divisible group attached to $\tilde{A}$. These are p-divisible groups over $\mathbb{Q}$ and $\mathbb{F}_p$ respectively. Let $\tilde{f} \in \mathrm{End}(\tilde{A})$. We wish to show that $\tilde{f}$ lifts to an endomorphism f of A. To prove this it suffices to show that the induced morphism $\tilde{f}(p)$ of $\tilde{A}(p)$ lifts to a morphism of $A(p)$. That is, it suffices to show that the induced map on p-divisible groups

$$\mathrm{End}(A(p)) \to \mathrm{End}(\tilde{A}(p)) \tag{2.5}$$

is a surjection (it is known to be an injection). In the setting of [14, Chapter IV, A.2.3–4] (where A is defined over $\mathbb{Z}_p$ to start with) the fact that (2.5) is surjective implies that (2.4) is surjective is attributed to an unpublished result of Tate. We now show, following a suggestion of Conrad, how the surjectivity of (2.4) follows from that of (2.5) in our setting where A is defined over $\mathbb{Q}$. Let ℓ denote an arbitrary prime and let $A(\ell)$ and $\tilde{A}(\ell)$ denote the associated ℓ-divisible groups over $\mathbb{Q}$ and $\mathbb{F}_p$ respectively. If (2.5) is an isomorphism then the natural map $\mathrm{End}(A(\ell)) \to \mathrm{End}(\tilde{A}(\ell))$ is an isomorphism for every prime ℓ since for $\ell \neq p$ it is automatically an isomorphism because both the ℓ-divisible groups involved are étale. On the other hand by the Tate conjecture for abelian varieties over number fields (due to Ribet in the case at hand, namely modular abelian varieties defined over $\mathbb{Q}$, but in general due to Faltings) we have

$$\mathrm{End}(A) \otimes \mathbb{Z}_\ell = \mathrm{End}_{G_L}(T_\ell(A)) = \mathrm{End}(A(\ell))$$

where L is a number field over which all the endomorphisms of A are defined and $G_L = \mathrm{Gal}(\bar{\mathbb{Q}}/L)$. Similarly (this time by a result of Tate himself) we have

$$\mathrm{End}(\tilde{A}) \otimes \mathbb{Z}_\ell = \mathrm{End}_H(T_\ell(\tilde{A})) = \mathrm{End}(\tilde{A}(\ell))$$

where H is a finite index subgroup of $\mathrm{Gal}(\bar{\mathbb{F}}_p/\mathbb{F}_p)$. Putting things together we see that if (2.5) is an isomorphism then (2.4) is an isomorphism when tensored with $\mathbb{Z}_\ell$ for all primes ℓ and therefore that (2.4) itself is an isomorphism.

We now prove that (2.5) is an isomorphism after possibly replacing A by an isogenous abelian variety. To do so we may replace A by the Néron model of A defined over $\mathbb{Z}_p$ which we continue to write as A.

Let $\mathcal{O}_f$ denote the ring of integers of K_f. By replacing A by an isogenous variety if necessary we may assume that $\mathcal{O}_f \subset \mathrm{End}(A)$. Consider the decomposition of p into prime ideals given by $p\mathcal{O}_f = \prod_{\wp|p} \wp^{e_\wp}$ where $e_\wp$ is the ramification index of $\wp|p$. For each prime $\wp$ of K_f lying over p let $A(\wp)$ denote the p-divisible group formed out of the commutative finite flat group schemes $A[\wp^{e_\wp n}]$ for $n = 1, 2, \ldots$. Then $A(\wp)$ has height $2e_\wp f_\wp$ where $f_\wp$ is the residue degree of $\wp|p$. Further the decomposition of the group scheme $A[p^n] = \prod_{\wp|p} A[\wp^{e_\wp n}]$ for each n induces the decomposition

$$A(p) = \prod_{\wp|p} A(\wp)$$

of the p-divisible group $A(p)$.

By hypothesis each $\rho_{f,\wp}$ splits at p. This means that $A(\wp)$, thought of as a p-divisible group over $\mathbb{Q}_p$, splits into two p-divisible groups over $\mathbb{Q}_p$ of height $e_\wp f_\wp$ each. By a theorem of Tate [19, Theorem 4] we deduce that $A(\wp)$ splits over $\mathbb{Z}_p$ itself:

$$A(\wp) = H_1(\wp) \times H_2(\wp)$$

where the $H_i(\wp)$ are p-divisible groups over $\mathbb{Z}_p$ of height $e_\wp f_\wp$. In particular $\tilde{A}(\wp) = \tilde{H}_1(\wp) \times \tilde{H}_2(\wp)$.

Since f is $\wp$-ordinary the characteristic polynomial of the crystalline Frobenius on the $\mathbb{Q}_p$-crystal attached to $\tilde{A}(\wp)$ namely

$$\mathrm{N}_{\mathbb{Q}_p}^{K_{f,\wp}}\left(x^2 - a(p,f)x + \chi(p)p^{k-1}\right) \tag{2.6}$$

has an ordinary Newton polygon. Since in the isogeny category of p-divisible groups over $\mathbb{F}_p$ the $\tilde{H}_i(\wp)$ are $K_{f,\wp}$-modules, one of the $\tilde{H}_i(\wp)$ must be étale and the other is therefore multiplicative. Say $\tilde{H}_1(\wp)$ is étale and $\tilde{H}_2(\wp)$ is multiplicative. If follows that $H_1(\wp)$ is étale and that $H_2(\wp)$ is multiplicative. In particular $A(p) = H_1 \times H_2$ where $H_1 = \prod H_1(\wp)$ is étale and $H_2 = \prod H_2(\wp)$ is multiplicative.

Since there are no maps between an étale p-divisible group and a multiplicative p-divisible group we are reduced to showing that if G is a 'pure' (that is either an étale or a multiplicative) p-divisible group over $\mathbb{Z}_p$ then

$$\mathrm{End}(G) = \mathrm{End}(\tilde{G}). \tag{2.7}$$

But by duality it suffices to prove (2.7) when G is étale in which case it is known to be true [14, Lemma, Chapter IV, A.2.3]. Summarizing we have shown that after a possible replacement of A by an isogenous abelian variety $\mathrm{End}(A(p)) = \mathrm{End}(\tilde{A}(p))$ and the map (2.4) is an isomorphism.

Recall that an abelian variety X defined over an arbitrary field is said to be of CM-type if $\mathrm{End}(X) \otimes \mathbb{Q}$ contains a semi-simple commutative $\mathbb{Q}$-subalgebra of dimension $2\dim(X)$ over $\mathbb{Q}$. Since $\tilde{A}$ is defined over a finite field it is an abelian variety of CM-type. After tensoring with $\mathbb{Q}$, (2.4) is an isomorphism for $A = A_f$, so A_f is of CM-type as well. By a result of Shimura [18, Proposition 1.6], f is a CM form. □

2.2. Case $p \mid N$

Let us turn our attention to the second case, namely, let us assume that $p|N$. Write N_p for the exponent of the exact power of p that divides N. Let C denote the conductor of χ and let C_p denote the exponent of the exact power of p that divides C. Then we have that either

1. (Principal series) $N_p = C_p$ in which case $|a(p, f)| = p^{(k-1)/2}$, or,
2. (Steinberg) $N_p = 1$ and $C_p = 0$ in which case $a(p, f)^2 = \chi(p)p^{k-2}$, or,
3. (Other) $N_p \geq 2$ and $N_p \neq C_p$ in which case $a(p, f) = 0$.

The last case contains the cases where π_p, the local factor at p in the automorphic representation π attached to f, is supercuspidal.

We now discuss each of the above cases in turn, mostly under the additional hypothesis that $k = 2$.

2.2.1. Principal series case

Proposition 5. *Suppose that f is a* CM *form of weight $k \geq 2$. Say $p|N$ and $N_p = C_p$. If f is $\wp$-ordinary then $\rho_{f,\wp}$ splits at p.*

Proof. The proof is similar to the proof of Proposition 3. The only difference worth noting is that the nebentypus character χ is now ramified at $\mathfrak{p}$. Nonetheless the equation (2.3) shows that at least one of ϕ or ϕ^σ is ramified at $\mathfrak{p}$ and the proof proceeds as before. Note also that the proof of Proposition 3 suggested by the referee applies verbatim in the present case. □

Again we can offer a converse similar to Proposition 4 when the weight of f is 2. Recall that complex conjugation induces an automorphism of K_f. When $k = 2$ we have $a(p, f)\overline{a(p, f)} = p$. In particular the sum of the $\wp$-adic valuation of $a(p, f)$ and the $\overline{\wp}$-adic valuation of $a(p, f)$ is non-zero.

Proposition 6. *Suppose that f has weight 2, and that $p|N$ with $N_p = C_p$. Suppose that for each prime $\wp|p$ of K_f either $a(p,f)$ is a $\wp$-adic unit and $\rho_{f,\wp}$ splits at p, or $a(p,f)$ is a $\overline{\wp}$-adic unit and $\rho_{f,\overline{\wp}}$ splits at p. Then f is a* CM *form.*

Proof. The proof is similar to that of Proposition 4 and so we only outline the differences here. This time the abelian variety $A = A_f$ has potentially good reduction at p. This may be seen by studying the ℓ-adic Galois representation $\rho : \mathrm{Gal}(\bar{\mathbb{Q}}/\mathbb{Q}) \to \mathrm{Aut}(T_\ell(A))$ on the ℓ-adic Tate module of A for a prime $\ell \neq p$. It turns out that $\rho(I_p)$ has finite image, so that ρ is unramified when restricted to a finite extension F of $\mathbb{Q}$. Further, the number field F is independent of ℓ and may be taken to be the field cut out by the p-part of the nebentypus character χ. By [15, Theorem 2], A has good reduction at $\mathfrak{p}$, the unique prime of F lying over p.

Let $B = A \times_{\mathbb{Q}} F$ and let $\tilde{B}$ denote the abelian variety over $\mathcal{O}_F/\mathfrak{p}$ which is the reduction of B at the prime $\mathfrak{p}$. The proof of Proposition 4 now goes through as before if we replace A by B. Indeed for each prime $\wp|p$ of K_f we must have $\wp \neq \overline{\wp}$ since the sum of the $\wp$-adic valuation of $a(p,f)$ and $\overline{\wp}$-adic valuation of $a(p,f)$ is non-zero whereas exactly one of these two valuations is zero by hypothesis. Thus, replacing B by an isogenous abelian variety if necessary, we may decompose the p-divisible group $B(p)$ as follows:

$$B(p) = \prod_{\wp|p} B(\wp) = G_1 \times G_2$$

where G_1 is the product of all the $B(\wp)$ with $a(p,f)$ a $\wp$-adic unit and $G_2 = \bar{G}_1$ is the product of the remaining $B(\wp)$.

Now by hypothesis $\rho_{f,\wp}$ is split when restricted to $H_{\mathfrak{p}}$, the decomposition group of F at $\mathfrak{p}$, for all primes $\wp$ of K_f such that $B(\wp)$ occurs in G_1. So for each factor $B(\wp)$ occurring in G_1 we may write $B(\wp) = H_1(\wp) \times H_2(\wp)$ as before. On the other hand the characteristic polynomial of the crystalline Frobenius on the $\mathbb{Q}_p$-crystal attached to $\tilde{B}(\wp)$ is

$$\mathrm{N}_{\mathbb{Q}_p}^{K_{f,\wp}}\left((x - a(p,f))(x - \chi'(p)\overline{a(p,f)})\right)$$

where χ' is the prime to p-part of χ (compare with (2.6)). Thus as before $\tilde{B}(\wp)$ is ordinary and is a product of an étale and a multiplicative p-divisible group. We conclude as before that say $H_1(\wp)$ is étale and $H_2(\wp)$ is multiplicative. In particular G_1 is a product of an étale and a multiplicative p-divisible group. Since $G_2 = \bar{G}_1$ we see that the same is true for G_2 and therefore also for G. The rest of the proof proceeds as before showing that (replacing B by an isogenous variety if necessary) $\mathrm{End}(B(p)) = \mathrm{End}(\tilde{B}(p))$ and B is an abelian variety of CM-type. We conclude that A is an abelian variety of CM-type and that f is a CM form. □

2.2.2. Steinberg case The following result of Ribet [13, Theorem 3.6.2 a)] completely answers Greenberg's question in the Steinberg case when $k = 2$. For completeness we will give the proof.

Theorem 2. *Suppose that f has weight 2 and that the p-part of χ is trivial. Suppose that $p||N$. Then f is ordinary at each prime $\wp$ of $\bar{\mathbb{Q}}$ and $\rho_{f,\wp}$ is not split at p for all such $\wp$.*

Proof. For the first statement note that f is ordinary at every prime $\wp$ of $\bar{\mathbb{Q}}$ simply because $a(p,f)$ is a root of unity. For the second, let $A = A_f$ be the abelian variety attached to f. First assume that $A = E$ is an elliptic curve with split multiplicative reduction at p (so $a(p,f) = 1$ and not -1). Then it is well known that the Galois representation attached to p is non-split when restricted to G_p. This follows from the theory of the Tate curve. Indeed, under the above hypotheses, $E_{/\mathbb{Q}_p}$ is isomorphic to a Tate curve defined over $\mathbb{Q}_p$, so that

$$E(\bar{\mathbb{Q}}_p) = \frac{\bar{\mathbb{Q}}_p^\times}{q^{\mathbb{Z}}}$$

for the Tate period $q \in \mathbb{Q}_p$ of E. A set of $q^{\mathbb{Z}}$-coset representatives of the p^n-torsion points of E is therefore given by $\zeta^l \alpha^m$ for $0 \leq l, m \leq n-1$ where ζ is a primitive $p^{n^{\text{th}}}$-root of 1 and α is a $p^{n^{\text{th}}}$-root of q. As a module for $G_p = \mathrm{Gal}(\bar{\mathbb{Q}}_p/\mathbb{Q}_p)$ we see that $E[p^n](\bar{\mathbb{Q}}_p)$ fits in an exact sequence

$$1 \to \mu_{p^n} \to E[p^n] \to \mathbb{Z}/p^n \to 1,$$

where the map $E[p^n] \to \mathbb{Z}/p^n$ is induced by $\zeta^l \alpha^m \mapsto m$. Thus the above sequence is not split as a sequence of $\mathbb{Z}[G_p]$-modules. If $V_p(E)$ is the Tate module of E tensored with $\mathbb{Q}_p$ we obtain a non-split exact sequence of $\mathbb{Q}_p[G_p]$-modules:

$$1 \to \mathbb{Q}_p(1) \to V_p(E) \to \mathbb{Q}_p \to 1$$

and thus ρ_f does not split at p in this case. If $A = E$ has non-split multiplicative reduction at p (so $a(p,f) = -1$), then E acquires split multiplicative reduction over an unramified quadratic extension K of $\mathbb{Q}_p$, and $E_{/K}$ is isomorphic to a Tate curve defined over K. Now the above argument goes through if one works over K, showing that ρ_f is not split on $\mathrm{Gal}(\bar{\mathbb{Q}}_p/K) \subset G_p$. It follows that ρ_f does not split at p.

Now say that $A = A_f$ has dimension $g \geq 1$. It is known that A has purely multiplicative (not necessarily split) reduction at p (see [3]). There is a general theory of p-adic uniformization of abelian varieties that are defined over finite extensions of $\mathbb{Q}_p$ and have purely multiplicative reduction at p. This is due to Tate, McCabe, Mumford and Raynaud (for references see [13]). This theory was used in [13] to show that $\rho_{f,\wp}$ restricted to G_p does not split at p for each prime $\wp$ of K_f with $\wp | p$. Let us recall the argument. It basically extends the argument for the case $g = 1$ outlined above. First assume that A has split multiplicative reduction at p. Then, working over $\mathbb{Q}_p$, one has

$$A(\bar{\mathbb{Q}}_p) = \frac{(\bar{\mathbb{Q}}_p^\times)^g}{L}$$

for a certain free subgroup $L = \langle q_1, q_2, \ldots, q_g \rangle$ of rank g generated by periods $q_i \in (\mathbb{Q}_p^\times)^g$. Write $q_i = (q_{i1}, q_{i2}, \ldots q_{ig})$ for $q_{ij} \in \mathbb{Q}_p^\times$. Let α_{ij} denote a $p^{n^{\text{th}}}$-root of q_{ij}. Then a set of L-coset representatives of the p^n-torsion points of A is

$$u(l_r, m_s) := (\zeta^{l_1}\alpha_{11}^{m_1}\alpha_{21}^{m_2}\cdots\alpha_{g1}^{m_g}, \zeta^{l_2}\alpha_{12}^{m_1}\alpha_{22}^{m_2}\cdots\alpha_{g2}^{m_g}, \ldots, \zeta^{l_g}\alpha_{1g}^{m_1}\alpha_{2g}^{m_2}\cdots\alpha_{gg}^{m_g})$$

where $0 \le l_r, m_s \le p^n - 1$ for $1 \le r, s \le g$. As a module for $G_p = \mathrm{Gal}(\bar{\mathbb{Q}}_p/\mathbb{Q}_p)$ we see that $A[p^n](\bar{\mathbb{Q}}_p)$ fits in the exact sequence

$$1 \to \mu_{p^n}^g \to A[p^n] \to (\mathbb{Z}/p^n)^g \to 1,$$

where the map $A[p^n] \to (\mathbb{Z}/p^n)^g$ is induced by

$$u(l_r, m_s) \quad \mapsto \quad (m_1, m_2, \ldots, m_g).$$

Again the above sequence is not split as a sequence of $\mathbb{Z}[G_p]$-modules. If $V_p(A)$ is the Tate module of A tensored with $\mathbb{Q}_p$ we obtain a non-split exact sequence of $\mathbb{Q}_p[G_p]$-modules:

$$1 \to \mathbb{Q}_p(1)^g \to V_p(A) \to \mathbb{Q}_p^g \to 1. \tag{2.8}$$

Now $V_p(A)$ is a free $K_f \otimes \mathbb{Q}_p$ module of rank 2. Note that

$$K_f \otimes \mathbb{Q}_p = \oplus_{\wp|p} K_{f,\wp}.$$

Since the action of K_f commutes with the action of G_p the module $\mathbb{Q}_p(1)^g$ above is a $K_f \otimes \mathbb{Q}_p$-module which in fact is free of rank 1. Thus, if $\wp|p$, we may take $\wp$-eigenspaces in the exact sequence (2.8) above, to obtain the exact sequence of $\mathbb{Z}[G_p]$-modules:

$$1 \to K_{f,\wp}(1) \to V_\wp(A) \to K_{f,\wp} \to 1. \tag{2.9}$$

Here $V_p(A) = \oplus_{\wp|p} V_\wp(A)$ and $V_\wp(A) = K_{f,\wp}^2$ as a vector space is the $\wp$-eigenspace of $V_p(A)$. An easy argument of Ribet now shows that the sequence (2.9) is still non-split. We conclude that $\rho_{f,\wp}$ restricted to G_p is non-split. If A does not have split multiplicative reduction at p, then, as in the elliptic curve case, we may extend scalars to a finite unramified extension K of $\mathbb{Q}_p$ over which A has split multiplicative reduction, and deduce that $\rho_{f,\wp}$ is non-split on $\mathrm{Gal}(\bar{\mathbb{Q}}_p/K)$. Hence $\rho_{f,\wp}$ is non-split on G_p. □

It is a consequence of the above theorem that if f has weight 2 and is Steinberg at p, that is $N_p = 1$ and $C_p = 0$, then f does not have CM. This can also be seen using ℓ-adic representations for $\ell \neq p$. Indeed, the inertia group I_p at p has finite image under the ℓ-adic representation attached to a CM form. On the other hand, it is a result of Langlands [11] that I_p has infinite image under the ℓ-adic representation attached to a form that is Steinberg at p.

2.2.3. Other cases If $N_p \ge 2$ and $N_p \neq C_p$ then $a(p, f) = 0$ and the cusp form f is not ordinary at any prime $\wp$ of $\bar{\mathbb{Q}}$. In particular Greenberg's question does not arise in this case.

3. Hida families

In the previous section we gave some partial answers to Greenberg's question (Question 1) under the assumption that f has weight $k = 2$. In this and the next section we deal with the case $k > 2$.

Ideally one should approach Question 1 in the higher weight case by studying the fine properties of the motive M_f attached to f since it is the étale cohomology of M_f which gives rise to the Galois representation ρ_f. However this approach has not yet borne fruit.

Instead we use Hida families and some Iwasawa theoretic arguments to study Question 1 in the higher weight case. While these tools do not allow us to answer Question 1 for any particular $\wp$-ordinary form f of weight $k > 2$, we are able to prove a result of the following form: all but finitely many of the Galois representations attached to the members of the Hida family containing f are non-split at p under a suitable non-splitting hypothesis on one member of the family.

In this section we recall some properties of Hida families and leave the statements of our results (and their proofs) to the next section. General references for the material presented below are [9] and [10].

Let $f \in S_k(N, \chi)$ be a primitive form of arbitrary weight $k \geq 2$. Fix an embedding ι_p of $\bar{\mathbb{Q}}$ into $\bar{\mathbb{Q}}_p$, equivalently, fix a prime $\wp$ of $\bar{\mathbb{Q}}$. Suppose that f is ordinary at $\wp$ (with respect to the embedding i_∞ of $\bar{\mathbb{Q}}$ into $\mathbb{C}$).

Assume for simplicity that

$$p \text{ is prime to } N. \tag{3.1}$$

Let $\{f_l \mid l \geq 1\}$ denote the Hida family to which f 'belongs'. It is a consequence of Hida's theory (more precisely an étaleness statement for the new part of Hida's ordinary Hecke algebra localized at a suitable height one prime of the Iwasawa algebra) that the forms f_l are uniquely determined by f. Now

$$f_l \in S_l^{\mathrm{ord}}(Np, \chi\omega_p^{k-l})$$

where $\omega_p : (\mathbb{Z}/p)^\times \to \mu_{p-1}$ is the Teichmüller character. The superscript ord means that for each $l \geq 1$, the cusp form f_l is ordinary at $\wp$ with respect to the embeddings ι_∞ and ι_p. Moreover f_k, the weight k member of this family, is related to f by the following process of 'p-stabilization'. Let α denote the unique $\wp$-adic unit root of $x^2 - a(p, f)x + \chi(p)p^{k-1}$ and let β be defined by $\alpha\beta = \chi(p)p^{k-1}$. Then $f(z)$ and $f(pz) \in S_k(Np, \chi)$ as old forms and $f_k = f(z) - \beta f(pz)$. Note that f_k is an old form that is a common eigenform of all the Hecke operators T_q for q prime to N with eigenvalue $a(q, f)$ and for the Hecke operators U_q for all primes $q|N$ with eigenvalue $a(q, f)$. It is also an eigenvector of U_p with eigenvalue α.

A reflection of the fact that f 'lives' in a family is that there is a 'large' Galois representation that interpolates the Galois representations $\rho_{f_l,\wp}$ attached to f_ℓ by Deligne for $l \geq 2$. Let us explain this in more detail. Let $\Lambda = \mathbb{Z}_p[[T]]$. This is a regular local domain of Krull dimension 2. Consider for each $l \geq 1$ the homomorphism ϕ_l of Λ given by

$$\begin{aligned} \phi_l : \Lambda &\rightarrow \mathbb{Z}_p \\ T &\mapsto (1+p)^l - 1. \end{aligned}$$

Let K be a sufficiently large field extension of finite degree of the quotient field of Λ. Let L be the integral closure of Λ in K. Let $\mathcal{O}$ be the ring of integers in a sufficiently large finite extension of $\mathbb{Q}_p$. Let $P_l : L \rightarrow \mathcal{O}$ denote an algebra homomorphism which extends the homomorphisms ϕ_l on Λ. Now Hida has constructed a Galois representation

$$\rho_\Lambda : \mathrm{Gal}(\bar{\mathbb{Q}}/\mathbb{Q}) \rightarrow \mathrm{GL}_2(K)$$

which has the following property: for each $l \geq 1$ there is a height one prime of L, which is the kernel of some homomorphism $P_l : L \rightarrow \mathcal{O}$ as above, such that the reduction of ρ_Λ at this prime yields a representation

$$\rho_\ell : \mathrm{Gal}(\bar{\mathbb{Q}}/\mathbb{Q}) \rightarrow \mathrm{GL}_2(\bar{\mathbb{Q}}_p)$$

which is equivalent to Deligne's representation $\rho_{f_l,\wp}$ when $l \geq 2$. In particular the representation ρ_k is equivalent to ρ_f.

The representation ρ_Λ above satisfies two important additional properties. Firstly, ρ_Λ restricted to a decomposition group is also 'upper triangular': there are characters δ_Λ and ϵ_Λ of G_p with values in $K^\times$ with ϵ_Λ unramified such that

$$\rho_\Lambda|_{G_p} \sim \begin{pmatrix} \delta_\Lambda & u_\Lambda \\ 0 & \epsilon_\Lambda \end{pmatrix}. \tag{3.2}$$

Again $u_\Lambda : G_p \rightarrow K$ is just a map, and not necessarily a homomorphism.

Secondly, the reductions of the Galois representations ρ_l for $l \geq 1$ are all mutually equivalent. More precisely let $\mathbb{F} = L/\mathfrak{m}_L$ denote the residue field of L and for $l \geq 1$ let

$$\bar{\rho}_l^{\mathrm{ss}} : \mathrm{Gal}(\bar{\mathbb{Q}}/\mathbb{Q}) \rightarrow \mathrm{GL}_2(\mathbb{F})$$

denote the semi-simplification of the reduction of ρ_l. Then

$$\bar{\rho}_{l_1}^{\mathrm{ss}} \sim \bar{\rho}_{l_2}^{\mathrm{ss}} \tag{3.3}$$

for any two weights $l_1 \geq 1$ and $l_2 \geq 1$. It is this last condition that is the key defining property of the Hida family.

4. Iwasawa theoretic arguments

Keep the notation of the previous section. In particular f is a $\wp$-ordinary form (of prime to p level N). We are interested in the behavior of ρ_f restricted to G_p. Recall that f_l for $l \geq 1$ are the members of the Hida family containing f.

Theorem 3. *Suppose that $\rho_{f_{l_0}}$ is non-split when restricted to G_p for one integer $l_0 \geq 1$. Then for all but finitely many $l \geq 1$ the representation ρ_{f_l} is non-split when restricted to G_p.*

Proof. Let $\mathfrak{m}_L$ denote the maximal ideal of L. By replacing L by a finite extension if necessary, we may assume that L contains a complete set of representatives of its residue field $\mathbb{F} = L/\mathfrak{m}_L$. Then

$$L^\times = \mu_{q-1} \times (1 + \mathfrak{m}_L) \tag{4.1}$$

where q denotes the order of $\mathbb{F}$. Note that μ_{q-1} has order prime to p and that $1 + \mathfrak{m}_L$ is a pro-p group.

For a character of $\psi : G_p \to L^\times$ write ψ^{t}, respectively ψ^{w} for the projection of ψ onto the first, respectively second, factor in (4.1) above.

Let E^{t} denote the union of the (finitely many) tamely ramified extensions of $\mathbb{Q}_p$ of degree dividing $q-1$. Let $\Delta = \mathrm{Gal}(E^{\mathrm{t}}/\mathbb{Q})$. Note that both $\delta_\Lambda^{\mathrm{t}}$ and $\epsilon_\Lambda^{\mathrm{t}}$ factor through Δ.

Let $\mathbb{Q}_p^{\mathrm{unr}}$ denote the maximal unramified $\mathbb{Z}_p$-extension of $\mathbb{Q}_p$ and let $\mathbb{Q}_p^{\mathrm{cycl}}$ denote the cyclotomic $\mathbb{Z}_p$-extension of $\mathbb{Q}_p$. Let E^{w} be the maximal abelian pro-p extension of $\mathbb{Q}_p$. It is well known that

$$E^{\mathrm{w}} = \mathbb{Q}_p^{\mathrm{unr}} \cdot \mathbb{Q}_p^{\mathrm{cycl}}$$

and that $\Gamma := \mathrm{Gal}(E^{\mathrm{w}}/\mathbb{Q}_p) \xrightarrow{\sim} \mathbb{Z}_p^2$. Note that both $\delta_\Lambda^{\mathrm{w}}$ and $\epsilon_\Lambda^{\mathrm{w}}$ factor through Γ.

Let $E = E^{\mathrm{t}} \cdot E^{\mathrm{w}}$ denote the compositum of E^{t} and E^{w}. Note that E is a Galois extension of $\mathbb{Q}_p$ with Galois group $G := \mathrm{Gal}(E/\mathbb{Q}_p) = \Delta \times \Gamma$. After the discussion above the characters δ_Λ and ϵ_Λ factor through G.

Let $H = \mathrm{Gal}(\bar{\mathbb{Q}}_p/E)$. Since δ_Λ and ϵ_Λ are trivial on H we have

$$\rho_\Lambda|_H = \begin{pmatrix} 1 & \lambda \\ 0 & 1 \end{pmatrix}$$

for a genuine homomorphism $\lambda : H \to L$.

We now claim that image$(\lambda) \neq 0$. Indeed if $\lambda = 0$ then ρ_Λ is split when restricted to H. This means that the specializations ρ_l for all $l \geq 1$ are all split when restricted to H. In particular ρ_{l_0} splits when restricted to H. In cohomological terms this means that the class of the cocycle c attached to $\rho_{l_0}|_{G_p}$ as in Section 1 has trivial image under the restriction map

$$\mathrm{H}^1(G_p, K_0(\chi)) \to \mathrm{H}^1(H, K_0(\chi)) \tag{4.2}$$

where $K_0 = \bar{K}_{f_{l_0},\wp} = \bar{\mathbb{Q}}_p$ and $\chi = \delta_{l_0}{\epsilon_{l_0}}^{-1}$. On the other hand the map (4.2) is injective. Accepting this for the moment we deduce that the class of c in $\mathrm{H}^1(G_p, K_0(\chi))$ is trivial. This contradicts the hypothesis of the theorem that ρ_{l_0} is not split when restricted to G_p, proving the claim.

It remains to show that (4.2) is injective. This can be seen as follows. By the inflation-restriction sequence the kernel of (4.2) is

$$\mathrm{H}^1(G, K_0(\chi)^H) = \mathrm{H}^1(G, K_0(\chi)) = \mathrm{H}^1(\Delta, K_0(\chi)^\Gamma) \oplus \mathrm{H}^1(\Gamma, K_0(\chi)^\Delta).$$

The first cohomology group on the right-hand side vanishes since Δ is a finite group. As for the second cohomology group it vanishes as well if χ is non-trivial on Δ, so we may assume that $K_0(\chi)^\Delta = K_0(\chi)$, in which case

$$\mathrm{H}^1(G, K_0(\chi)^H) = \mathrm{H}^1(\Gamma, K_0(\chi)).$$

Now $\Gamma = \Gamma_1 \times \Gamma_2$ with $\Gamma_i = \mathbb{Z}_p$ for $i = 1, 2$. Consider the exact sequence

$$0 \to \mathrm{H}^1(\Gamma_2, K_0(\chi)^{\Gamma_1}) \to \mathrm{H}^1(\Gamma, K_0(\chi)) \to \mathrm{H}^1(\Gamma_1, K_0(\chi)).$$

Note that $\chi \neq 1$ on Γ. Without loss of generality we may assume that χ restricts to a non-trivial character on Γ_1. Then $K_0(\chi)^{\Gamma_1} = 0$ and

$$\mathrm{H}^1(\Gamma_1, K_0(\chi)) = K_0/(\chi(\gamma) - 1)K_0 = 0$$

where γ is a topological generator of Γ_1. This shows that $\mathrm{H}^1(\Gamma, K_0(\chi)) = 0$ and that the map (4.2) is injective.

Let I denote the (non-zero) ideal of L generated by the image of λ. Recall that P_l is an extension to L of the weight l specialization map ϕ_l of Λ. Let $\mathfrak{p}_l := \ker P_l \subset L$. It is a height one prime ideal of L. Clearly ρ_l is split at G_p if and only if image$(\lambda) \subset \mathfrak{p}_l$, that is, if and only if $I \subset \mathfrak{p}_l$. Since the intersection of infinitely many height one primes in L is 0, the ideal I can only be contained in finitely many $\mathfrak{p}_l$ and we conclude that ρ_l is split when restricted to G_p for finitely many weights $l \geq 1$. This proves the theorem. □

We now try and get a better handle on the image of the homomorphism $\lambda : H \to L$. This will allow us to describe the number of generators of the ideal I of L generated by this image. We keep the notation introduced in the proof of Theorem 3.

Let $\mu = \delta_\Lambda {\epsilon_\Lambda}^{-1}$ as a character of G with values in $L^\times$. Let $\Lambda_G = \mathbb{Z}_p[[G]]$ be the completed group algebra of G. We write μ again for the induced homomorphism of algebras:

$$\mu : \Lambda_G \to L.$$

Write $L(\mu)$ for L considered as a Λ_G-module via μ.

Let M denote the maximal pro-p abelian extension of E and let

$$X := \mathrm{Gal}(M/E).$$

Since X is both a $\mathbb{Z}_p$-module and a G-module it is a module for Λ_G. Moreover since L is a pro-p group, the homomorphism $\lambda : H \to L$ factors through X. We have:

Lemma 2. *The homomorphism $\lambda : X \to L(\mu)$ is a morphism of Λ_G-modules.*

Proof. This follows from the following identity

$$\begin{pmatrix} \delta_\Lambda & 0 \\ 0 & \epsilon_\Lambda \end{pmatrix} \cdot \begin{pmatrix} 1 & \lambda \\ 0 & 1 \end{pmatrix} \cdot \begin{pmatrix} {\delta_\Lambda}^{-1} & 0 \\ 0 & {\epsilon_\Lambda}^{-1} \end{pmatrix} = \begin{pmatrix} 1 & \mu \cdot \lambda \\ 0 & 1 \end{pmatrix}.$$

□

We may write $\Lambda_G = \mathbb{Z}_p[[G]] = \mathbb{Z}_p[[\Gamma]][\Delta]$. Since Δ has order prime to p, X decomposes into eigenspaces for the action of Δ on X:

$$X = \bigoplus_{\chi \in \widehat{\Delta}} X^{\chi}. \tag{4.3}$$

Each eigenspace is a $\mathbb{Z}_p[[\Gamma]]$-module which is essentially free of rank 1. More precisely, the $\mathbb{Z}_p[[\Gamma]]$-module structure of the eigenspaces is given by the following proposition. We shall assume for simplicity that

$$L = \Lambda \text{ so that } q = p. \tag{4.4}$$

Proposition 7. *Under (4.4), X^{χ} is a projective $\mathbb{Z}_p[[\Gamma]]$-module of rank 1. More precisely we have:*

1. *If $\chi \neq \omega_p$ then $X^{\chi} \xrightarrow{\sim} \mathbb{Z}_p[[\Gamma]]$.*
2. *If $\chi = \omega_p$ then $X^{\chi} \xrightarrow{\sim} \mathfrak{a}$ where $\mathfrak{a}$ is the augmentation ideal in $\mathbb{Z}_p[[\Gamma]]$, considered as a power series ring in two variables.*

Proof. The proof follows from local class field theory. The argument is similar to the proof of [7, Theorem 3.11]. □

Consider μ^{t} as a character on Δ. By Lemma 2 we see that $\lambda : X \to L(\mu)$ factors through $X^{\mu^{\mathrm{t}}}$. Let M^{λ} denote the fixed field of the kernel of λ and let

$$X(\lambda) := \mathrm{Gal}(M^{\lambda}/E).$$

Then there are natural surjections $X \twoheadrightarrow X^{\mu^{\mathrm{t}}} \twoheadrightarrow X(\lambda)$.

Let $\Lambda_\Gamma = \mathbb{Z}_p[[\Gamma]]$. By the above proposition $X^{\mu^{\mathrm{t}}}$ is isomorphic to Λ_Γ or to $\mathfrak{a}$ as a module over Λ_Γ. The character μ^{w} of Γ induces a homomorphism $\mu^{\mathrm{w}} : \Lambda_\Gamma \to R$. Therefore, by restricting to $\mathfrak{a}$ if necessary, we may consider μ^{w} as a homomorphism

$$\mu^{\mathrm{w}} : X^{\mu^{\mathrm{t}}} \to L.$$

Let M^{μ} denote the fixed field of $\ker(\mu^{\mathrm{w}}) \subset X^{\mu^{\mathrm{t}}}$ and let

$$X(\mu) := \mathrm{Gal}(M^{\mu}/E).$$

Thus we have natural surjections $X \twoheadrightarrow X^{\mu^{\mathrm{t}}} \twoheadrightarrow X(\mu)$.

Lemma 3. *We have $M^{\lambda} \subset M^{\mu}$.*

Proof. To prove the lemma it suffices to show that $\ker(\mu^{\mathrm{w}}) \subset \ker(\lambda)$ as subgroups of $X^{\mu^{\mathrm{t}}}$. Suppose that $r \in \ker(\mu^{\mathrm{w}})$. Then by Lemma 2 we have

$$\lambda(r \cdot x) = \mu^{\mathrm{w}}(r) \cdot \lambda(x) = 0$$

for all $x \in X^{\mu^{\mathrm{t}}}$. If $X^{\mu^{\mathrm{t}}} = \Lambda_\Gamma$, then taking $x = 1$ we see that $\lambda(r) = 0$ as desired. If $X^{\mu^{\mathrm{t}}} = \mathfrak{a}$ then we still have that

$$\mu^{\mathrm{w}}(x) \cdot \lambda(r) = \lambda(x \cdot r) = \lambda(r \cdot x) = 0$$

for all $x \in \mathfrak{a}$. But L is an integral domain. Since $\mu^{\mathrm{w}}(x) \neq 0$ for some $x \in \mathfrak{a}$ we see that $\lambda(r) = 0$, as desired. □

By Lemma 3 we have $X(\mu) \twoheadrightarrow X(\lambda)$. In sum we see that there is a map of Λ_Γ-modules $X^{\mu^t} \twoheadrightarrow \mathrm{image}(\lambda)$.

Since X^{μ^t} is generated by at most two elements as a Λ_Γ-module (cf. Proposition 7) we see that image(λ) is generated by less than two elements as a Λ_Γ-module. In particular I, the ideal generated by image(λ) in L, is generated by at most two elements. More precisely we have:

Corollary 1. *Assume that $L = \Lambda$. Then the ideal $I \subset L$ is generated by at most two elements, and, if $\mu^t \neq \omega_p$ then it is generated by one element. In particular if $\mu^t \neq \omega_p$ then there is a power-series $g(T) \in \Lambda = \mathbb{Z}_p[[T]]$ such that the finite set of weights $l \geq 1$ for which ρ_l is split are zeros of $g(T)$ of the form $T = (1+p)^l - 1$.*

Proof. The first statement follows from the discussion above. For the second say that I is generated by the power series $g(T) \in \Lambda$. Recall that $\rho_l|_{G_p}$ is split if and only if $I \subset \mathfrak{p}_l$ and in the present context this translates into the condition that $g(T)$ has a zero at $T = (1+p)^l - 1$. This proves the corollary. □

We remark that in the situation $L = \Lambda$ and $\mu^t \neq \omega_p$ another way to see that the set of weights $l \geq 1$ for which $\rho_l|_{G_p}$ is split is finite is to note that the non-zero power series $g(T)$ has only finitely many zeros by the Weierstrass preparation theorem.

5. The Δ-function

Let $\Delta = \sum \tau(n) q^n$ be the unique primitive cusp form of level $N = 1$ and weight $k = 12$. Let p be a prime which is ordinary for Δ and let $\rho_\Delta : \mathrm{Gal}(\bar{\mathbb{Q}}/\mathbb{Q}) \to \mathrm{GL}_2(\mathbb{Q}_p)$ denote the associated Galois representation. In this section we illustrate how one can obtain some information towards Greenberg's question for ρ_Δ using the results proved in this paper. The main thrust of the argument is to combine the weight 2 results in Section 2 with Theorem 3 taking $l_0 = 2$. We obtain, under an assumption of total ordinariness, a criterion for whether $\rho_\Delta|_{G_p}$ is split in terms of the vanishing of certain power series.

Recall that Δ has Fourier coefficients $\tau(n) \in \mathbb{Z}$. Moreover $\tau(n)$ is coprime to n for all primes n smaller than 10^6 except for $n = 2, 3, 5, 7$ and 2411 (see Gouvêa [6]). In particular $p \geq 11$. Note Δ is automatically ordinary at every prime $\wp$ of $\bar{\mathbb{Q}}$ lying over p since $\tau(p) \in \mathbb{Z}$. Fix a prime $\wp$ of $\bar{\mathbb{Q}}$ lying over p. Let $\Delta_l \in S_l(p, \omega_p^{12-l})$ denote the weight $l \geq 1$ member of the corresponding Hida family which contains Δ. Recall that here the word contains means that the level p form Δ_{12} is the p-stabilized modular form associated to (the level one form) Δ by the process described in Section 3.

In this case we may take $L = \Lambda = \mathbb{Z}_p[[T]]$. Moreover one may check that

$$\det(\rho_\Lambda) = \kappa \cdot \omega_p^k \cdot \nu^{-1}$$

where $\kappa : \mathrm{Gal}(\bar{\mathbb{Q}}/\mathbb{Q}) \to \Lambda^\times$ is the character induced by composing the projection $\mathrm{Gal}(\bar{\mathbb{Q}}/\mathbb{Q}) \twoheadrightarrow \mathrm{Gal}(\mathbb{Q}_\infty/\mathbb{Q}) = 1 + p\mathbb{Z}_p$ where $\mathbb{Q}_\infty$ is the cyclotomic $\mathbb{Z}_p$-extension of $\mathbb{Q}$ with the character $1+p\mathbb{Z}_p \to \Lambda^\times$ which takes $(1+p)^s$ for $s \in \mathbb{Z}_p$ to $(1+T)^s \in \Lambda^\times$

and $\nu : \mathrm{Gal}(\bar{\mathbb{Q}}/\mathbb{Q}) \twoheadrightarrow \mathbb{Z}_p^\times \hookrightarrow \Lambda^\times$ is the p-adic cyclotomic character. Recall that $\mu = \delta_\Lambda {\epsilon_\Lambda}^{-1} = \det(\rho_\Lambda) \cdot {\epsilon_\Lambda}^{-2}$. Hence $\mu^{\mathrm{t}} = \omega_p^{k-1} \cdot (\epsilon_\Lambda^{\mathrm{t}})^{-2}$. Since $\epsilon_\Lambda^{\mathrm{t}}$ is unramified it is not hard to see that if $p > 11$ then $\mu^t \neq \omega_p$. Thus assuming $p \neq 11$ puts us in the setting of Corollary 1 and we obtain a power series $g(T) \in \mathbb{Z}_p[[T]]$. Applying Theorem 3 we get:

Proposition 8. *Let $p > 11$ be ordinary for Δ. Assume that $\rho_{l_0}|_{G_p}$ is not split for some weight $l_0 \geq 1$. Then $\rho_\Delta|_{G_p}$ splits if and only if $g(T)$ vanishes at $T = (1+p)^{12} - 1$.*

We are interested in applying the above proposition when $l_0 = 2$. The form $\Delta_2 \in S_2(p, \omega_p^{10})$. The associated nebentypus ω_p^{10} has order $\frac{p-1}{\gcd(10,p-1)}$. Thus if $p > 11$ then $\omega_p^{10} \neq 1$ and the component at p of the automorphic representation corresponding to Δ_2 is in the (ramified) principal series. On the other hand we have:

Lemma 4. *Let $p \geq 11$. Then Δ_2 does not have* CM.

Proof. This follows from the general fact that the specializations of a non-CM Λ-adic modular form are non-CM forms in all weights $l \geq 2$ (see, for instance, [5]). In this particular case the lemma can also be seen directly. If $p = 11$ then Δ_2 is Steinberg at p and so cannot have CM. So assume that $p > 11$. Recall (cf. [17, Lemma 3]) that a general cusp form g has CM if g is of the form $g = \sum_{\mathfrak{a}} \lambda(\mathfrak{a}) q^{N(\mathfrak{a})}$ where λ is a Hecke character of an imaginary quadratic field M and $\mathfrak{a}$ varies through all integral ideals of M. Further there is an integer $\nu \geq 1$ and an integral ideal $\mathfrak{c}$ of M such that $\lambda((\alpha)) = \alpha^\nu$ for all $\alpha \equiv 1 \mod \mathfrak{c}$. If M has discriminant D then g is a cusp form of level $\mathrm{N}(\mathfrak{c})|D|$, weight $\nu + 1$ and nebentypus $\lambda|_{\mathbb{Q}} \cdot \omega_{M/\mathbb{Q}}$ where $\lambda|_{\mathbb{Q}}$ is the Dirichlet character of level $\mathrm{N}(\mathfrak{c})$ taking a to $\lambda((a))/a^\nu$ for all $a \in \mathbb{Z}$ with $(a, \mathrm{N}(\mathfrak{c})) = 1$ and $\omega_{M/\mathbb{Q}}$ is the quadratic character corresponding to $M/\mathbb{Q}$. A necessary condition for λ as above to exist is

$$\zeta^\nu = 1 \text{ for all roots of unity } \zeta \in K \text{ with } \zeta \equiv 1 \mod \mathfrak{c}. \tag{5.1}$$

If $\Delta_2 = g$ for some g as above then we must have $\nu = 1$ and $\mathfrak{c} = 1$ (and $D = -p$ where p must satisfy $p \equiv 3 \mod 4$ and $p > 11$). Taking $\zeta = -1$ we see that the condition (5.1) fails and so Δ_2 does not have CM.

Let K_2 denote the Hecke field of Δ_2. The lemma shows that K_2 is a CM field when $p > 11$. Assume from now on that $p > 11$. As in the introduction we shall say that p is totally ordinary for Δ_2 if for each prime $\wp$ of K_2 lying above p either $a(p, \Delta_2)$ is a $\wp$-adic unit or it is a $\overline{\wp}$-adic unit. We have the following theorem.

Theorem 4. *Assume that $p > 11$ and that Δ_2 is totally ordinary at p. Then ρ_Δ is split at p if and only if $g(T)$ has a zero at $T = (1+p)^{12} - 1$.*

Proof. Since Δ_2 does not have CM then by Proposition 6 there must be a place $\wp$ of $\bar{\mathbb{Q}}$ lying above p such that $\rho_{2,\wp}|_{G_p}$ is non-split. By conjugating with an element of $\mathrm{Gal}(\bar{\mathbb{Q}}/\mathbb{Q})$ if necessary we may assume that $\wp$ is the place of $\bar{\mathbb{Q}}$ that was earlier fixed in order to speak of the Hida family containing Δ. The theorem now follows from Proposition 8. □

Note that the condition of total ordinariness on Δ_2 is easy to check numerically. It is true if $p = 13$ for instance.

6. Filtered modules

We conclude with an equivalent formulation by Breuil of Greenberg's question, in the case that N is prime to p, in terms of filtered modules.

Let f be a $\wp$-ordinary primitive form of level N prime to p. Fontaine has associated a filtered (ϕ, N)-module $D = D_{\mathrm{st}}(\rho_f|_{G_p})$ to the restriction of $\rho_f : \mathrm{Gal}(\bar{\mathbb{Q}}/\mathbb{Q}) \to \mathrm{GL}_2(K_{f,\wp})$ to the decomposition group at p. This is a finite-dimensional $\mathbb{Q}_p$-vector space which is equipped with a linear map $\phi : D \to D$, a decreasing exhaustive filtration $\mathrm{Fil}^i(D)$ of $\mathbb{Q}_p$-subspaces of D, and a linear operator $N : D \to D$ called the monodromy operator, which, because of our assumption that p is prime to the level, is trivial. Moreover D has an action of $K_{f,\wp}$ which commutes with all the structures on D mentioned above and makes D into a two-dimensional $K_{f,\wp}$-module.

Let μ_1 and μ_2 be defined by $\mu_1\mu_2 = \chi(p)$ and $p^{k-1}\mu_1 + \mu_2 = a(p, f)$. Let $K'_{f,\wp}$ be a sufficiently large finite extension of $K'_{f,\wp}$ and consider the following filtered (ϕ, N)-modules of dimension two over $K'_{f,\wp}$ with basis e_1, e_2 defined by

$$D_{\text{split}} : \begin{cases} \phi(e_1) = p^{k-1}\mu_1 e_1 \\ \phi(e_2) = \mu_2 e_2 \\ \mathrm{Fil}^{k-1}(D) = K'_{f,\wp} \cdot e_1 \\ N = 0 \end{cases} \qquad D_{\text{non-split}} : \begin{cases} \phi(e_1) = p^{k-1}(\mu_1 e_1 + e_2) \\ \phi(e_2) = \mu_2 e_2 \\ \mathrm{Fil}^{k-1}(D) = K'_{f,\wp} \cdot e_1 \\ N = 0 \end{cases} .$$

Let D^* denote the filtered (ϕ, N)-module corresponding to the representation which is dual to $\rho_f|_{G_p}$. It turns out that $D^* \sim D_{\text{split}}$ if ρ_f is split at p and $D^* \sim D_{\text{non-split}}$ if ρ_f is not split at p. Breuil has asked the following question (cf. [1]). It is the exact analogue of Greenberg's question in the setting of filtered modules (in the case that p is prime to N).

Question 2. *Suppose that f is a $\wp$-ordinary primitive modular cusp form of weight at least 2. Is the associated filtered module D^* isomorphic to D_{split} or to $D_{\text{non-split}}$?*

Acknowledgements. We wish to thank R. Greenberg for posing the question investigated in this paper during a visit to TIFR in January 2001, and for very helpful discussions related to it. We also thank C. Breuil, A. Brown, O. Bültel, B. Conrad, D. Prasad and the referee for some helpful remarks.

References

[1] C. Breuil. p-adic Hodge theory, Deformations and Local Langlands. *CRM Barcelona Lecture Notes*, 2001.

[2] R. Coleman and J. F. Voloch. Companion forms and Kodaira-Spencer theory. *Invent. Math.*, 110(2):263–281, 1992.

[3] P. Deligne and M. Rapoport. Les schémas de modules de courbes elliptiques. In *Modular functions of one variable, II (Proc. Internat. Summer School, Univ. Antwerp, Antwerp, 1972)*, volume 349 of *Lecture Notes in Math.*, pages 143–316. Springer, Berlin, 1973.

[4] M. Emerton. A p-adic variational Hodge conjecture and modular forms with complex multiplication. *Preprint.*

[5] E. Ghate and V. Vatsal. On the local behaviour of ordinary Λ-adic representations. To appear.

[6] F. Gouvêa. Non-ordinary primes: a story. *Experiment. Math.*, 6(3):195–205, 1997.

[7] R. Greenberg. Introduction to Iwasawa theory of elliptic curves. *IAS/Park City Mathematics Institute Lecture Notes*, 1999.

[8] B. Gross. A tameness criterion for Galois representations associated to modular forms (mod p). *Duke. Math. J.*, 61(2):445–517, 1990.

[9] H. Hida. Iwasawa modules attached to congruences of cusp forms. *Ann. Sci. École Norm. Sup.*, 19(2):231–273, 1986.

[10] H. Hida. Galois representations into $\mathrm{GL}_2(\mathbf{Z}_p[[X]])$ attached to ordinary cusp forms. *Invent. Math.*, 85:545–613, 1986.

[11] R. Langlands. Modular forms and ℓ-adic representations. In *Modular functions of one variable, II (Proc. Internat. Summer School, Univ. Antwerp, Antwerp, 1972)*, volume 349 of *Lecture Notes in Math.*, pages 361–500. Springer, Berlin, 1973.

[12] B. Mazur and A. Wiles. On p-adic analytic families of Galois representations. *Compositio Math.*, 59:231–264, 1986.

[13] K. Ribet. Galois action on division points of Abelian varieties with real multiplications. *Amer. J. Math.*, 98(3):751–804, 1976.

[14] J.-P. Serre. *Abelian l-adic representations and elliptic curves. Second edition.* Advanced Book Classics. Addison-Wesley Publishing Company, Redwood City, CA, 1989.

[15] J.-P. Serre and J. Tate. Good reduction of abelian varieties. *Ann. of Math*, 88:492–517, 1968.

[16] G. Shimura. *Introduction to the arithmetic theory of automorphic functions.* Princeton Univ. Press, Princeton, 1971.

[17] G. Shimura. On elliptic curves with complex multiplication as factors of the Jacobians of modular function fields. *Ngoya Math. J.*, 43:199–208, 1971.

[18] G. Shimura. Class fields over real quadratic fields and Hecke operators. *Ann. of Math*, 95(2):130–190, 1972.

[19] J. Tate. p-divisible groups. In *Proc. Conf. Local Fields (Driebergen, 1966)*, pages 158–183. Springer, Berlin, 1967.

[20] A. Wiles. On ordinary λ-adic representations associated to modular forms. *Inv. Math.*, 94(3):529–573, 1988.

Eknath Ghate
School of Mathematics
Tata Institute of Fundamental Research
Homi Bhabha Road
Mumbai, 400 005, India
e-mail: eghate@math.tifr.res.in

Progress in Mathematics, Vol. 224, 125–139

Arithmetic of $\mathbb{Q}$-curves

Josep González, Joan-Carles Lario, and Jordi Quer

Abstract. This paper is a survey on the arithmetic of $\mathbb{Q}$-curves: the elliptic curves defined over number fields which are isogenous to all their Galois conjugates. Our purpose is to review some results concerning their basic properties such as: the moduli classification, fields of definition, relationship with abelian varieties of GL_2-type, and optimal quotients. Most of the results were separately published before (see [4, 5, 9]), and our aim here is to reinforce a global presentation. What we call "central" $\mathbb{Q}$-curves will constitute the leitmotif.

1. Introduction

Mazur asked in [7] which elliptic curves defined over number fields are modular, i.e., admit a covering from a modular curve $X_1(N)$. The elliptic curves satisfying this property are easily seen to be isogenous to all their Galois conjugates. Elliptic curves defined over number fields that are isogenous to all conjugates were called $\mathbb{Q}$-curves by Ribet (who borrowed the term from Gross [6]) and he conjectured in [11] that this property is enough for ensuring modularity. Since then several efforts have been taken to prove that fact (an account is given in the paper by Ellenberg in this same volume).

An abelian variety A defined over $\mathbb{Q}$ is said to be of GL_2-type if the $\mathbb{Q}$-algebra of endomorphisms defined over $\mathbb{Q}$ is a number field of degree equal to the dimension of the variety. The $\mathbb{Q}$-curves can be characterized as the elliptic curves arising as one-dimensional factors of abelian varieties of GL_2-type (Ribet [11]). The varieties A_f attached by Shimura to weight 2 newforms f for the group $\Gamma_1(N)$ are of GL_2-type. Ribet conjectured [11] that every variety of GL_2-type is isogenous to some A_f, and he proved that this conjecture would follow from Serre's conjecture on the modularity of odd irreducible Galois representations in positive characteristic.

Our research focuses on arithmetic properties of $\mathbb{Q}$-curves, without a special emphasis on modularity aspects or assuming it when necessary. The plan of the paper is as follows. In the following four sections the covered topics include:

(i) Moduli of $\mathbb{Q}$-curves. The action of the absolute Galois group on the isogeny classes of $\mathbb{Q}$-curves gives a rich symmetry governed by the Atkin-Lehner involutions. The rational points of the modular curves $X^*(M)$, for square-free levels M, parametrize the isogeny classes of $\mathbb{Q}$-curves. Section 2 gives an

Research partially supported by grants BFM-2000-0794-C02-02 and 2002SGR 00148.

assortment of results on: rational and elliptic cases, exotic $\mathbb{Q}$-curves, and reduction of their Néron models.

(ii) Fields of definition. In Section 3 we recall the main facts concerning the number fields over which the $\mathbb{Q}$-curves are defined, either as elliptic curves and as $\mathbb{Q}$-curves (i.e., including the isogenies between conjugates), and we also give some results that clarify the relation between $\mathbb{Q}$-curves and abelian varieties of GL_2-type.

(iii) Optimal quotients. In Sections 4 and 5 we show how to obtain the $\mathbb{Q}$-curves packaged by modular abelian varieties, and regard them as optimal quotients. We also introduce the Manin ideal, and generalize the Manin-Stevens conjecture to the setting of $\mathbb{Q}$-curves.

2. Moduli of $\mathbb{Q}$-curves

Let C be an elliptic curve over $\overline{\mathbb{Q}}$, and denote by $\mathcal{C}$ its isogeny class.

Definition 2.1. *We say that C is a $\mathbb{Q}$-curve if, equivalently:*

(i) *For every $\sigma \in \mathrm{Gal}(\overline{\mathbb{Q}}/\mathbb{Q})$, there is an isogeny $\phi_\sigma\colon {}^\sigma C \to C$;*
(ii) $\mathrm{Gal}(\overline{\mathbb{Q}}/\mathbb{Q})$ *acts on $\mathcal{C}$.*

From now on, we will assume that all elliptic curves in this paper are without complex multiplication. The CM elliptic curves deserve a particular treatment, and it is partially covered in the classical literature of class field theory. For analog results in the CM case to those discussed here the reader is refereed to [6] and to the paper by Nakamura in this same volume.

Associated with the isogeny class $\mathcal{C}$ there is a natural graph whose vertices represent isomorphism classes of the elliptic curves belonging to $\mathcal{C}$ and edges represent isogenies. When C is a $\mathbb{Q}$-curve, the action of $G_\mathbb{Q}$ on $\mathcal{C}$ yields a certain symmetry on it. In particular, there are some special vertices (j-invariants) that we call *central* and that play an important role in the arithmetic phenomena (including modularity) of $\mathbb{Q}$-curves. To be more precise, let M be a square-free integer and consider the modular curve $X^*(M) = X_0(M)/B(M)$, where $X_0(M)$ is the modular curve associated to the congruence subgroup $\Gamma_0(M)$ and $B(M)$ denotes the group of Atkin-Lehner involutions of level M. By considering the natural projection $\pi\colon X_0(M) \to X^*(M)$, it is immediate that the pre-image of non-cuspidal non-CM points in $X^*(M)(\mathbb{Q})$ under π gives rise to sets of $\mathbb{Q}$-curves. Indeed, if $P = \pi((j(\tau), j(M\tau)) \in X^*(M)(\mathbb{Q})$ then $\pi^{-1}(P) = \{(j(d\tau), j(M\tau/d))\}_{d|M}$ is a Galois stable set of isogenous elliptic curves.

Definition 2.2. *We say that C is a central $\mathbb{Q}$-curve if it admits isogenies to all its Galois conjugates of square-free degrees. Their isomorphism classes are obtained from $\pi^{-1}(P)$ where $P \in X^*(M)(\mathbb{Q})$ is non-cusp non-CM point and M runs over the square-free integers.*

Notice that central $\mathbb{Q}$-curves can be defined over polyquadratic number fields. To be more precise, let M be the l.c.m. of the square-free degrees of the isogenies linking a central $\mathbb{Q}$-curve to all its conjugate curves and let G be the subgroup of the Atkin-Lehner involutions corresponding to these degrees. Then, the field of moduli of this curve is a polyquadratic field whose Galois group is isomorphic to G. In fact, up to Galois conjugation, each central class gives rise to $[B(M) : G]$ classes of $\mathbb{Q}$-curves and each of these provides a rational point on the intermediate curve $X_0(M)/G$.

The following result was first obtained by Elkies in [2]:

Theorem 2.3 (Elkies). *Every isogeny class of a $\mathbb{Q}$-curve contains a central $\mathbb{Q}$-curve.*

By making use of modular units in $X_0(M)$, the first two authors gave a method to parametrize the central $\mathbb{Q}$-curves in the genus zero and genus one cases [4]. It can be seen that $X^*(M)$ is a rational or an elliptic curve exactly for the following levels:

$\text{genus}(X^*(M)) = 0$	$2, 3, 5, 6, 7, 10, 11, 13, 14, 15, 17, 19, 21, 22, 23, 26, 29, 30, 31, 33, 34, 35, 41, 42, 46, 47, 51, 55, 59, 62, 66, 69, 70, 71, 78, 87, 94, 95, 105, 110, 119.$
$\text{genus}(X^*(M)) = 1$	$37, 43, 53, 57, 58, 61, 65, 74, 77, 79, 82, 83, 86, 89, 91, 101, 102, 111, 114, 118, 123, 130, 131, 138, 142, 143, 145, 155, 159, 174, 182, 190, 195, 210, 222, 231, 238.$

Each one of these 81 cases gives rise to infinitely many central $\mathbb{Q}$-curves.

Example 2.4. When $M = 6$, the corresponding central j-invariants are

$$j = 1728\,\frac{\left(3+3\,t+5\,\sqrt{1+2\,t}-2\,\sqrt{t}\,(3+\sqrt{1+2\,t})\right)\left(5\,t+5\,(1+\sqrt{1+2\,t})-2\,\sqrt{t}\,(5+\sqrt{1+2\,t})\right)^3}{\left(2+\sqrt{t}\right)^6\left(2-\sqrt{t}\right)^2}$$

with $t \in \mathbb{Q}$. Generically, one obtains biquadratic $\mathbb{Q}$-curves over $\mathbb{Q}(\sqrt{t}, \sqrt{1+2\,t})$. For future reference, we consider the specific curve corresponding to $X^*(6)$ obtained by taking the value of the parameter $t = -4/7$. We get four quadratic $\mathbb{Q}$-curves whose j-invariants are:

$$\begin{cases} j_A = 27\,(13171 + 2759\,\sqrt{-7})/128 \\ j_B = 27\,(13171 - 2759\,\sqrt{-7})/128 \end{cases} \qquad \begin{cases} j_C = 27\,(8889 - 3751\,\sqrt{-7})/128 \\ j_D = 27\,(8889 + 3751\,\sqrt{-7})/128 \end{cases}$$

In the diagram below $A, B, \ldots, G, H, \ldots$ denote isomorphism classes of $\mathbb{Q}$-curves. The ones circled correspond to the central $\mathbb{Q}$-curves in the isogeny graph for this particular example.

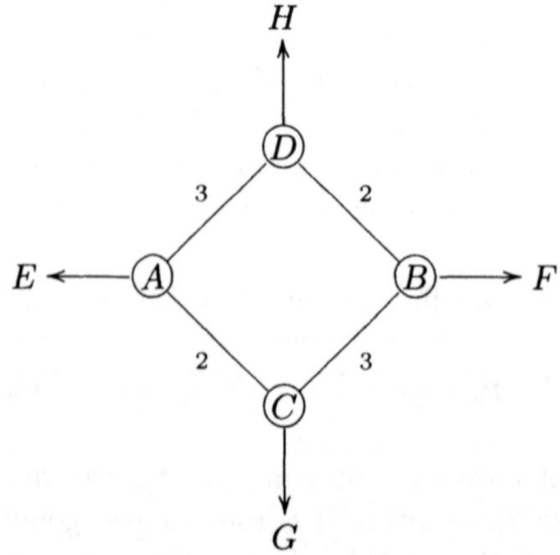

Note that these ones are quadratic $\mathbb{Q}$-curves of degree 6, which provide two rational points on the curve $X_0(6)/\langle W_6\rangle$.

Exotic $\mathbb{Q}$-curves. The central $\mathbb{Q}$-curves arising from $X^*(M)$ of genus greater than one are called exotic. The only 5 known examples when M is a prime are due to Elkies and Galbraith and listed in the table below.

M	j
73	$5/2\big(3(-26670989-15471309\sqrt{\Delta})/2^{25}\big)^3$
103	$19\big(3.2^4(1623826405+30228849\sqrt{\Delta})\big)^3$
137	$\big(2(52655804415199-91887305813\sqrt{\Delta})\big)^2\big(3(1814829135-35283128\sqrt{\Delta})/(2^{47}137^2)\big)^3$
191	$(724537954586714121+16056976492100\sqrt{\Delta})\big(480(7725788647437+95942438\sqrt{\Delta})/191^2\big)^3$
311	$(86643784170081+434217620\sqrt{\Delta})\big(240(22278331350442515+32106727993\sqrt{\Delta})/96721\big)^3$

The corresponding quadratic discriminants in factored form are

$$\Delta = \quad -127, \quad 5\cdot 577, \quad -31159, \quad 61\cdot 229\cdot 145757, \quad 11\cdot 17\cdot 9011\cdot 23629.$$

Conjecturally, one expects finiteness of exotic exemplars.

Properties of quadratic $\mathbb{Q}$-curves. As for the quadratic case, we have the following results given in [3]. The first concerns the fields of definition and the second the arithmetical properties of j-invariants of central $\mathbb{Q}$-curves.

Theorem 2.5. *If there exists a quadratic $\mathbb{Q}$-curve of degree M defined over a quadratic field K, then each divisor $M_1|M$ such that*

$$M_1 \equiv 1 \pmod 4 \quad \text{or} \quad M_1 \ \text{even and} \ M/M_1 \equiv 3 \pmod 4$$

is a norm of K.

Theorem 2.6. *Let C/K be a quadratic $\mathbb{Q}$-curve of degree M. Let j denote its invariant, and σ denote the non-trivial automorphism of K. Then, we have*

i) *If* $M \equiv 1 \pmod 2$, *then* $N_{K/\mathbb{Q}}(j - 1728) \in \mathbb{Q}^2$.
ii) *If* $M \equiv -1 \pmod 3$, *then* $N_{K/\mathbb{Q}}(j) \in \mathbb{Q}^3$.
iii) *If* $M \equiv 1 \pmod 3$, *then* $j/{}^{\sigma}j \in K^3$.

When the degree M is prime, we also have

iv) *If C has multiplicative reduction at a prime $\mathfrak{L}$, then $\ell = \mathfrak{L}\,{}^{\sigma}\mathfrak{L}$ splits in K and C has also multiplicative reduction at ${}^{\sigma}\mathfrak{L}$. Moreover, if $v_{\mathfrak{L}}(j) \le v_{{}^{\sigma}\mathfrak{L}}(j)$ then $v_{\mathfrak{L}}(j) = M\, v_{{}^{\sigma}\mathfrak{L}}(j)$.*
v) *If C has ordinary reduction at $\mathfrak{M} \mid M$, then $M = \mathfrak{M}\,{}^{\sigma}\mathfrak{M}$ splits in K.*
vi) *If C is supersingular at $\mathfrak{M}$ and $M = \mathfrak{M}\,{}^{\sigma}\mathfrak{M}$, then* $j \equiv 0, 1728 \pmod{\mathfrak{M}}$ *and* $M \not\equiv 1 \pmod{12}$.
vii) *If M is inert in K, then* $j \equiv 1728 \pmod M$ *and* $M \not\equiv 1 \pmod 4$.

From this result, one sees the resemblance between j-invariants of $\mathbb{Q}$-curves and j-invariants of CM elliptic curves. Moreover, it can be thought as a first step in attempting to prove the finiteness conjecture.

3. From $\mathbb{Q}$-curves to abelian varieties of GL_2-type

Let $\mathcal{C}$ be the isogeny class of a $\mathbb{Q}$-curve. In this section we consider several properties of the curves in that isogeny class: the fields of moduli of the curves $C \in \mathcal{C}$, both as elliptic curves and as $\mathbb{Q}$-curves, and the curves defined over a number field K such that the L-series $L(C/K, s)$ is a product $\prod L(f, s)$ of L-series of weight 2 cuspidal newforms f.

We say that a $\mathbb{Q}$-curve C is *completely defined* over a number field K (or that is defined over K as a $\mathbb{Q}$-curve) if all the conjugates of C and the isogenies between them are defined over the number field K. It is easily seen that the smallest field over which a $\mathbb{Q}$-curve is completely defined (its field of moduli as a $\mathbb{Q}$-curve) is a finite Galois extension of $\mathbb{Q}$. The main tool for investigating matters concerning fields of definition is a 2-cocyle built from isogenies between the conjugate curves.

Let C be a $\mathbb{Q}$-curve completely defined over a field K, finite Galois over $\mathbb{Q}$. For every $\sigma \in \mathrm{Gal}(K/\mathbb{Q})$ choose an isogeny $\phi_\sigma : {}^{\sigma}C \to C$, and define the map

$$c_K\colon \mathrm{Gal}(K/\mathbb{Q}) \times \mathrm{Gal}(K/\mathbb{Q}) \to \mathrm{End}(C)^* \simeq \mathbb{Q}^*\,, \qquad c_K(\sigma,\tau) = \phi_\sigma\, {}^{\sigma}\phi_\tau\, \phi_{\sigma\tau}^{-1}.$$

This map is a 2-cocycle of the group $\mathrm{Gal}(K/\mathbb{Q})$ with values in $\mathbb{Q}^*$, viewed as a trivial module; its cohomology class $[c_K] \in H^2(K/\mathbb{Q}, \mathbb{Q}^*)$ depends only on the K-isogeny class of the curve C. For every $\sigma \in G_{\mathbb{Q}}$ we can consider the isogeny ϕ_σ corresponding to the action of σ on K; then we have a locally constant set of isogenies $\{\phi_\sigma\}_{\sigma \in G_{\mathbb{Q}}}$ and we can define a continuous cocycle c of $G_{\mathbb{Q}}$ in the same way, whose cohomology class $[c] \in H^2(G_{\mathbb{Q}}, \mathbb{Q}^*)$ is the inflation of $[c_K]$ and is an invariant of the isogeny class $\mathcal{C}$. For every σ let $d(\sigma)$ denote the degree of the isogeny ϕ_σ. The map $d\colon G_{\mathbb{Q}} \to \mathbb{Q}^*/\mathbb{Q}^{*2}$ is then a group homomorphism depending only on the isogeny class $\mathcal{C}$. Since the square $c(\sigma,\tau)^2 = d(\sigma)d(\tau)d(\sigma\tau)^{-1}$ is a coboundary, the cohomology classes $[c_K]$ and $[c]$ belong to the 2-torsion of the corresponding cohomology groups.

The decomposition $\mathbb{Q}^* = \{\pm 1\} \times P$, where P is the multiplicative group of positive rationals, induces a natural isomorphism

$$H^2(G_{\mathbb{Q}}, \mathbb{Q}^*)[2] \simeq H^2(G_{\mathbb{Q}}, \{\pm 1\}) \times \operatorname{Hom}(G_{\mathbb{Q}}, \mathbb{Q}^*/\mathbb{Q}^{*2}). \tag{1}$$

Under this isomorphism the class of the cocycle c corresponds to the pair formed by the cohomology class of the 2-cocycle $\operatorname{sgn}(c)$ and the degree map d.

Let K_d denote the fixed field of $\ker(d)$, which is an abelian extension of $\mathbb{Q}$ of exponent 2, that can be written as $K_d = \mathbb{Q}(\sqrt{a_1}, \ldots, \sqrt{a_m})$, and let $d_1, \ldots, d_m$ be squarefree integers generating $\operatorname{im}(d)$ that are a Kummer dual basis for the a_i. Then the element $[\operatorname{sgn}(c)] \in H^2(G_{\mathbb{Q}}, \{\pm 1\}) \simeq \operatorname{Br}_2(\mathbb{Q})$ can be expressed as a product of quaternion algebras in the following way (cf. Quer [9, Theorem 4.2])

$$[\operatorname{sgn}(c)] = \prod_{i=1}^{m} (a_i, d_i). \tag{2}$$

The following result due to Ribet [11, Section 8] characterizes the fields of definition of $\mathbb{Q}$-curves, as elliptic curves, up to isogeny.

Theorem 3.1 (Ribet). *Assume $\mathcal{C}$ is an isogeny class of a $\mathbb{Q}$-curve. Let K be a number field. Then, there is C in $\mathcal{C}$ defined over K if and only if*

$$[c] \in \ker\left(\operatorname{Res}\colon H^2(G_{\mathbb{Q}}, \mathbb{Q}^*) \to H^2(G_K, \mathbb{Q}^*)\right)$$

where Res *denotes the restriction map on cohomolgy.*

Now from the decomposition (1) and the formula (2) it is immediate that the smallest field of definition of a $\mathbb{Q}$-curve up to isogeny is the field K_d cut-out by the degree map, so that there exist $\mathbb{Q}$-curves in the isogeny class $\mathcal{C}$ with j-invariant belonging to that field. In this way we recover Elkies' result (Theorem 2.3 of Section 2).

In general the curve cannot be completely defined over that field. Indeed, the characterization of the fields of moduli as $\mathbb{Q}$-curves of the elements of $\mathcal{C}$ can be also obtained from the cohomology class $[c]$.

Theorem 3.2 (Quer, [9, Proposition 2.3]**).** *Assume $\mathcal{C}$ is an isogeny class of a $\mathbb{Q}$-curve. Let K be a Galois number field. Then, there is C in $\mathcal{C}$ that is completely defined over K if and only if*

$$[c] \in \operatorname{im}\left(\operatorname{Inf}\colon H^2(\operatorname{Gal}(K/\mathbb{Q}), \mathbb{Q}^*) \to H^2(G_{\mathbb{Q}}, \mathbb{Q}^*)\right)$$

where Inf *denotes the inflation map on cohomolgy.*

The situation here is quite different from the previous case: in general there is no smallest field of definition as $\mathbb{Q}$-curve up to isogeny, but only minimal such fields. For example, it is easily seen that the cohomology class $[c]$ can always be inflated from the polyquadratic field $\mathbb{Q}(\{\sqrt{a_i}\}, \{\sqrt{d_i}\}) = K_d(\sqrt{d_1}, \ldots, \sqrt{d_m})$. As we shall see in a moment there are also representatives in $\mathcal{C}$ completely defined over fields of the type $K_\varepsilon \cdot K_d$ with K_ε a cyclic extension of $\mathbb{Q}$. In fact the proof of the previous theorem shows how to construct a curve whose attached cohomology class $[c_K]$ is any given class on $H^2(\mathrm{Gal}(K/\mathbb{Q}), \mathbb{Q}^*)$ whose inflation is $[c]$: one starts with a curve $C \in \mathcal{C}$ which is already defined over the field K, and then one finds a curve completely defined over K as a twist of the curve C over a quadratic extension of K which is a solution of a certain Galois embedding problem.

As mentioned before, Ribet [11, Theorem 6.1] proves that every $\mathbb{Q}$-curve is a factor of some abelian variety of GL_2-type. In fact he finds such a variety as a factor over $\mathbb{Q}$ of the variety obtained by restriction of scalars $\mathrm{Res}_{K/\mathbb{Q}}(C/K)$ from a sufficiently large number field K. Now we consider a more precise question, that of finding curves $C \in \mathcal{C}$ completely defined over a number field K and such that the variety obtained by restriction of scalars not only contains a varietiey of GL_2-type but factors over $\mathbb{Q}$ as a product of such varieties. We remark that the interesting fact in this situation is that the L-series of the curve over the number field K is then a product of L-series of cuspidal newforms of weight 2 for $\Gamma_1(N)$.

Given a character $\varepsilon\colon G_\mathbb{Q} \to \overline{\mathbb{Q}}^*$, the map

$$\theta_\varepsilon(\sigma,\tau) = \sqrt{\varepsilon(\sigma)}\sqrt{\varepsilon(\tau)}\sqrt{\varepsilon(\sigma\tau)}^{-1}$$

is a 2-cocycle of $G_\mathbb{Q}$ with values in $\{\pm 1\}$. Let K_ε be the fixed field of ε.

Theorem 3.3 (Quer, [9, Theorem 4.2]**).** *Let $\mathcal{C}$ be an isogeny class of a $\mathbb{Q}$-curve. Then, for every Galois character $\varepsilon\colon G_\mathbb{Q} \to \overline{\mathbb{Q}}^*$ such that*

$$[\theta_\varepsilon] = [\mathrm{sgn}(c)] \quad \textit{in} \quad \mathrm{Br}_2(\mathbb{Q})$$

there is a $\mathbb{Q}$-curve C in $\mathcal{C}$ defined over $L = K_\varepsilon \cdot K_d$ such that $\mathrm{Res}_{L/\mathbb{Q}}(C/L)$ is a product of abelian varieties of GL_2-type. Moreover, the $\mathbb{Q}$-curve C/L is central.

The local component at a prime p of the element $[\theta_\varepsilon] \in \mathrm{Br}_2(\mathbb{Q})$ is just the parity of corresponding local character. Hence every element of $\mathrm{Br}_2(\mathbb{Q})$, and in particular the class $[\mathrm{sgn}(c)]$, can be realized as a $[\theta_\varepsilon]$ for some Galois character ε and every isogeny class $\mathcal{C}$ of $\mathbb{Q}$-curves contains representatives as in the previous theorem. For more information on the number of factors of $\mathrm{Res}_{L/\mathbb{Q}}(C/L)$ and the structure of their endomorphism algebras the reader is refered to [9].

Example 3.4. Consider the elliptic curve over $\mathbb{Q}(\sqrt{-7})$ given by the Weierstrass equation

$$C : Y^2 = X^3 + 3\,(3 + 4\sqrt{-7})\,X - 2\,(23 + 6\sqrt{-7})$$

obtained from the j-invariant j_A of Example 2.4. It is isogenous to its Galois conjugate with isogeny of degree 6. The field K_d is $\mathbb{Q}(\sqrt{-7})$ and the curve is defined over this field. Using formula (2) one computes $[\mathrm{sgn}(c)] = (-7, 6)$, which is a quaternion division algebra. Since this element of $\mathrm{Br}_2(\mathbb{Q})$ cannot be obtained by inflation from an element of $H^2(\mathrm{Gal}(K_d/\mathbb{Q}), \{\pm 1\})$ there is no curve in the isogeny class of C that is completely defined over K_d. In fact the curve C itself is completely defined over the biquadratic field $\mathbb{Q}(\sqrt{-7}, \sqrt{6})$. The restriction of scalars of the curve C from any number field can never be a product of varieties of GL_2-type (cf. [9, Proposition 5.2]).

The simplest character satisfying the hypothesis of Theorem 3.3 is the quadratic character ε whose fixed field is the quadratic field $K_\varepsilon = \mathbb{Q}(\sqrt{21})$. Then, $[\theta_\varepsilon] = (-1, 21) = [\mathrm{sgn}(c)]$. From Theorem 3.3 it follows that we can find a nice representative in $\mathcal{C}$ completely defined over the field $L = \mathbb{Q}(\sqrt{-3}, \sqrt{-7})$ whose restriction of scalars from this field is a product of abelian varieties of GL_2-type. The explicit resolution of the appropriate Galois embedding problem over L produces the element $\gamma = -3\,(-3 - 3\sqrt{-3} + 3\sqrt{-7} + \sqrt{21})/4 \in L$. The twist of the curve C corresponding to the quadratic extension $L(\sqrt{\gamma})/L$ is the curve with equation

$$C_\gamma : Y^2 = X^3 + 3\,\gamma^2(3 + 4\sqrt{-7})\,X - 2\,\gamma^3(23 + 6\sqrt{-7})$$

and its Weil restriction $\mathrm{Res}_{L/\mathbb{Q}}(C_\gamma/L)$ is an abelian variety of GL_2-type with algebra of endomorphisms defined over $\mathbb{Q}$ isomorphic to the number field $\mathbb{Q}(\sqrt{-1}, \sqrt{6})$.

4. From abelian varieties of GL_2-type to $\mathbb{Q}$-curves

In the previous section we studied how to obtain abelian varieties of GL_2-type starting from a $\mathbb{Q}$-curve; in fact, starting from the j-invariant of a $\mathbb{Q}$-curve. In this section we consider the other way. We start with an abelian variety of GL_2-type and consider the problem of deciding whether it factors as a power of a $\mathbb{Q}$-curve, what is the smallest field where this happens, and how to get an explicit equation of such a curve.

Let A be an abelian variety of GL_2-type, and let $E = \mathrm{End}_{\mathbb{Q}}(A)$, which is a number field of degree $[E : \mathbb{Q}] = g = \dim A$. Especially important is the case where $f = \sum a_n q^n$ is a (normalized) newform in $S_2(N, \varepsilon)$ and A is the variety A_f attached to it by Shimura, in which case the field E is $\mathbb{Q}(\{a_n\})$. Conjecturally all varieties A of GL_2-type up to $\mathbb{Q}$-isogeny are of this form.

Shimura proved that the variety A contains some simple variety of CM-type if, and only if, it is isogenous to a power of a CM elliptic curve; this is also equivalent to the fact that A has CM over $\overline{\mathbb{Q}}$. From now on we assume that A is not such a CM variety. In the papers by Ribet (Propositions 3.6 and 5.2) and of Pyle (Theorem 1.2 and Proposition 1.7) in this same volume, one can find two different proofs of the following fact.

Theorem 4.1 (Ribet, Pyle). *Let $A/\mathbb{Q}$ be an abelian variety of* GL_2*-type containing no CM-type simple subvariety. Then* $\mathrm{End}(A)$ *is a central simple algebra over a totally real field $F \subseteq E$ with Schur index* 1 *or* 2*; the field E is a maximal subfield of* $\mathrm{End}(A)$ *and is an abelian extension of F.*

Let t be the Schur index of $\mathrm{End}(A)$. Then A factors up to isogeny over $\overline{\mathbb{Q}}$ as a power of a simple variety B whose endomorphism algebra is equal to the field F if $t = 1$ or is a quaternion division algebra over F otherwise.

Since $\mathrm{End}(A)$ is a simple algebra and the elements of the center $F \subseteq E$ are defined over $\mathbb{Q}$, by the Skolem-Noether theorem the Galois action is performed by inner automorphisms

$$^{\sigma}\psi = \alpha(\sigma)\psi\alpha(\sigma)^{-1}, \qquad \sigma \in G_{\mathbb{Q}}, \qquad \psi \in \mathrm{End}(A).$$

Since the endomorphisms of the maximal subfield E are defined over $\mathbb{Q}$, the automorphisms $\alpha(\sigma)$ belong to E^*, and by definition they are determined by σ only up to elements of F^*. The map $c(\sigma,\tau) = \alpha(\sigma)\alpha(\tau)\alpha(\sigma\tau)^{-1}$ is a 2-cocycle of $G_{\mathbb{Q}}$ with values in F^*. Restricting the cohomology class $[c] \in H^2(G_{\mathbb{Q}}, F^*)$ to the subgroup G_F and viewing it with values in $\overline{\mathbb{Q}}^*$, one obtains an element $[c] \in \mathrm{Br}(F) \simeq H^2(G_F, \overline{F}^*)$. Ribet [11, Theorem 5.6] proves that this element $[c]$ is the Brauer class of $\mathrm{End}(A)$. In Quer [8] a formula for $[c]$ is given that, for modular varieties A_f, permits the computation just knowing a few Fourier coefficients of the modular form f.

For every $s \in \mathrm{Gal}(E/F)$ let $\chi_s : G_{\mathbb{Q}} \to E^*$ be the Galois character defined by $\chi_s(\sigma) = {}^s\alpha(\sigma)\alpha(\sigma)^{-1}$, that we view also as a Dirichlet character; when $A = A_f$ these Dirichlet characters are just the inner twists of the form f in the sense of [10, Section 3]. Let $L/\mathbb{Q}$ be the abelian extension cut out by these characters, $L = \overline{\mathbb{Q}}^{\cap \ker \chi_s}$. It is the smallest field of definition of all the endomorphisms of A and the smallest field over which the decomposition $A \sim B^n$ of A into its building blocks takes place.

The full endomorphism algebra $\mathrm{End}(A)$ is isomorphic to the crossed product algebra $(E/F, j)$ corresponding to the Jacobi cocycle $j(s,t) = J(\chi_s^{-1}, {}^s\chi_t^{-1})$, where J denotes the Jacobi sum (cf. Chi [1, Corollary 5.11] and Ribet [11, Section 5]). Let η_s be an E-basis of $\mathrm{End}(A)$ according to this isomorphism, such that

$$\eta_s\, a = {}^s a \eta_s, \qquad \eta_s\eta_t = j(s,t)\eta_{st}.$$

In the case $A = A_f$ such a basis of endomorphisms has been explicitly constructed by Shimura (cf. [5, Section 2]). If the Brauer class of $\mathrm{End}(A)$ is trivial, i.e., the Schur index is $t = 1$, then these elements can be used to split the variety A in the following way.

Theorem 4.2 (González-Lario, [5, Theorem 2.1]**).** *Let $s \mapsto \beta_s$ be a map* $\mathrm{Gal}(E/F) \to E^*$ *such that $j(s,t) = \beta_s {}^s\beta_t \beta_{st}^{-1}$, and let $w \in \mathrm{End}(A)$ be the endomorphism defined by*

$$w = \sum_{s \in \mathrm{Gal}(E/F)} \beta_s^{-1}\, \eta_s.$$

Then the abelian variety $B = w(A)$ is a building block of A and all $\overline{\mathbb{Q}}$-simple abelian subvarieties of A are the varieties of the form $a(B)$ with $a \in E^*$.

In the case of abelian varieties A_f the pull-back under w of the differentials $H^0(B, \Omega^1_{B/\mathbb{C}})$ can be easily computed, as is also done in [5, Theorem 2.1]. In the particular case that B is a $\mathbb{Q}$-curve, the corresponding direction is attained by the cusp form

$$h_w = \frac{1}{\lambda} \sum_{s\in \mathrm{Gal}(E/F)} \frac{g(\chi_s^{-1})}{\beta_s}\, {}^s f = \sum_{n=1}^{\infty} \lambda_n q^n \in S_2(\Gamma_1(N)), \tag{3}$$

where g denotes the corresponding Gauss sum and $\lambda \in \overline{\mathbb{Q}}$ is a normalizing factor ensuring that the first Fourier coefficient is $\lambda_1 = 1$. The form h_w has its Fourier coefficients in the field L. Note that when $L \neq \mathbb{Q}$ there are infinitely many directions in $H^0(A_f, \Omega^1_{A_f/\mathbb{C}})$ which are pull-back of the vector space of regular differentials of elliptic quotients, the Hecke operators acting transitively on them. The following result provides information about the optimal elliptic quotients.

Theorem 4.3 (González-Lario, [5, Lemma 3.2]**).** *Assume that B is an elliptic curve. Let $\mathcal{D} = \{d_1, \dots, d_k\}$ be the set of the square-free positive integer representatives of the set $\{a_p^2/\varepsilon(p) \mod \mathbb{Q}^{*2}\}$ where p runs over the primes not dividing N and such that $a_p \neq 0$. Set $M = \mathrm{lcm}(d_1, \dots, d_k)$. Then, all optimal elliptic quotients of A_f of level N are the central $\mathbb{Q}$-curves which are the preimage of a rational point of $X^*(M)$. In particular, there are exactly $|\mathrm{Div}(M)|/|\mathcal{D}|$ optimal elliptic quotients, up to Galois conjugation.*

By using the previous facts it is possible to construct a table of non-CM modular $\mathbb{Q}$-curves which are optimal quotients of the Jacobian of $X_1(N)$, similar to Birch's and Cremona's tables of modular elliptic curves over the rational numbers. The process is as follows: for every pair N, ε of level and Nebentypus character one determines the non-CM newforms $f = \sum a_n q^n \in S_2(N, \varepsilon)$ using the modular symbols method; then the field F is computed from the values $a_p^2/\varepsilon(p)$ and the forms with $F \neq \mathbb{Q}$ are discarded. Next one computes the Brauer class of $\mathrm{End}(A_f)$ using the formula in [8]. Then every form f with trivial class gives rise to conjugation classes of $\mathbb{Q}$-curves; in order to obtain equations for them one computes the direction h_w from some splitting map β and integrates the differential form $h_w(dq/q)$ over the homology $H_1(X_1(N), \mathbb{Z})$ thereby obtaining a lattice $\Lambda \subset \mathbb{C}$ whose co-invariants c_4 and c_6 are defined over L and correspond to a $\mathbb{Q}$-curve that is a quotient of $X_1(N)$. In case of $|\mathcal{D}| < |\mathrm{Div}(M)|$, we determine the other classes as follows. We take a prime $l \mid M$ such that $l \notin \mathcal{D}_1 := \mathcal{D}$ and we look for an integer $d_i \in \mathcal{D}_1$ multiple of l and a prime p such that $a_p^2 = m^2\varepsilon(p)d_i$ with $m \in \mathbb{Z}\backslash\{0\}$. Then, the Hecke operator $T := l + T_p/m$ gives the direction $h' := T^*(h_w)/c$, where $T^*(h_w) = cq + O(q^2)$, which provides another class of optimal elliptic quotient. Next, set $\mathcal{D}_2 := \mathcal{D}_1 \cup \{ld/\gcd(l,d)^2 \mid d \in \mathcal{D}_1\}$ and repeat the process until getting $\mathcal{D}_i = \mathrm{Div}(M)$.

Example 4.4. Let $N = 42$ and let ε be the quadratic character of conductor 21. The space $S_2^{\mathrm{new}}(42, \varepsilon)$ has dimension 4 and is generated by the Galois conjugates of a newform with Fourier coefficients in $E = \mathbb{Q}(i, \sqrt{6})$ starting as

$$f = q - i\,q^2 - \tfrac{1}{2}(1+i)\sqrt{6}\,q^3 - q^4 + \sqrt{6}\,q^5 - \tfrac{1}{2}(1-i)\sqrt{6}\,q^6 - (1 - i\sqrt{6})\,q^7 + \cdots$$

All the Galois conjugates of the form f correspond to twisting by Dirichlet characters: the complex conjugation s corresponds to twisting by (the inverse of) the Nebentypus character $\chi_s = \varepsilon$, and the conjugation t that is the involution of $\mathbb{Q}(\sqrt{6})$ and fixes $\mathbb{Q}(i)$ corresponds to the quadratic Dirichlet character χ_t of conductor 7; in particular the field $F = \mathbb{Q}(\{a_p^2/\varepsilon(p)\})$ is $\mathbb{Q}$. Using the formula in [8] one computes the Brauer class of $\operatorname{End}(A_f)$ obtaining the quaternion algebra $(-7,6)[c_\varepsilon] = (-7,6)(-1,21) = 1$ in $\operatorname{Br}_2(\mathbb{Q})$. Hence A_f factors up to $\overline{\mathbb{Q}}$-isogeny as the fourth power of a $\mathbb{Q}$-curve. The field L over which the splitting is accomplished is the field $\mathbb{Q}(\sqrt{21}, \sqrt{-7})$.

Let a_n be the nth Fourier coefficient in the q-expansion of f. The numbers $\beta_1 = 1, \beta_s = a_{42}, \beta_t = a_{14}, \beta_{st} = a_3$ are a splitting map for the Jacobi cocycle defined from the twisting Dirichlet characters. Using this splitting map the formula (3) produces the cusp form with coefficients in L starting as

$$h_w = q + \tfrac{1}{4}(-3-\sqrt{-3}-\sqrt{-7}-\sqrt{21})\,q^2 + \tfrac{1}{4}(-3+3\sqrt{-3}+3\sqrt{-7}-\sqrt{21})\,q^3 - q^4 + \cdots$$

Now we will obtain the corresponding curve as an optimal quotient of the curve $X_\varepsilon(42)$ associated to the congruence subgroup $\Gamma_\varepsilon(N) \subset \Gamma_0(N)$ consisting of matrices with upper-left entry belonging to $\ker\varepsilon$. It is a curve of genus 9 whose Jacobian contains A_f as a factor up to isogeny. The period lattice

$$\Lambda = \left\{2\pi i \int_\gamma h_w\, d\tau \,\middle|\, \gamma \in H_1(X_\varepsilon(42), \mathbb{Z})\right\}$$

is computed and one obtains

$$\Lambda = (3.93832148\cdots - 0.29887963\ldots i)\mathbb{Z} \oplus (-0.05266443\cdots + 4.94519255\ldots i)\mathbb{Z}.$$

Then, computing the co-invariants of this lattice with enough precision, the following algebraic integers are identified

$$c_4 = -\frac{3}{4}(69 + 43\sqrt{-3} + 29\sqrt{-7} + 17\sqrt{21}),$$
$$c_6 = -3(207 - 84\sqrt{-3} - 54\sqrt{-7} + 46\sqrt{21}).$$

These invariants correspond to a curve isomorphic over the field $L = \mathbb{Q}(\sqrt{-7}, \sqrt{21})$ to the curve C_γ of Example 3.4.

5. The Manin ideal

Keep the notations as in the previous section. Given a $\mathbb{Q}$-curve which is an optimal quotient attached to a weight 2 newform f as above, we introduce an ideal of L associated with modular parametrizations $\pi\colon X_1(N) \to C$ induced by f. This ideal plays a similar role as the Manin constant for the case of rational elliptic curves. For this reason, in the general setting of $\mathbb{Q}$-curves we call it the *Manin ideal.*

Fix a splitting map β for the Jacobi cocycle defined from the inner twists of f. This induces a non-constant morphism $\pi\colon X_1(N) \to C$ such that the pull-back of any regular differential on the $\mathbb{Q}$-curve C is a multiple of $h_w\, dq/q$, where h_w is the cusp form corresponding to β (see (3) in the previous section.)

Notice that π is defined over L, the abelian extension of $\mathbb{Q}$ cut out by the inner twists of f. Let $\mathcal{N}_C$ be the Néron model of C over the ring of integers $\mathcal{O}_L$. Recall that $H^0(\mathcal{N}_C, \Omega^1_{\mathcal{N}_C/\mathcal{O}_L})$ is a locally free $\mathcal{O}_L$-module of rank one inside the one-dimensional L-vector space $H^0(C, \Omega^1_{C/L})$.

Definition. The Manin ideal $\mathfrak{c}(\pi)$ attached to the parametrization (C, π) is the (fractional) ideal of L satisfying

$$\pi^* H^0(\mathcal{N}_C, \Omega^1_{\mathcal{N}_C/\mathcal{O}_L}) = \mathfrak{c}(\pi) \left(\pi^* H^0(C, \Omega^1_{C/L}) \cap \mathcal{O}_L[[q]]\, dq \right) .$$

Alternatively, the Manin ideal attached to (C, π) can be defined as follows. Let $X_1(N)^0$ be the smooth locus of the minimal proper regular model of $X_1(N)$ over $S = \operatorname{Spec}(\mathcal{O}_L)$. By the Néron property, π extends to a morphism $\pi\colon X_1(N)^0 \to \mathcal{N}_C$ over $\mathcal{O}_L$. Then, the Manin ideal $\mathfrak{c}(\pi)$ is the pull-back under the cuspidal section $\infty\colon S \to X_1(N)^0$ of the ramification divisor (*branch locus*) of the morphism $\pi\colon X_1(N)^0 \to \mathcal{N}_C$.

There is a fairly easy way to carry out the computation of the Manin ideal in particular instances. For it, let ω be a non-zero differential in $H^0(C, \Omega^1_{C/L})$. With regard to the modular parametrization, we consider the fractional ideal attached to (C, π, ω) by

$$\mathfrak{m}_\omega(\pi) = \{\beta \in L\colon \beta\, \pi^*(\omega) \in \mathcal{O}_L[[q]]\, dq\} .$$

The class of $\mathfrak{m}_\omega(\pi)$ in the class group $\operatorname{Pic}(\mathcal{O}_L)$ is invariant by L-isogenies. Attached to (C, ω) we also consider its Weierstrass ideal δ_ω; it is the fractional ideal of L defined by the equality

$$\delta_\omega = \prod \mathfrak{p}^{\operatorname{ord}_\mathfrak{p}(\omega_\mathfrak{p}/\omega)} ,$$

where $\omega_\mathfrak{p}$ is the invariant differential corresponding to a minimal Weierstrass equation for C at $\mathfrak{p}$. The Weierstrass ideal measures the obstruction of C having a global minimal equation over L and its class $[\delta_\omega]$ in $\operatorname{Pic}(\mathcal{O}_L)$ is invariant of the L-isomorphism class of C. Then, one checks that the ideal $\mathfrak{m}_\omega(\pi)\, \delta_\omega$ is independent of the choice for ω, and that the Manin ideal $\mathfrak{c}(\pi)$ is equal to the inverse of $\mathfrak{m}_\omega(\pi)\, \delta_\omega$.

As in the classical case (Manin constant for elliptic curves over $\mathbb{Q}$), we have the following properties.

Theorem 5.1 (González-Lario, [5]).

(i) *The Manin ideal* $\mathfrak{c}(\pi)$ *is an integral ideal of* L.
(ii) *If* $\mathfrak{p}$ *is a prime ideal in* L *not dividing* $2N$, *then* $\operatorname{ord}_\mathfrak{p}(\mathfrak{c}(\pi)) = 0$.
(iii) *For all* $\nu \in \operatorname{Gal}(L/\mathbb{Q})^2$, *we have* ${}^\nu\mathfrak{c}(\pi) = \mathfrak{c}(\pi)$.

Evidence from experimental computations lead us to the following conjectures, which generalize the analogous conjectures of Manin and Stevens respectively.

Conjecture 5.2. *Let $\pi: X_1(N) \to C$ be an optimal parametrization defined over L. Then, the Manin ideal $\mathfrak{c}(\pi)$ is (1).*

Conjecture 5.3. *Let (C, π) be an optimal parametrization defined over L. Denote by $\mathcal{C}_L$ the L-isogeny class of C. For every $\mathbb{Q}$-curve C' in $\mathcal{C}_L$, there exists a central $\mathbb{Q}$-curve C'' (in $\mathcal{C}_L$ but not necessarily the same C) with an étale isogeny $C'' \to C'$.*

Recall that an isogeny between elliptic curves defined over a number field is said to be étale if its extension to Néron models is an étale morphism. It is easy to see that étale isogenies are characterized by the equality between the Weierstrass ideals attached to an invariant differential and its pull-back under the isogeny.

Example 5.4. Finally, we go back again to our example. Let $f(q) = \sum a_n q^n \in S_2(42, \varepsilon)$, the splitting map β, and $h_w = \sum \lambda_n q^n$ be as in Example 4.4. The lattice Λ obtained by integration of h_w on the singular homology of $X_\varepsilon(42)$ corresponds to the $\mathbb{Q}$-curve C_γ in Example 4.4. For future reference, denote by $42\,A$ this elliptic curve. Following the steps explained above, one checks that the Manin ideal $\mathfrak{c}(\pi)$ attached to this modular parametrization is trivial.

There is a natural Hecke action on lattices induced by the action on cusp forms. Let Λ_{a_5} be the lattice obtained by integration of $T_5^* h_w = a_5^* h_w = \lambda_5 \bar{h}_w$ where the bar denotes complex conjugation. Here,

$$\lambda_5 = \frac{1}{4}\left(3 + \sqrt{-3} - 3\sqrt{-7} + \sqrt{21}\right).$$

It turns out that $\Lambda_{a_5} = \lambda_5 \bar{\Lambda}$ is a sublattice of Λ of index 6, and we denote the corresponding central $\mathbb{Q}$-curve by $42\,B$. Likewise, the sublattices

$$\Lambda_{2+a_5}, \Lambda_{3+a_5} \subseteq \Lambda$$

provide the other two isomorphism classes of optimal quotients of A_f. Denote these new $\mathbb{Q}$-curves by $42\,C$ and $42\,D$, respectively.

The L-isogeny class of $\mathbb{C}/\Lambda$ contains (at least) 8 elliptic curves. The following table gives the corresponding co-invariants:

	c_4	c_6
$42\,A$	$3/4\,(-69, -43, 29, 17)$	$3\,(-207, 84, -54, 46)$
$42\,B$	$3/4\,(-69, -43, -29, -17)$	$3\,(-207, 84, 54, -46)$
$42\,C$	$3/4\,(-149, 37, -19, 33)$	$33\,(27, 28, -18, -6)$
$42\,D$	$3/4\,(-149, 37, 19, -33)$	$33\,(27, 28, 18, 6)$
$42\,E$	$3/4\,(2411, -283, 109, -543)$	$3\,(30033, 2436, -1566, -6674)$
$42\,F$	$3/4\,(2411, -283, -109, 543)$	$3\,(30033, 2436, 1566, 6674)$
$42\,G$	$3/4\,(891, -5483, 3421, -47)$	$3\,(174177, 122612, -78822, -38706)$
$42\,H$	$3/4\,(891, -5483, -3421, 47)$	$3\,(174177, 122612, 78822, 38706)$

where $r\,(a,b,c,d)$ stands for $r\,(a+b\sqrt{-3}+c\sqrt{-7}+d\sqrt{21})$. As for the j-invariants, we have:

	j
$42\,A$	$27/2^7\,(13171+2759\sqrt{-7})$
$42\,B$	$27/2^7\,(13171-2759\sqrt{-7})$
$42\,C$	$27/2^4\,(8889-3751\sqrt{-7})$
$42\,D$	$27/2^4\,(8889+3751\sqrt{-7})$
$42\,E$	$27/2^{31}\,(152709070611-49352194649\sqrt{-7})$
$42\,F$	$27/2^{31}\,(152709070611+49352194649\sqrt{-7})$
$42\,G$	$-27/2^{16}\,(4154888461+3566479289\sqrt{-7})$
$42\,H$	$-27/2^{16}\,(4154888461-3566479289\sqrt{-7})$

Finally, the graph of L-isogenies looks like

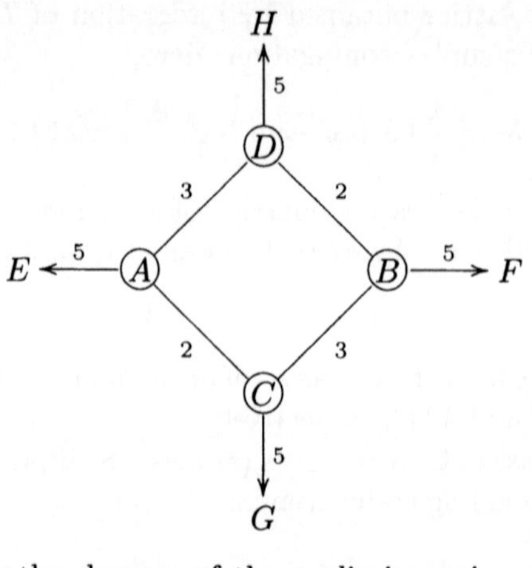

The numbers indicate the degrees of the cyclic isogenies, and the edges denote étale isogenies with origin in the optimal quotients A, B, C, and D.

References

[1] W. Chi, *Twists of Central Simple Algebras and Endomorphism Algebras of Some Abelian Varieties.* Math. Ann. 276 (1987), 615–632.

[2] N. Elkies, *Remarks on elliptic k-curves.* Preprint, 1993. Revised version published in this volume.

[3] J. González, *On the j-invariants of the quadratic* $\mathbb{Q}$*-curves.* J. London Math. Soc. (2) 63 (2001), no. 1, 52–68.

[4] J. González, J.C. Lario, *Rational and elliptic parametrizations of $\mathbb{Q}$-curves.* J. Number Theory 72 (1998), no. 1, 13–31.

[5] J. González, J.C. Lario, *$\mathbb{Q}$-curves and the Manin ideal.* Amer. J. Math. 123 (2001), no. 3, 475–503.

[6] B.H. Gross, *Arithmetic on elliptic curves with complex multiplication.* Lecture Notes in Mathematics 776, Springer, Berlin, 1980.

[7] B. Mazur, *Number Theory as gadfly,* Amer. Math. Monthly 98 (1991), 593–610.

[8] J. Quer, *La classe de Brauer de l'algèbre d'endomorphismes d'une variété abélienne modulaire,* C. R. Acad. Sci. Paris Sér. I Math. 327 (1998), no. 3, 227–230.

[9] J. Quer, *$\mathbb{Q}$-curves and abelian varieties of* GL_2*-type,* Proc. London Math. Soc. (3) 81 (2000), no. 2, 285–317.

[10] K. Ribet, *Twists of Modular Forms and Endomorphisms of Abelian Varieties,* Math. Ann. 253 (1980), 43–62.

[11] K. Ribet, *Abelian varieties over $\mathbb{Q}$ and modular forms.* In *Algebra and Topology 1992 (Taejŏn),* 53–79. Korea Adv. Inst. Sci. Tech., (Taejŏn), 1992. Reprinted in this volume.

Josep González, Joan-Carles Lario and Jordi Quer
Facultat de Matemàtiques i Estadística
Universitat Politècnica de Catalunya
Pau Gargallo 5,
08028-Barcelona, Spain
e-mail: `josepg@mat.upc.es`
e-mail: `Joan.Carles.Lario@upc.es`
e-mail: `Jordi.Quer@upc.es`

Progress in Mathematics, Vol. 224, 141–149

Serre's Conjecture for mod 7 Galois Representations

Jayanta Manoharmayum

Abstract. We give an account of Serre's conjecture for Galois representations with values in $GL_2(\mathbb{F}_7)$. For this, we construct elliptic curves over totally real soluble extensions with given mod 7 representation, and use base change results to obtain modularity over $\mathbb{Q}$.

1. Introduction

Let k be a finite field, and let

$$\overline{\rho} \; : \; \mathrm{Gal}(\mathbb{Q}^{ac}/\mathbb{Q}) \longrightarrow GL_2(k)$$

be a continuous absolutely irreducible odd representation of the absolute Galois group of $\mathbb{Q}$. A well-known conjecture of Serre (see [11]) states that such a representation is modular. The conjecture is fully understood if either the coefficient field k is $\mathbb{F}_3$, or the projective image of $\overline{\rho}$ is dihedral (see Theorem 3.14 of [4]). The result, in these cases, essentially is the work of Langlands and Tunnell.

Serre's conjecture is also known to be true, modulo restrictions at certain decomposition groups, for a few other coefficient fields k (in fact, when $|k| \leq 9$). For $GL_2(\mathbb{F}_4)$ representations, the result is due to Shepherd-Barron and Taylor (see [12]). The case of $GL_2(\mathbb{F}_5)$ with cyclotomic determinant is essentially equivalent to the Shimura-Taniyama-Weil conjecture; and, this is now known (see [2]) following on from the pioneering work of Andrew Wiles on Fermat's Last Theorem. More recently, Ellenberg (see [7]) has established the conjecture for a large class of Galois representations with values in $GL_2(\mathbb{F}_9)$.

In this article, we give an account of the conjecture when the coefficient field is $\mathbb{F}_7$. The precise statement is the following:

Theorem 1.1. *Let*

$$\overline{\rho} \; : \; Gal(\mathbb{Q}^{ac}/\mathbb{Q}) \longrightarrow GL_2(\mathbb{F}_7)$$

be a continuous, absolutely irreducible, odd representation of $Gal(\mathbb{Q}^{ac}/\mathbb{Q})$. Further, suppose that:

- *The determinant of $\overline{\rho}$ has even order when restricted to the inertia group at* 7.
- *The projective image of inertia at* 3 *has odd order.*

Then $\overline{\rho}$ is modular.

The author, in [9], showed the above result under the added hypothesis that $\overline{\rho}|_{D_7}$ is reducible; the mechanisms involved in the proof there are the same for the new cases presented here. The additional result required to handle these cases (see Section 2.2 of this article) is covered by the author and A.F. Jarvis in [8]. Also, details of the proof of Theorem 1.1 skipped in this article can be found in [8].

We now set out some notation. As is customary, we denote the decomposition and inertia groups at a place v of a number field F by D_v and I_v respectively. We denote by $\omega_{2,v}$ the second fundamental character of I_v. Thus, for a rational prime p,

$$\omega_{2,p} : I_p \longrightarrow \mathbb{F}_{p^2}^{\times} \quad \text{sends} \quad \sigma \longrightarrow \frac{\sigma(p^{1/(p^2-1)})}{p^{1/(p^2-1)}}$$

(and we have made an identification of $\mathbb{F}_{p^2}^{\times}$ with the group of (p^2-1)th roots of unity). We denote by ω_n the mod n cyclotomic character. The algebraic closure of a field L is denoted by L^{ac}. For an elliptic curve E and a positive integer n, the $GL_2(\mathbb{Z}/n\mathbb{Z})$ representation of the (relevant) Galois group is denoted by $\overline{\rho}_{E,n}$.

I would like to thank the organizers for a wonderful and stimulating conference, and for the support I received.

2. Modularity of Galois representations

Let F be a totally real number field, Galois over $\mathbb{Q}$. Fix an odd prime ℓ, and consider odd, continuous irreducible representations

$$\rho : G_F \longrightarrow GL_2(\mathbb{Q}_\ell^{ac})$$

satisfying the following conditions:

- ρ is unramified outside finitely many primes, and
- the determinant of ρ is the ℓ-adic cyclotomic character.

We now describe some conditions which will guarantee 'modularity' of such a representation. Modularity, in this context, is taken to mean a representation which is associated to a Hilbert modular cuspidal eigenform of parallel weight 2. For a precise description, see [13].

2.1.

We fix a continuous, odd, irreducible representation

$$\overline{\rho} : G_F \longrightarrow GL_2(\mathbb{F}_\ell^{ac})$$

satisfying the following conditions:

- the determinant of $\overline{\rho}$ is the mod ℓ cyclotomic character, and
- for each place $v|\ell$, we have

$$\overline{\rho}|_{D_v} \simeq \begin{pmatrix} \delta_v^{(1)} & * \\ 0 & \delta_v^{(2)} \end{pmatrix} \quad \text{with} \quad \delta_v^{(1)} \neq \delta_v^{(2)}.$$

Let us call a representation $\rho : G_F \longrightarrow GL_2(\mathbb{Q}_\ell^{ac})$ as in the beginning of this section to be a *nearly ordinary lift* of $\overline{\rho}$ if additonally, we have

- the reduction modulo ℓ of ρ is equivalent to $\overline{\rho}$, and
- for each place $v|\ell$, we have

$$\rho|_{D_v} \simeq \begin{pmatrix} \psi_v^{(1)} & * \\ 0 & \psi_v^{(2)} \end{pmatrix}$$

with $\psi_v^{(2)}|_{I_v}$ having finite order and $\delta_v^{(2)} = \psi_v^{(2)} \pmod{\ell}$.

We then have the following important result of Skinner and Wiles (see Theorem 5.1 of [13]).

Theorem 2.1. *Let $\overline{\rho}$ be as above, and let*

$$\rho : G_F \longrightarrow GL_2(\mathbb{Q}_\ell^{ac})$$

be a nearly ordinary lift of $\overline{\rho}$. Then ρ is modular, provided there is a nearly ordinary lift of $\overline{\rho}$ which is modular.

2.2.

Assume now that the ramification index of F at every prime above ℓ is less than or equal to $\ell - 1$. Consider a continuous, odd, irreducible representation

$$\overline{\rho} : G_F \longrightarrow GL_2(\mathbb{F}_\ell^{ac})$$

with determinant the mod ℓ cyclotomic character, and satisfying the following conditions:

- the restriction of $\overline{\rho}$ to the absolute Galois group of $F(\sqrt{(-1)^{\ell-1/2}\ell})$ is irreducible; and,
- for each place $v|\ell$, we have

$$\overline{\rho}|_{I_v} \sim \begin{pmatrix} \omega_{2,\ell} & 0 \\ 0 & \omega_{2,\ell}^{\ell} \end{pmatrix}\Bigg|_{I_v}.$$

Such a representation is necessarily finite at primes above ℓ (i.e., for every $v|\ell$, there is a finite flat group scheme defined over the ring of integers of F_v whose generic fibre corresponds to $\overline{\rho}|_{D_v}$).

Now let $\rho : G_F \longrightarrow GL_2(\mathbb{Q}_\ell^{ac})$ be a representation as in the beginning of the section. The representation ρ takes values in $GL_2(\mathcal{O})$ with $\mathcal{O}$ the ring of integers in some finite extension of $\mathbb{Q}_\ell$. We say that ρ is a *crystalline lift* of $\overline{\rho}$ if we also have:

- the residual representation of ρ is equivalent to $\overline{\rho}$, and
- the representations

$$\rho \pmod{\lambda^n} : G_F \longrightarrow GL_2(\mathcal{O}/\lambda^n\mathcal{O})$$

(λ a uniformizer of $\mathcal{O}$) are finite at primes above ℓ for every $n \geq 0$.

For such Galois representations, the following theorem is proved in [8] by Frazer Jarvis and the author.

Theorem 2.2. *Let F and $\overline{\rho}$ be as above. A crystalline lift*

$$\rho : G_F \longrightarrow GL_2(\mathbb{Q}_\ell^{ac})$$

of $\overline{\rho}$ is modular, provided $\overline{\rho}$ has a modular crystalline lift.

3. Key proposition

Throughout this section, we let

$$\overline{\rho} : \mathrm{Gal}(\mathbb{Q}^{ac}/\mathbb{Q}) \longrightarrow GL_2(\mathbb{F}_7)$$

be as in the statement of Theorem 1.1. Thus $\overline{\rho}$ is a continuous, absolutely irreducible, odd representation; the order of $\mathrm{Proj}\overline{\rho}(I_3)$ is odd and $\det\overline{\rho}$ has even order when restricted to the inertia group at 7.

One checks that if $\overline{\rho}|_{D_7}$ is irreducible, there are integers m, n such that

$$\overline{\rho}|_{I_7} \sim \omega_{2,7}^{n+8m} \oplus \omega_{2,7}^{n+8m} \quad \text{with} \quad n = 1, 3, 5.$$

Furthermore, the projective image of $\overline{\rho}$ in this case is either dihedral or insoluble. We can thus assume the projective image to be insoluble if $\overline{\rho}|_{D_7}$ is irreducible.

Proposition 3.1. *There is a finite totally real, soluble extension $F/\mathbb{Q}$, an elliptic curve E/F satisfying the following conditions.*

- *$\overline{\rho}_{E,7} \sim \overline{\rho}$ (up to twist by a character).*
- *$\overline{\rho}_{E,3} : Gal(F^{ac}/F) \longrightarrow GL_2(\mathbb{F}_3)$ is surjective.*
- *Ramification properties at 7:*
  - *If $\overline{\rho}|_{D_7}$ is reducible, then F has odd ramification index at 7 and E has good ordinary or potentially multiplicative reduction at every place above 7.*
  - *If $\overline{\rho}|_{D_7}$ is irreducible, the ramification index of F at 7 is less than or equal to 6, and E has good supersingular reduction at primes above 7. Furthermore, we have $\overline{\rho}_{E,7}|_{I_v} \sim (\omega_{2,7} \oplus \omega_{2,7}^7)|_{I_v}$ for any place $v|7$.*
- *F has odd ramification index at 3, and E has good ordinary reduction at primes above 3.*

The deduction of the theorem, given the above proposition, is given in [9]. The main modification required when $\overline{\rho}|_{D_7}$ is absolutely irreducible is the existence of a lift

$$\rho : \mathrm{Gal}(\mathbb{Q}^{ac}/\mathbb{Q}) \longrightarrow GL_2(\mathbb{Z}_7)$$

of $\overline{\rho}$ which is a crystalline lift over F. The relevant local cohomology information needed to be able to apply Ravi Ramakrishna's result (see [10], and [14] for an axiomatic set up) is known from the work of Conrad, [3]. The existence of a crystalline lift when

$$\overline{\rho}|_{I_7} \sim \omega_{2,7} \oplus \omega_{2,7}^7$$

can be found in [14]; we will only prove the above proposition under this added hypothesis in the next section.

One deduces the modularity of ρ as follows: applying the result of Skinner and Wiles to the 3-adic representation given by E, we get 'E is modular'. Therefore, the 7-adic representation given by E will be a nearly ordinary/crystalline lift of $\overline{\rho}$ over F. Applying Theorem 2.1/Theorem 2.2, we get the desired modularity of ρ.

There is however a subtlety involved. In order to apply Theorem 2.1 to the 3-adic representation given by E, we need to be sure that $\overline{\rho}_{E,3}$ has a nearly ordinary lift – and, this does not seem to be so obvious.

We will also give a slightly stronger version of the above proposition in the last section of this article where we insist that $\overline{\rho}_{E,3}$ is reducible and decomposable at every prime above 3. The existence of a nearly ordinary modular lift is widely believed in this case, and seems to be well known to the experts. A forthcoming work of Ellenberg and the author will address this issue.

4. Constructing elliptic curves with prescribed mod 7 representation

We shall now give an indication of the proof of Proposition 3.1. We deal with the case when the mod 7 representation restricted to a decomposition group at 7 is absolutely irreducible. (As mentioned before, the reducible case is dealt in [9].) For simplicity, we shall assume from here onwards that

$$\overline{\rho} : \mathrm{Gal}(\mathbb{Q}^{ac}/\mathbb{Q}) \longrightarrow GL_2(\mathbb{F}_7)$$

satisfies, in addition to the hypotheses of theorem, the following:

- the order of inertia at 3 is odd,
- the determinant of $\overline{\rho}$ is the mod 7 cyclotomic character, and
- $\overline{\rho}|_{I_7} \sim \omega_{2,7} \oplus \omega_{2,7}^7$.

In making an identification of the determinant with the mod 7 cyclotomic character, we have implicitly fixed a primitive 7th root of unity. We can think of $\overline{\rho}$ as an $\mathbb{F}_7$-vector space scheme of dimension 2 defined over $\mathbb{Q}$ (it is, in fact, defined over $\mathbb{Z}[1/N]$ where N is the product of primes where $\overline{\rho}$ ramifies). Denoting the resulting vector space scheme by $\mathcal{W}$, the determinant condition gives an alternating, perfect pairing $\mathcal{W} \times \mathcal{W} \longrightarrow \boldsymbol{\mu}_7$. We then have a fine moduli space classifying pairs (E, i), defined over appropriate bases, where

- E is an elliptic curve, and
- $i : \mathcal{W} \xrightarrow{\sim} E[7]$ is an isomorphism which sends the pairing on $\mathcal{W}$ to the Weil pairing on $E[7]$.

We denote the resulting fine moduli space by $Y(\overline{\rho})$, and write $X(\overline{\rho})$ for its compactification (which also has a moduli interpretation in terms of generalized elliptic curves).

We now recall some facts about the above moduli space (see for example [6]). It is known that $X(\overline{\rho})$ is a geometrically irreducible, smooth, non-hyperelliptic curve of genus 3 defined over $\mathbb{Q}$. The canonical divisor thus embeds $X(\overline{\rho})$ as a smooth quartic in $\mathbb{P}^2$. We fix such an embedding, and *we identify* $X(\overline{\rho})$ *with a quartic in* $\mathbb{P}^2/\mathbb{Q}$ *throughout this section.*

There is a morphism $j : X(\overline{\rho}) \longrightarrow_{/\mathbb{Q}} \mathbb{P}^1$ which is obtained by 'taking j-invariants of elliptic curves'. The degree of j is 168, and it is ramified at 0, 1728 and ∞. As for $Y(\overline{\rho})$, it is the affine curve given by $X(\overline{\rho}) - j^{-1}(\infty)$.

Let us say that a geometric point on $X(\overline{\rho})$ has property P if the corresponding elliptic curve possesses property P (assuming it all makes sense).

Lemma 4.1. *There is a finite Galois extension* $F_7/\mathbb{Q}_7$ *with ramification index less than or equal to* 4 *and a line* L_7 *in* $\mathbb{P}^2$ *defined over* F_7 *with the property that* L_7 *intersects* $X(\overline{\rho})_{/F_7}$ *at four distinct supersingular points defined over* F_7.

Proof. Over $\mathbb{C}$, the curve $X(\overline{\rho})$ is isomorphic to the Klein quartic (it is defined by the equation $x^3y + y^3z + z^3x = 0$). There are 84 points in $j^{-1}(1728)$, and these can be partitioned into 21 sets of 4 collinear points. Given a point in $j^{-1}(1728)$, there is a unique line in the projective plane which passes through the given point and intersecting the Klein quartic at points with j-invariant 1728.

Now let $E/\mathbb{Q}_7$ be the elliptic curve with Weierstrass equation $y^2 = x^3 + x$. It has j-invariant 1728 and good supersingular reduction. The mod 7 representation given by E is equivalent to $\overline{\rho}$ when we restrict to inertia at 7. However, it is not at all clear if we can conclude $(E, E[7]) \in X(\overline{\rho})(\mathbb{Q}_7^{\mathrm{nr}})$.

To see this, consider the twisted modular curve $X(E[7])$, where the specified pairing on $E[7]$ is the Weil pairing

$$\langle\ ,\ \rangle : E[7] \times E[7] \longrightarrow \boldsymbol{\mu}_7.$$

For each $i \in (\mathbb{Z}/7\mathbb{Z})^\times$, let $X(E[7]_i)$ be the twisted modular curve given by taking the pairing on $E[7]$ to be the composite

$$E[7] \times E[7] \xrightarrow{\langle\ ,\ \rangle} \boldsymbol{\mu}_7 \xrightarrow{\times i} \boldsymbol{\mu}_7.$$

Let α be an automorphism of $E[7]$, and consider the embedding

$$E[7] \xrightarrow{\alpha} E[7] \longrightarrow E$$

obtained by composing α with the natural one. With this choice, we will get a point on $X(E[7]_{\det\alpha})(\mathbb{Q}_7)$ where the determinant is the one given by the Weil pairing.

The problem is that there may be very few automorphisms! Indeed, we know that $\overline{\rho}_{E,7}$ is an absolutely irreducible representation, and so the automorphisms of $E[7]$ over $\mathbb{Q}_7$ are given by scalar multiplication – their determinants are always a square modulo 7.

But when we go up to $\mathbb{Q}_7^{\mathrm{nr}}$, we can certainly find extra automorphisms. The image $\overline{\rho}_{E,7}(I_7)$ is a cyclic subgroup of $GL_2(\mathbb{F}_7)$ of order 48 – we can find automorphisms so that E gives a point on $X(E[7]_i)(\mathbb{Q}_7^{\mathrm{nr}})$ for each $i \in (\mathbb{Z}/7\mathbb{Z})^\times$.

We can thus conclude that $X(\overline{\rho})(\mathbb{Q}_7^{\mathrm{nr}})$ contains a point P which is supersingular with j-invariant 1728. Now take the unique line through this point which cuts $X(\overline{\rho})$ at three other points with j-invariant 1728. It can be checked, using the description given in [6] for instance, that one of the three points is already supersingular point defined over $\mathbb{Q}_7^{\mathrm{nr}}$ since it is 2-isogenous to P. One then checks that the remaining points are elliptic curves with good supersingular reduction defined over $\mathbb{Q}_7(7^{1/4})$ by analysing the automorphisms of X which stabilises the unique line through these four points. For details, see Lemma 9.3 of [8]. □

Lemma 4.2. *There is a finite extension F_3 of $\mathbb{Q}_3$ having odd ramification index and a line L_3 in $\mathbb{P}^2$ defined over F_3 which cuts $X(\overline{\rho})$ at four distinct good ordinary reduction points, each defined over F_3.*
There is a line L_∞ defined over $\mathbb{R}$ which intersects $X(\overline{\rho})$ at four distinct real points.

We refer to [9] for the proof of the above lemma (but see also the next section).

Proof of Proposition 3.1. There is a totally real soluble extension F' such that its completion at primes above 3 (resp. 7) is isomorphic to F_3 (resp. F_7). Take a line in $\mathbb{P}^2$ defined over F' which is close enough to L_∞, L_3 and L_7 for the respective completions. Then take F to be the Galois closure of the extension given by intersecting this line with $X(\overline{\rho})$, and take the elliptic curve given by an intersection point. Generically, we should get surjectivity of the mod 3 representation – we will expand on this a bit more in the next section. □

5. More on the geometry of $X(7)$ and a modular correspondence

Working over the complex numbers, we have an identification of the $X(7)$ as a quotient of the upper half-plane (along with cusps) by $\Gamma(7)$, the principal congruence subgroup of level 7. The correspondence associated to the double coset

$$\Gamma(7)\eta\begin{pmatrix}2 & 0\\ 0 & 1\end{pmatrix}\Gamma(7) \quad \text{where} \quad \eta \equiv \begin{pmatrix}2 & 0\\ 0 & 4\end{pmatrix} \bmod 7,\ \eta \in SL_2(\mathbb{Z})$$

is studied by Adler in [1]. Following Theorem 5 of [1], we denote this correspondence by Ξ. Observe that for a point P, the elliptic curves given by $\Xi(P)$ are the three curves isogenous to the one determined by P by an isogeny of degree 2.

Let K be the twisted sextic curve in $\mathbb{P}^3$ birational to $X(7)$ defined in Section 8 of [1]. Up to a linear change of coordinates, K is the image of the Klein quartic under the rational map $\mathbb{P}^2 - \to \mathbb{P}^3$ given by

$$(x : y : z) \longrightarrow \left(\frac{x}{y} : \frac{y}{z} : \frac{z}{x} : 1\right).$$

Then $\Xi(P)$ is a trisecant of K for every point P, and all trisecants arise in this way. A plane Π through $\Xi(P)$ determines an effective divisor $\Pi \cap K - \Xi(P)$ of degree 3; the divisor

$$\Pi \cap K - \Xi(P) + P$$

is in the canonical class (see page 160 of [5]).

Through any point on K, there are exactly three trisecants passing through it. A point P lies on a trisecant $\Xi(Q)$ if and only if Q lies on the trisecant $\Xi(P)$.

These can be easily seen from the moduli point of view. It follows that the trisecants passing through P are the trisecants determined by the points of $\Xi(P)$.

Proposition 5.1. *Let P be a point on $X(7)(\mathbb{C})$, and suppose $\Xi(P) = P_1 + P_2 + P_3$. Then the divisor $\Xi(P_1) - P + P_2 + P_3$ is in the canonical class.*

Proof. The plane through $\Xi(P_1)$ and $\Xi(P_2)$ intersects K at a sixth point – and this point is P_3 (see page 160 of [5]). The lemma follows (from couple of paragraphs above). □

The corollary that now follows originates from a conversation the author had with Bjorn Poonen during the MCAV conference. The (content of the) corollary was already known to Poonen; the statement given here is a consequence of the author's attempt at reconstructing the conversation.

Corollary 5.2. *Let M be a field of characteristic zero containing a seventh root of unity. Let E/M be an elliptic curve with all of its 4-division points are defined over M, and denote by $I(E)$ the set of degree 2 or 4 isogenies of E. If E gives a point in $X(7)(M)$, then there is a line in $\mathbb{P}^2$ which intersects the Klein quartic at 4 distinct points defined over M, with the corresponding elliptic curves in the set $I(E)$.* □

Proposition 5.3. *In Proposition 3.1, we can insist that the mod 3 representation given by E is reducible and decomposable at every prime above 3.*

Proof. Start with an elliptic curve $\mathbb{Q}_3$ with good ordinary reduction and whose mod 3 representation is reducible and decomposable. Take the line given by the above corollary (after going up to an appropriate extension of $\mathbb{Q}_3$), and carry on as before. □

Now for those words on the surjectivity part of the proposition: one way of doing it is to impose further local restrictions. For example, take a prime p such that $\bar{\rho}|_{D_p}$ is trivial. The Tate curve over $\mathbb{Q}_p$ with parameter p^{28} gives a line L_p defined over $\mathbb{Q}_p$ adjoined 4th and 7th roots of unity. If we insist in the proof of the proposition that the global approximating line is p-adically close enough to L_p, the image of the mod 3 representation will certainly contain elements of order 3.

How do we make sure that the image of the mod 3 representation contains elements of order 4? For this, let q be a prime such that the representation

$$\begin{pmatrix} \omega_4 & 0 \\ 0 & 1 \end{pmatrix} \times \begin{pmatrix} \omega_3 & 0 \\ 0 & 1 \end{pmatrix} \times \bar{\rho} : \mathrm{Gal}(\mathbb{Q}^{ac}/\mathbb{Q}) \longrightarrow GL_2(\mathbb{Z}/4\mathbb{Z}) \times GL_2(\mathbb{F}_3) \times GL_2(\mathbb{F}_7)$$

is trivial at q. Let $\psi : \mathrm{Gal}(\mathbb{Q}^{ac}/\mathbb{Q}) \longrightarrow GL_2(\mathbb{F}_3)$ be a continuous representation with determinant ω_3 such that $\psi|_{D_q}$ is unramified and has order 4 (i.e., the order of $\psi(D_q)$ is 4). Finally, let

$$\phi \stackrel{\text{def}}{=} \begin{pmatrix} \omega_4 & 0 \\ 0 & 1 \end{pmatrix} \times \psi \times \bar{\rho} : \mathrm{Gal}(\mathbb{Q}^{ac}/\mathbb{Q}) \longrightarrow GL_2(\mathbb{Z}/4\mathbb{Z}) \times GL_2(\mathbb{F}_3) \times GL_2(\mathbb{F}_7).$$

The corresponding moduli space $X(\phi)$ has good reduction at q. For large enough q, we will thus get a $\mathbb{Q}_q$ rational point on $X(\phi)$. An application of Corollary 5.2

to the resulting elliptic curve then gives us a line over $\mathbb{Q}_q$. The rest is as in the previous paragraph.

Alternatively, one can deduce the surjectivity if there is a line in $\mathbb{P}^2$ defined over some number field such that the elliptic curves given by the intersection points all have mod 3 representations with 'big' image (e.g., the image contains $SL_2(\mathbb{F}_3)$). For this, let N be the extension of $\mathbb{Q}$ through which $\overline{\rho}$ factors. A general elliptic curve over N will have linearly disjoint mod 3, mod 4 and 7 representations with 'big' images. Going up to the extension obtained by adjoining the 7 and 4 torsion points, we will get our desired line by Corollary 5.2.

References

[1] A. Adler, *Characterization of modular correspondences by geometric properties,* Pacific J. Math., **155** (1992), no. 1, 1–27.

[2] C. Breuil, B. Conrad, F. Diamond and R. Taylor, *On the modularity of elliptic curves over $\mathbb{Q}$: wild 3-adic exercises.,* J. Amer. Math. Soc., **14** (2001), no. 4, 843–939.

[3] B. Conrad, *Ramified deformation problems,* Duke Math. J., **97** (1999), no. 3, 439–513.

[4] H. Darmon, F. Diamond and R. Taylor, *Fermat's last theorem,* in *Elliptic curves, modular forms & Fermat's last theorem (Hong Kong, 1993),* Internat. Press, Cambridge, MA, 1997, 2–140

[5] W.L. Edge, *The Klein quartic in three dimensions,* Acta Math., **79** (1947), 153–223.

[6] N. Elkies, *The Klein quartic in number theory,* in *The eightfold way,* Math. Sci. Res. Inst. Publ., **35** (1999), Cambridge Univ. Press, Cambridge, 51–101.

[7] J. Ellenberg, *Serre's conjecture over* $\mathbb{F}_9$, preprint.

[8] A.F. Jarvis and J. Manoharmayum, *Modularity of elliptic curves over totally real fields,* preprint.

[9] J. Manoharmayum, *On the modularity of certain* $\mathrm{GL}_2(\mathbb{F}_7)$ *Galois representations,* Math. Res. Lett., **8** (2001), no. 5–6, 703–712.

[10] R. Ramakrishna, *Deforming Galois representations and the conjectures of Serre and Fontaine-Mazur, preprint.*

[11] J.-P. Serre, *Sur les représentations de degré 2 de Gal($\overline{\mathbb{Q}}/\mathbb{Q}$),* Duke Math. J., **54** (1987), 179–230.

[12] N. I. Shepherd-Barron and R. Taylor, mod2 *and* mod5 *icosahedral representations,* J. Amer. Math. Soc. 6, **10** (1997), no. 2, 283–298.

[13] C. Skinner, A. Wiles, *Nearly ordinary deformations of irreducible residual representations,* Ann. Fac. Sci. Toulouse Math. (6), **10** (2001), no. 1, 185–215

[14] R. Taylor, *On icosahedral Artin representations II,* to appear in the American Journal of Mathematics. Also available at `http://www.math.harvard.edu/~rtaylor`

Jayanta Manoharmayum
Department of Pure Mathematics
University of Sheffield
Sheffield S3 7RH
United Kingdom
e-mail: `j.manoharmayum@sheffield.ac.uk`

Progress in Mathematics, Vol. 224, 151–163

Pairings in the Arithmetic of Elliptic Curves

Barry Mazur and Karl Rubin

Introduction

The recent work of Cornut, Vatsal, Bertolini, Darmon, Nekovár and others on the Mordell-Weil and Shafarevich-Tate groups of elliptic curves over anticyclotomic towers has made it timely to organize, and possibly sharpen, the collection of yet unresolved conjectures regarding the finer structure we expect to be true about this piece of arithmetic. That was the general aim of two of our recent preprints ([MR1], [MR2]) the former being the text of an address given at the International Congress of Mathematicians in Beijing in August 2002 by the second author. It was also the aim of the lecture delivered by the first author at the conference in Barcelona in July 2002. Since there was some overlap in our two lectures (Beijing, Barcelona), we felt it reasonable not to repeat things, but rather, in this write-up for the proceedings of the Barcelona conference, to concentrate only on the following part of the story which was not really covered in any of the other accounts.

In a recent preprint [N2], Nekovár developed a *derived complex* version of the theory of Selmer modules and expressed some of the standard dualities enjoyed by Selmer modules in the language of derived complexes. The advantage of this is its naturality and generality: the cohomological machinery directly yields such a structure, and we are well advised to keep that structure in view. The disadvantage is the unavoidable largeness of the technique, which makes it slightly inconvenient, for example, to explain it in an hour. Nevertheless, if one is to understand, and build upon, the collection of outstanding unsolved problems regarding our anticyclotomic context, it might be a good idea to hold in one's head all these dualities at the same time. As an expository tool, it occurred to us to hypothesize a certain rigidification of Nekovár's theory (a single skew-Hermitian module) from which one might obtain the Selmer module of interest, and all of its attendant height pairings, and Cassels pairings, and upon which one can impose some of the finer structure carried, in some contexts, by Selmer modules (see for example the discussion of Heegner points and L-functions in [MR1]). This hypothesized skew-Hermitian module has the virtue of being very easy to explain, making use of only low-tech methods. The great disadvantage of our theory is that we have no natural way of constructing the relevant skew-Hermitian matrices in any nontrivial case. Our hope that perhaps we may eventually be able to find such natural

Both authors partially supported by the National Science Foundation.

constructions, at least in some cases, was inspired by the vague analogy it has to Seifert's classical results which express the algebraic topology of abelian covers of knot complements in terms of skew-symmetric matrices with coefficients in an appropriate localization of the group ring of the first homology group of the knot complement. The Cassels pairing in arithmetic has a formal structure similar to the Blanchfield pairing in knot theory. The height pairings, however, which play such a prominent role in arithmetic also have analogues in knot theory, but, to our knowledge, these analogues have not been explicitly used in either classical knot theory, or in the theory of knots in higher dimensions.

We offer Part I of this article as a possibly helpful expository warm-up for any student interested in learning Nekovář's "Selmer complexes." Part II discusses the arithmetic of elliptic curves over anticyclotomic towers from the viewpoint sketched in Part I.

Part I. The structure of Selmer modules

1. Selmer groups of elliptic curves

Consider the structure "carried by" the *p-Selmer group* $S_p(E,K)$ of an elliptic curve E over a number field K. Recall that this Selmer group is the $\mathbf{Z}_p$-module of finite type defined as the Pontrjagin dual of:

$$\ker(H^1(K, E[p^\infty]) \to \prod_v H^1(K_v, E)),$$

where $E[p^\infty]$ is the Galois module of p-power torsion on E, and the product is over all places v of K. We define $S_p(E,K)_{\mathrm{tors}}$ to be the torsion subgroup of $S_p(E,K)$ and $S_p(E,K)_{\mathrm{free}}$ the canonical torsion-free quotient $S_p(E,K)/S_p(E,K)_{\mathrm{tors}}$.

Let the superscripts $^{\mathcal{D}}$ and $^{\diamond}$ denote Pontrjagin dual and $\mathbf{Z}_p$-dual, respectively, so that $M^{\mathcal{D}} := \mathrm{Hom}(M, \mathbf{Q}/\mathbf{Z})$ and $M^{\diamond} := \mathrm{Hom}(M, \mathbf{Z}_p)$.

For simplicity we will assume the Shafarevich-Tate conjecture, so $S_p(E,K)_{\mathrm{tors}}$ is canonically isomorphic to the Pontrjagin dual of the p-primary subgroup of the Shafarevich-Tate group of E over K. It follows that there are canonical isomorphisms

$$S_p(E,K)_{\mathrm{free}} \cong E(K)^{\diamond}, \qquad S_p(E,K)_{\mathrm{tors}} \cong Ш(E,K)[p^\infty]^{\mathcal{D}},$$

$$(E(K)/E(K)_{\mathrm{tors}}) \otimes \mathbf{Z}_p \cong S_p(E,K)^{\diamond}_{\mathrm{free}}.$$

The Cassels pairing induces a nondegenerate skew-symmetric pairing

$$S_p(E,K)^{\mathcal{D}}_{\mathrm{tors}} \otimes S_p(E,K)^{\mathcal{D}}_{\mathrm{tors}} \longrightarrow \mathbf{Q}_p/\mathbf{Z}_p.$$

Since this pairing is nondegenerate, it induces an isomorphism

$$S_p(E,K)_{\mathrm{tors}} \cong S_p(E,K)^{\mathcal{D}}_{\mathrm{tors}},$$

and hence a nondegenerate skew-symmetric pairing

$$S_p(E,K)_{\text{tors}} \otimes S_p(E,K)_{\text{tors}} \longrightarrow \mathbf{Q}_p/\mathbf{Z}_p. \tag{1'}$$

Assume now that every place v of K of residual characteristic p is a prime of good, ordinary reduction for E. By a $\mathbf{Z}_p$-*power extension* of K let us mean a Galois extension with Galois group isomorphic to $\mathbf{Z}_p^d$ for some natural number d, and by $\Gamma := \Gamma(K)$ let us mean the Galois group of the maximal $\mathbf{Z}_p$-power extension of K in an algebraic closure of K (assumed chosen and fixed). Recall that we have a canonical symmetric, bilinear, p-adic height pairing

$$S_p(E,K)^{\diamond}_{\text{free}} \otimes S_p(E,K)^{\diamond}_{\text{free}} \longrightarrow \Gamma(K) \otimes \mathbf{Q}_p \tag{2'}$$

and we may even guarantee that the values of this pairing are nearly integers, in the sense that they are contained in

$$\frac{1}{\delta(E,K)} \cdot \Gamma(K) \subset \Gamma(K) \otimes \mathbf{Q}_p$$

where the denominator $\delta(E,K)$ is a very controllable number, often 1 (cf. [MT]). The p-adic height pairing is conveniently functorial in K, and in particular when $K/\mathbf{Q}$ is Galois the pairing is $\mathrm{Gal}(K/\mathbf{Q})$-equivariant in an evident sense.

We will be studying, later, more machinery than (1′) and (2′) with which Selmer groups are naturally endowed, but for the moment let us take a step backwards and repeat the algebraic structure discussed so far.

Let A be a discrete valuation ring and a $\mathbf{Z}_p$-algebra (A was simply $\mathbf{Z}_p$ in the previous paragraph). Denote by $\mathcal{K}$ the field of fractions of A. Let Γ be an abelian pro-p-group, and let S be an A-module of finite type. Form the canonical exact sequence of A-modules

$$0 \longrightarrow T \longrightarrow S \longrightarrow F \longrightarrow 0,$$

where T is a torsion A-module and F free. We assume that we have a nondegenerate bilinear skew-symmetric pairing

$$T \otimes_A T \longrightarrow \mathcal{K}/A, \tag{1}$$

and a bilinear symmetric pairing

$$F^{\diamond} \otimes F^{\diamond} \rightarrow \Gamma \otimes_{\mathbf{Z}_p} \mathcal{K} \tag{2}$$

where the superscript $^{\diamond}$ now denotes A-duality, i.e., $F^{\diamond} := \mathrm{Hom}_A(F,A)$. Also, it is good to bear in mind (without formally including this feature in our axiomatic set-up above) that in the prototype, the symmetric pairing (2) takes values in

$$\frac{1}{\delta} \cdot \Gamma \otimes_{\mathbf{Z}_p} A \subset \Gamma \otimes_{\mathbf{Z}_p} \mathcal{K},$$

where δ is something computable.

How might one naturally come across an A-module S with the two types of pairings as sketched above?

2. Hermitian and skew-Hermitian modules

Fix a commutative ring R endowed with an involution $\iota : R \to R$.

If M is an R-module, let M^ι denote the R-module having the same underlying abelian group M, but with R-module structure obtained from that of M by composition with ι. Clearly $(M^\iota)^\iota = M$. By a *semi-linear* (R,ι)*-module* we mean an R module W endowed with an involution $i : W \to W$ such that $i(rw) = \iota(r) \cdot i(w)$ for all $r \in R$ and $w \in W$. Equivalently, we may think of the involution i as an R-module isomorphism $i : W \to W^\iota$ such that $i^\iota \cdot i : W \to (W^\iota)^\iota = W$ is the identity. We refer to such a pair (W, i) as a *semi-linear module*, for short. The involution ι of the free R-module R endows that module with a natural semi-linear structure. If M is an R-module and W is a semi-linear R-module, the R-module $\mathrm{Hom}_R(M, W)$ inherits a semi-linear structure as follows. For $f \in \mathrm{Hom}(M, W)$ let $i(f) \in \mathrm{Hom}_R(M, W)$ be given by $i(f) := i \circ f$. For a free R-module Φ of finite rank, by the *semi-linear R-dual* Φ^* of Φ we mean the R-module $\Phi^* := \mathrm{Hom}_R(\Phi^\iota, R)$ with the semi-linear structure as given above.

If $I \subset R$ is an ideal which is stable under the action ι then the quotient R/I inherits an involution compatible with ι; we denote it again ι. Let us call such an ideal I a *switching ideal* if ι induces the identity involution on R/I and the automorphism "multiplication by -1" on I/I^2.

Example 2.1. A principal example for us is the following. Suppose B is a commutative $\mathbf{Z}_p$-algebra and G is a commutative pro-p group. Let R be the completed group ring $B[[G]]$, and let ι be the involution which operates as the identity on B and by $g \mapsto g^{-1}$ on G.

If $H \subset G$ is a closed subgroup, let $I_H \subset R$ be the closed ideal generated by all elements of the form $h - 1 \in R$ for $h \in H$. That is, I_H is the kernel of the natural projection $B[[G]] \to B[[G/H]]$. We have a canonical isomorphism of $B[[G/H]]$-modules

$$H \otimes_{\mathbf{Z}_p} B[[G/H]] \cong I_H/I_H^2$$

characterized by the property that the element $h \otimes 1$ is sent to $h - 1$ modulo I_H^2 for all $h \in H$. The augmentation ideal $I_G \subset B[[G]]$ is a switching ideal.

If Φ is an R-module, and W a semi-linear R-module, a pairing

$$\pi : \Phi \otimes_R \Phi^\iota \to W$$

is called *Hermitian* if

$$\pi(a \otimes_R b) = +i(\pi(b \otimes_R a)),$$

and *skew-Hermitian* if

$$\pi(a \otimes_R b) = -i(\pi(b \otimes_R a)).$$

Remark 2.2. When we talk of Hermitian or skew-Hermitian pairings with values in the ring R we will always mean to view the ring R *with semi-linear R-module structure given by its involution ι*. Note however, that if $i : W \to W$ is a semi-linear involution of an R-module W, then so is $-i$, and if we switch to considering

$i' = -i$, then Hermitian pairings with values in the semi-linear R-module (W, i) are skew-Hermitian pairings with values in (W, i'), and vice versa.

3. Derived pairings

Suppose from now on that Φ is free over R of finite rank, and $W = R$. Suppose further that we have a nondegenerate R-valued skew-Hermitian pairing $\Phi \otimes \Phi^\iota \to R$. Such a pairing corresponds to an injective R-homomorphism

$$h : \Phi \longrightarrow \Phi^*$$

and the skew-Hermitian property of the pairing is then equivalent to the fact that the induced map

$$\Phi^\iota = \mathrm{Hom}(\Phi^*, R) \xrightarrow{h^*} \mathrm{Hom}(\Phi, R) = (\Phi^*)^\iota$$

is identified with $-h$ under the canonical isomorphism

$$\mathrm{Hom}_R(\Phi, \Phi^*) = \mathrm{Hom}_R(\Phi^\iota, (\Phi^*)^\iota).$$

Let S denote the cokernel of h, so that

$$0 \longrightarrow \Phi \xrightarrow{h} \Phi^* \longrightarrow S \longrightarrow 0 \tag{3}$$

is a free resolution of the R-module S, giving, in particular that the A-modules $\mathrm{Tor}^i_R(S, A)$ and $\mathrm{Ext}^i_R(S, A)$ vanish for every R-algebra A and every $i > 1$. If A is an R-algebra, put

$$M(A) := \mathrm{Tor}^1_R(S, A) = \ker(h \otimes_R A),$$
$$S(A) := S \otimes_R A = \mathrm{coker}(h \otimes_R A)$$

(the letter M is chosen to remind us of *Mordell-Weil*, while the letter S is chosen to remind us of *Selmer*; see §5). These definitions give us an exact sequence of A-modules

$$0 \longrightarrow M(A) \longrightarrow \Phi \otimes_R A \xrightarrow{h \otimes A} \Phi^* \otimes_R A \longrightarrow S(A) \longrightarrow 0. \tag{4}$$

Now let us pass to the special case where A is an R-algebra with involution, compatible with the involution ι of R. We use the same letters ι and i to refer to the involutions of A and of $\Phi^* \otimes_R A$. We have that $h^* \otimes_R A = -h \otimes_R A$, and using this along with (4) (for the upper exact sequence) and (3) (for the lower exact sequence) gives a commutative diagram of A-modules,

$$\begin{array}{ccccccccc}
0 \longrightarrow & M(A)^\iota & \longrightarrow & \Phi^\iota \otimes A & \xrightarrow{-h \otimes A} & (\Phi^*)^\iota \otimes A & \longrightarrow & S(A)^\iota & \longrightarrow 0 \\
 & & & \downarrow \cong & & \downarrow \cong & & & \\
0 \to & \mathrm{Hom}_R(S, A) & \to & \mathrm{Hom}(\Phi^*, A) & \xrightarrow{h^* \otimes A} & \mathrm{Hom}(\Phi, A) & \to & \mathrm{Ext}^1_R(S, A) & \to 0.
\end{array}$$

Thus we obtain canonical isomorphisms

$$M(A)^\iota \cong \mathrm{Hom}_R(S, A), \tag{5}$$

$$S(A)^\iota \cong \mathrm{Ext}^1_R(S, A). \tag{6}$$

Assume now that A is a quotient R-algebra with compatible involution, that is, $A = R/I$ where $I \subset R$ is stable under ι. Tensoring the exact sequence

$$0 \longrightarrow I \longrightarrow R \longrightarrow A \longrightarrow 0$$

with S gives a canonical injection

$$0 \longrightarrow \mathrm{Tor}^1_R(S, A) \longrightarrow I \otimes_R S$$

and composing this with the natural pairing

$$(I \otimes_R S) \otimes_R \mathrm{Hom}_R(S, A) \longrightarrow I \otimes_R A = I/I^2$$

we get the pairing

$$\mathrm{Tor}^1_R(S, A) \otimes_A \mathrm{Hom}_R(S, A) \longrightarrow I/I^2.$$

Now, using the definition of $M(A)$ and (5), we obtain the pairing:

$$M(A) \otimes_A M(A)^\iota \longrightarrow I/I^2. \tag{7}$$

Note that if ι induces multiplication by -1 on I/I^2, then (7) is an A-bilinear *symmetric* pairing.

Remark 3.1. Here is a more direct description of the pairing (7). Let $\langle\ ,\ \rangle$ denote the skew-Hermitian pairing corresponding to h, and if $m \in M(A) \subset \Phi/I\Phi$ let $\tilde{m} \in \Phi$ denote any choice of lifting of m. Then, from the definition of $M(A)$, $\langle \tilde{m}, x\rangle \in I \subset R$ for every $x \in \Phi$. If $m_1, m_2 \in M(A)$ we see that the value $\langle \tilde{m}_1, \tilde{m}_2\rangle \in I$, when taken modulo I^2, is dependent only upon the elements $m_1, m_2 \in M(A)$ and independent of the choices of liftings $\tilde{m}_1, \tilde{m}_2 \in \Phi$. Then the A-bilinear pairing (7) is defined by the rule

$$m_1 \otimes m_2 \ \mapsto\ \langle \tilde{m}_1, \tilde{m}_2\rangle \pmod{I^2}\ \in\ I/I^2.$$

Now let us suppose, for simplicity, that A is a principal ideal domain, and denote its field of fractions by $\mathcal{K}$. The natural exact sequence

$$0 \longrightarrow S(A)_{\mathrm{tors}} \longrightarrow S(A) \longrightarrow S(A)_{\mathrm{free}} \longrightarrow 0,$$

decomposing the A-module $S(A)$ into its torsion A-submodule and torsionfree quotient, has a (noncanonical) splitting. Combining this with the natural long exact sequence coming from applying the functor $\mathrm{Hom}_R(S,\ \cdot\ )$ to the exact sequence of R-modules $0 \to A \to \mathcal{K} \to \mathcal{K}/A \to 0$, we obtain an exact sequence

$$0 \longrightarrow \mathrm{Hom}_A(S(A)_{\mathrm{tors}}, \mathcal{K}/A) \longrightarrow \mathrm{Ext}^1_R(S, A) \longrightarrow \mathrm{Ext}^1_R(S, \mathcal{K}),$$

from which we see that the A-torsion submodule of $\mathrm{Ext}^1_R(S, A)$ may be identified with $\mathrm{Hom}_A(S(A)_{\mathrm{tors}}, \mathcal{K}/A)$. Therefore, restricting the isomorphism (6) to $S(A)^\iota_{\mathrm{tors}}$

we obtain a pairing

$$S(A)_{\mathrm{tors}} \otimes_A S(A)^{\iota}_{\mathrm{tors}} \longrightarrow \mathcal{K}/A \tag{8}$$

which is A-bilinear and nondegenerate.

Remark 3.2. Here is a more direct description of the pairing (8). Suppose $s \in S(A)_{\mathrm{tors}}$, say $as = 0$ with $a \in A$. From the definition (4) of $S(A)$, we can choose $\tilde{s} \in \Phi \otimes A$ and $\tilde{s}^* \in \Phi^* \otimes A$ such that $\tilde{s}^*$ lifts s (under (4)) and $\tilde{s}$ lifts $a\tilde{s}^*$. Similarly, if $t \in S(A)^{\iota}_{\mathrm{tors}}$ and $bt = 0$ we can lift to $\tilde{t} \in \Phi^{\iota} \otimes A$ whose image in $(\Phi^*)^{\iota} \otimes A$ is b times a lift of t.

Let $\langle\ ,\ \rangle_A$ denote the skew-Hermitian pairing $(\Phi \otimes A) \otimes (\Phi^{\iota} \otimes A) \to A$ induced by h. Then the pairing (8) is given by

$$s \otimes t \mapsto (ab)^{-1}\langle \tilde{s}, \tilde{t}\rangle_A \pmod{A} \in (ab)^{-1}A/A \subset \mathcal{K}/A.$$

This is independent of all the choices that were made.

In summary, given a skew-Hermitian module Φ over R, with the hypotheses above, for every ideal $I \subset R$ stable under the action of ι with $A := R/I$ a principal ideal domain, we get an A-bilinear pairing (7) on $M(A)$ with values in I/I^2 and a nondegenerate A-bilinear pairing (8) on $S(A)_{\mathrm{tors}}$ with values in $\mathcal{K}/A$.

In the special case where $I \subset R$ is a switching ideal, the pairing (7) is symmetric and the pairing (8) is skew-symmetric.

4. Discriminants, "L-functions", and regulators

Let Φ be a skew-Hermitian module over R as in §3, and continue to suppose that $A = R/I$ is a principal ideal domain. Denote by $D \in R$ the discriminant of the skew-Hermitian pairing on Φ (well defined up to an element uu^{ι} with $u \in R^{\times}$). Let r denote the rank of the free A-module $M(A)$, and denote by $d \in I^r/I^{r+1}$ the discriminant of the symmetric pairing (7) (well defined up to an element vv^{ι} with $v \in A^{\times}$). Let $s \in A$ denote a generator of the Fitting ideal of the A-torsion module $S(A)_{\mathrm{tors}}$.

Proposition 4.1. *The discriminant D lies in I^r, and if $D^{(r)} \in I^r/I^{r+1}$ denotes the image of D in I^r/I^{r+1} then we have (up to a unit in $A^{\times}$)*

$$D^{(r)} = d \cdot s.$$

Remark 4.2. One might think of this formula as analogous to the standard (conjectured, or sometimes proved) formulas linking a special value of the rth derivative of an L-function to the product of a height-pairing regulator times a term which reflects torsion in the Selmer group.

Proof of Proposition 4.1. Form the free A-modules $\Psi := \Phi/I\Phi$ and $\Psi^* := \Phi^*/I\Phi^*$. Then $\Psi^* = \mathrm{Hom}(\Psi^{\iota}, A)$, and both Ψ and Ψ^* are free over A of rank $g := \mathrm{rank}_R(\Phi)$. The A-module Ψ inherits an A-bilinear skew-Hermitian pairing from the skew-Hermitian pairing on Φ.

Since $M(A)$ is the kernel of $h \otimes A : \Psi \to \Psi^*$, $M(A)$ is a direct summand of Ψ. Fix an R-basis of Φ whose first r elements reduce modulo I to an A-basis of $M(A)$, and let N be the submodule of Ψ generated by the reductions of the last $g - r$ basis elements. Then $\Psi = M(A) \oplus N$, and with respect to this basis the pairing on Φ is given by a matrix

$$B = \begin{pmatrix} B_1 & B_2 \\ B_3 & B_4 \end{pmatrix}$$

where B_1, B_2, and B_3 are $r \times r$, $r \times (g-r)$, and $(g-r) \times r$ matrices with entries in I, and B_4 is a $(g-r) \times (g-r)$ matrix with entries in R. Thus we see directly that $D = \det(B) \in I^r$ and

$$D^{(r)} \equiv \det(B_1) \det(B_4) \pmod{I^{r+1}}. \tag{9}$$

By Remark 3.1, the matrix giving the pairing (7) on $M(A)$ with respect to our chosen basis is $B_1 \pmod{I^2}$, so $d \equiv \det(B_1) \pmod{I^{r+1}}$.

Let B' and B_4' be the reductions of B and B_4 modulo I, so their entries lie in A and

$$B' = \begin{pmatrix} 0 & 0 \\ 0 & B_4' \end{pmatrix}.$$

Then B' is the matrix describing the map $h \otimes A : \Psi \to \Psi^*$. Thus B' has rank $g - r$, so the $(g-r) \times (g-r)$-matrix B_4' has nonzero determinant in A. Further, we have

$$S(A) = \operatorname{coker}(h \otimes A) = \Psi^*/B'\Psi = M(A)^* \oplus (N^*/B_4'N)$$

so $S(A)_{\mathrm{tors}} = N^*/B_4'N$ has Fitting ideal $\det(B_4')$. Now the proposition follows from (9). □

Part II. Skew-Hermitian modules and arithmetic

5. The arithmetic of (E, K, p)

Let us fix an elliptic curve E over a number field K, having good ordinary reduction at all primes of K above p. We keep the notation as in §1. Let $\tilde{K}$ be the maximal $\mathbf{Z}_p$-power extension of K, and $\Gamma(K) = \mathrm{Gal}(\tilde{K}/K)$. Let R be a localization of the completed group ring $\mathbf{Z}_p[[\Gamma(K)]]$, viewed as $\mathbf{Z}_p[[\Gamma(K)]]$-algebra, endowed with the (unique) involution $\iota : R \to R$ compatible with the standard involution of $\mathbf{Z}_p[[\Gamma(K)]]$ (given by inversion of group elements). By an *intermediate field extension* L we mean an extension field L of K contained in $\tilde{K}$. Put $H_L := \mathrm{Gal}(\tilde{K}/L) \subset \Gamma(K)$, and note that the ideal I_L of R generated by I_{H_L} is stable under the involution ι and I_K is a switching ideal. Define the quotient R-algebra

$$A_L := R/I_L = R \otimes_{\mathbf{Z}_p[[\Gamma(K)]]} \mathbf{Z}_p[[\mathrm{Gal}(L/K)]].$$

For arbitrary intermediate extensions L define the Selmer A_L-module

$$S(L) = S_p(E, L; R) := \varprojlim_{K'} (S_p(E, K') \otimes_{\mathbf{Z}_p[[\Gamma(K)]]} R),$$

the projective limit being taken over all finite extensions K' of K in L. Denote the R-module $S(\tilde{K})$ simply S.

Assume the following:

Perfect Control axiom. *The natural A_L-homomorphism*

$$S \otimes_R A_L \longrightarrow S(L)$$

is an isomorphism for all intermediate fields L.

Remark 5.1. Although the Perfect Control axiom does not universally hold, the homomorphism $S \otimes_R A_L \to S(L)$ usually has rather small kernel and cokernel bounded independently of L (as implied by the "Control Theorem", which holds in a fairly general context) and it is often possible to choose localizations R of $\mathbf{Z}_p[[\Gamma(K)]]$ so as to avoid the support of these kernels and cokernels, and thereby force this axiom to hold (see [M], [G]). An alternate strategy is to simply replace the classical Selmer groups $S(L)$ by $S'(L) := S \otimes_R A_L$ relying on the control theorem, when it applies, to guarantee that these $S'(L)$ are not very different from the true $S(L)$. In what follows we just assume the Perfect Control axiom.

Put

$$M(L) := \mathrm{Hom}_R(S, A_L)^\iota = \mathrm{Hom}_{A_L}(S(L), A_L)^\iota,$$

the latter equality holding because of the Control axiom. If $L \subset L'$ is an inclusion of intermediate fields, using the first equality above we have a natural induced homomorphism $M(L') \to M(L)$. Moreover, again using the first equality, for any intermediate field L we have

$$M(L) \cong \varprojlim_{K'} M(K')$$

where K' ranges through any increasing sequence of intermediate fields whose union is L. In particular, for any intermediate field L, $M(L)$ can be computed as a projective limit of $M(K')$ for intermediate fields K' of finite degree over K.

If $L \subset L'$ is an inclusion of intermediate fields of finite degree over K, let $E(L') \to E(L)$ denote the natural trace (also called "norm") mapping on Mordell-Weil groups. For any intermediate field L, of finite degree or not, consider the A_L-module

$$U(L) := \varprojlim_{K'} (E(K')/E(K')_{\mathrm{tors}} \otimes \mathbf{Z}_p) \otimes_{\mathbf{Z}_p[[\Gamma(K)]]} A_L$$

where K' ranges through any increasing sequence of intermediate fields of finite degree over K whose union is L.

Proposition 5.2. *Let L be an intermediate field.*

(i) *If $[L : K]$ is finite and $G = \mathrm{Gal}(L/K)$, then there is a natural isomorphism of A_L-modules*

$$M(L) \cong (E(L)/E(L)_{\mathrm{tors}}) \otimes_{\mathbf{Z}[G]} A_L.$$

(ii) *For arbitrary L there is a natural isomorphism of A_L-modules*

$$M(L) \cong U(L).$$

Proof. We will prove (i), and then (ii) follows by passing to the projective limit. So we assume that L/K is finite. Recall that we have also assumed the Shafarevich-Tate conjecture and the Perfect Control axiom.

We have

$$M(L)^{\iota} = \mathrm{Hom}_{A_L}(S(L), A_L) = \mathrm{Hom}_{\mathbf{Z}_p[G]}(S_p(E,L), \mathbf{Z}_p[G]) \otimes A_L.$$

The maximal $\mathbf{Z}_p$-torsion-free quotient

$$S_p(E,L)_{\mathrm{free}} := S_p(E,L)/S_p(E,L)_{\mathrm{tors}}$$

of $S_p(E,L)$ satisfies

$$S(L)_{\mathrm{free}} \cong \mathrm{Hom}(E(L), \mathbf{Z}_p).$$

Therefore we get isomorphisms

$$\begin{aligned} \mathrm{Hom}_{\mathbf{Z}_p[G]}(S_p(E,L), \mathbf{Z}_p[G]) &\cong \mathrm{Hom}_{\mathbf{Z}_p[G]}(S_p(E,L)_{\mathrm{free}}, \mathbf{Z}_p[G]) \\ &\cong \mathrm{Hom}_{\mathbf{Z}_p[G]}(\mathrm{Hom}(E(L), \mathbf{Z}_p), \mathbf{Z}_p[G]) \\ &\cong \mathrm{Hom}_{\mathbf{Z}_p}(\mathrm{Hom}(E(L), \mathbf{Z}_p), \mathbf{Z}_p) \\ &\cong (E(L)/E(L)_{\mathrm{tors}}) \otimes \mathbf{Z}_p \end{aligned}$$

using (for example) [Br] Proposition VI.3.4 for the third isomorphism. Tensoring with A_L proves (i). □

Let us now examine whether we have analogues of the two basic isomorphisms (5), (6) of §3. Specifically (assuming the Perfect Control axiom) are there natural isomorphisms of A_L-modules

$$M(L) \overset{?}{\cong} \mathrm{Tor}_R(S, A_L), \qquad S(L)^{\iota} \overset{?}{\cong} \mathrm{Ext}^1_R(S, A_L) \tag{10}$$

for every intermediate extension L?

In [N2] (cf. §6.6, and passim) Nekovár constructs what he calls the *Selmer complex* of R-modules. Applying his construction to our triple (E, K, p) we get that the associated Selmer complex lies in the derived category of bounded R-modules, and is supplied with a skew-Hermitian self-pairing, which, if one assumes the Perfect Control axiom and the vanishing of what Nekovár calls "error terms", produces the isomorphisms (10). Given a skew-Hermitian R-module Φ let us view the corresponding R-homomorphism $\mathbf{\Phi} = \{\Phi \to \Phi^*\}$ as being a complex in the derived category of bounded R-modules (putting Φ^* in degree 1) and note that the skew-Hermitian pairing gives an isomorphism

$$\eta : \mathbf{\Phi} \xrightarrow{\sim} \mathrm{Hom}_R(\mathbf{\Phi}^{\iota}, R)[-3].$$

Definition 5.3. Let (E, K, p) be as above, and R a localization of the completed p-adic group ring of $\Gamma(K)$ with involution ι which extends the natural involution on the group ring. Let us say that the skew-Hermitian R-module Φ *organizes the arithmetic of* (E, K, p) if the corresponding R-homomorphism $h : \Phi \to \Phi^*$ is injective, the Perfect Control axiom holds, and the complex $\mathbf{\Phi}$ is quasi-equivalent to Nekovár's Selmer complex, the quasi-equivalence being compatible with skew-Hermitian pairings.

If Φ organizes the arithmetic of (E, K, p), then for every intermediate extension L/K we have an exact sequence of A_L-modules

$$0 \longrightarrow U(L) \longrightarrow \Phi \otimes_R A_L \longrightarrow \Phi^* \otimes_R A_L \longrightarrow S_p(E, L) \otimes A_L \longrightarrow 0.$$

Further, the induced pairing on $U(L)$ described in §3 coincides with the p-adic height pairing on $U(L)$, and if $[L : K]$ is finite and A_L is a principal ideal domain then the induced pairing on $(S_p(E, L) \otimes A_L)_{\mathrm{tors}}$ described in §3 coincides with the Cassels pairing.

6. The classical anticyclotomic setting

In this section we assume all the hypotheses and notation of Section 5 and, furthermore, that E is an elliptic curve over $\mathbf{Q}$ without complex multiplication, K is a quadratic imaginary field, and p is an odd prime number. We have that $\Gamma(K)$ is free of rank two over $\mathbf{Z}_p$ and decomposes canonically as the product

$$\Gamma(K) = \Gamma^{\mathrm{anti}} \times \Gamma^{\mathrm{cycl}}$$

where $\Gamma^{\mathrm{anti}} := \mathrm{Gal}(K^{\mathrm{anti}}/K)$ and $\Gamma^{\mathrm{cycl}} := \mathrm{Gal}(K^{\mathrm{cycl}}/K)$. Here K^{anti}/K and K^{cycl}/K are the classical anticyclotomic and cyclotomic $\mathbf{Z}_p$-extensions of K, respectively.

We will be particularly interested in the anticyclotomic $\mathbf{Z}_p$-extension, so put

$$L := K^{\mathrm{anti}}.$$

As in [MR1], [MR2] let us consider the sign of $\chi_K(-N)$ where χ_K is the quadratic Dirichlet character attached to the field K, and N is the conductor of E. Conditional upon the hypotheses that we have made, this sign controls two things. First, thanks to a result of Nekovár [N1], if the sign is "+" then the rank of $E(K)$ is even, and if the sign is "−" then that rank is odd. Thanks also to recent results of Vatsal, Cornut, Kolyvagin and others, if the sign is "+" then $U(L)$ is trivial, while if it is "−" then $U(L)$ is free of rank one over A_L. In the case where the sign is "−" we have the height pairing

$$h_L : U(L) \otimes_{A_L} U(L)^\iota \longrightarrow I_L/I_L^2 = \Gamma^{\mathrm{cycl}} \otimes_{\mathbf{Z}_p} A_L.$$

Under the above hypotheses, in [MR1] we made the following conjecture.

Conjecture 6.1. *If* $\chi_K(-N) = -1$ *then the height-pairing* h_L *is an isomorphism of free, rank-one* A_L*-modules.*

Put another way, Conjecture 6.1 says that the *regulator* of the height pairing h_L on $U(L)$ is a unit.

Now suppose that the skew-Hermitian R-module Φ organizes the arithmetic of (E, K, p), as in Definition 5.3, with $(x, y) \mapsto \langle x, y\rangle$ denoting the R-valued skew-Hermitian pairing on Φ. Choose an element $u \in \Phi$ that projects to a generator of the A_L-module $U(L)$ in $\Phi/I_L\Phi$. Then $\langle u, v\rangle \in I_L$ for all $v \in \Phi$, and Conjecture 6.1 would imply that $\langle u, u\rangle$ is a generator of the ideal $I_L \subset R$.

Under these conditions, the Gram-Schmidt projector $\pi : \Phi \to \Phi$ defined by

$$v \mapsto \pi(v) = v - \frac{\langle v, u\rangle}{\langle u, u\rangle} u$$

splits the skew-Hermitian module Φ into the direct sum of the free, rank-one skew-Hermitian R-module Ru and its orthogonal complement $\pi(\Phi)$.

Remark 6.2. If the Conjecture 6.1 holds, and if there is a skew-Hermitian organizer of the arithmetic of (E, K, p), then under the above conditions when $\chi_K(-N) = -1$ the Selmer complex in the derived category splits as a direct sum of a complex quasi-isomorphic to the complex $R \to R$ (the mapping sending 1 to an "imaginary" generator of I_L) plus another complex. The former complex would account for the universal norms in Mordell-Weil compiled by ascending the anticyclotomic tower; the latter complex cannot vanish if the rank of the Mordell-Weil group of E over K is greater than 1 and indeed the R-rank g of an organizer Φ must be at least equal to the rank of the Mordell-Weil group of E over K.

Remark 6.3. If Φ organizes the arithmetic of (E, K, p), then the characteristic ideal of the Selmer R-module S is the discriminant D of the pairing $h : \Phi \to \Phi^*$. In this situation the "two-variable main conjecture" is equivalent to the following.

Conjecture 6.4. *The discriminant of h is equal, up to a unit in R, to the two-variable p-adic L-function of Haran, Hida, and Perrin-Riou* [Ha], [Hi], [PR1], [PR2].

For more about the relation between our setup and other anticyclotomic conjectures, see [MR1].

References

[Br] K. Brown, Cohomology of groups, *Grad. Texts in Math.* **87**, Springer, New York (1982).

[Co] C. Cornut, Mazur's conjecture on higher Heegner points, *Invent. math.* **148** (2002) 495–523.

[G] R. Greenberg, Galois theory for the Selmer group of an abelian variety. To appear.

[Ha] S. Haran, p-adic L-functions for elliptic curves over CM fields, thesis, MIT 1983.

[Hi] H. Hida, A p-adic measure attached to the zeta functions associated with two elliptic modular forms. I, *Invent. Math.* **79** (1985), 159–195.

[M] B. Mazur, Rational points of abelian varieties with values in towers of number fields, *Invent. Math.* **18** (1972) 183–266.

[MR1] B. Mazur, K. Rubin, Elliptic curves and class field theory. In: *Proceedings of the International Congress of Mathematicians, ICM 2002, Beijing,* Ta Tsien Li, ed., vol II. Beijing: Higher Education Press (2002) 185–195.

[MR2] ———, Studying the growth of Mordell-Weil. To appear in a special issue of *Documenta Math.* dedicated to K. Kato.

[MT] B. Mazur, J. Tate, Canonical height pairings via biextensions. In: *Arithmetic and Geometry,* Progr. Math. **35**, Birkhäuser, Boston (1983) 195–237.

[N1] J. Nekovář, On the parity of ranks of Selmer groups. II, *C. R. Acad. Sci. Paris Sér. I Math.* **332** (2001) 99–104.

[N2] ———, Selmer complexes. Preprint available at `http://www.math.jussieu.fr/~nekovar/pu/`.

[PR1] B. Perrin-Riou, Fonctions L p-adiques, théorie d'Iwasawa et points de Heegner, *Bull. Soc. Math. France* **115** (1987), 399–456.

[PR2] ———, Points de Heegner et dérivées de fonctions L p-adiques, *Invent. Math.* **89** (1987), 455–510.

[V] V. Vatsal, Uniform distribution of Heegner points, *Invent. math.* **148** (2002) 1–46.

Barry Mazur
Department of Mathematics
Harvard University
Cambridge, MA 02138 USA
e-mail: `mazur@math.harvard.edu`

Karl Rubin
Department of Mathematics
Stanford University
Stanford, CA 94305 USA
e-mail: `rubin@math.stanford.edu`

Progress in Mathematics, Vol. 224, 165–179

Explicit Parametrizations of Ordinary and Supersingular Regions of $X_0(p^n)$

Ken McMurdy

Abstract. In this paper, we explore the geometry of the modular curves $X_0(p^n)$, over $\mathbb{C}_p$ in a few non-trivial examples. In particular, we give explicit rigid-analytic parameterizations of the ordinary and supersingular regions. This is done by first building towers of models for $X_0(p^i)$, $0 \le i \le n$, and then pulling back the ordinary and supersingular regions of $X(1)$ via the forgetful maps.

1. Introduction

On $X(1)$, as a rigid analytic curve over $\mathbb{C}_p$, define the supersingular region to be the rigid subspace whose points correspond to elliptic curves over $\mathbb{C}_p$ with good supersingular reduction. This is a finite disjoint union of rigid-analytic discs, one for each supersingular j-invariant in characteristic p. Now let

$$\pi_1 : X_0(N) \to X(1)$$

be the "forgetful" map, the moduli-theoretic map which takes the point corresponding to (E, C) to the point corresponding to E. We then define the supersingular region of $X_0(N)$ to be the inverse image under π_1 of the supersingular region of $X(1)$. The complement is called the ordinary region.

While these regions have been defined rigid-analytically, they can also be studied by looking at the special fibers in various scheme-theoretic models for $X_0(N)$ over extensions of $\mathbb{Z}_p$. A natural starting point is the Katz-Mazur model of $X_0(N)$ as the compactified coarse moduli scheme representing the $[\Gamma_0(N)]$ moduli problem [KM, 3.4, 5.1]. For $X_0(p^n)$, the irreducible components of the special fiber of this model are explicitly described in [KM, 13.5.6], with crossings at the supersingular points as described in [KM, 13.4.7]. From this information alone, however, it is impossible even to conclude the number of (rigid) connected components of the ordinary region. In the simplest case of $X_0(p)$ ($p \neq 2, 3$) the minimal resolution over $\mathbb{Z}_p$ has been worked out in [DR, VI.6.16] and provides a fairly complete description of both the ordinary and supersingular regions. In particular, the ordinary region must have two trivial connected components (genus 0 affinoids as defined below), reducing to the two "vertical" copies of P^1 in the DR model. The supersingular region must consist of ss wide-open annuli, the width of each being determined by the number (if any) of "horizontal" copies of P^1 in the resolution of the corresponding supersingular point. A similar description of $X_0(p^2)$ follows from the special fiber of Edixhoven's model [E, 2.1.2]. From this model we know that the ordinary region of $X_0(p^2)$ must have four connected components, reducing to the four vertical curves. The connected components of the supersingular region reduce to the ss horizontal curves for which Edixhoven provides equations.

Unfortunately, for $X_0(p^n)$ with $n > 2$ there are no such complete and general results analogous to these of Deligne-Rapoport and Edixhoven. The connected components of the ordinary region can now be completely described in moduli-theoretic terms by recent results of Coleman ([C2]). This is done by further analysing the affinoids $X_{a,b}$ reducing to the irreducible components in [KM, 13.5.6] via the theory of the canonical subgroup. In [C2] Coleman also provides the following moduli-theoretic interpretation of the horizontal components in $X_0(p^2)$. Coleman shows that points of the horizontal components of $X_0(p^2)$ correspond to pairs (E, C) such that E/pC is too-supersingular in the sense of [B, 3.3]. This moduli-theoretic description of the horizontal components, however, has not yet been generalized to $X_0(p^n)$ for $n > 2$.

In this paper we will take a different approach toward understanding the ordinary and supersingular regions of $X_0(p^n)$. The starting point is to build very explicit models for $X_0(p)$, $X_0(p^2)$, and even $X_0(p^3)$ (in the $p = 5$ example), always determining the formula for the forgetful map in terms of the chosen parameters. The ordinary and supersingular regions can then be described quite explicitly by simply pulling back the corresponding regions of $X(1)$. It then becomes a rigid-analysis problem to determine whatever data is desired. In this paper, we will focus for example on the number of connected components of the ordinary region as well as their genera in the following sense. Define the genus of a smooth one-dimensional affinoid X to be the minimal genus of a smooth proper curve containing X (which exists). We show in each example, using (a souped-up version of) Hensel's Lemma, that each affinoid connected component X of the ordinary region is contained in a rigid subspace W such that $W - X$ is a wide-open annulus (as in [C1, II]). Attaching a disc D to W along this annulus (as in the proof of [C1, 3.3]) gives $X \cup D$ the structure of a proper curve whose genus can be readily calculated and is the genus of X. While our approach toward understanding $X_0(p^n)$ will most likely not result directly in a general theorem, it could conceivably provide a great deal of relevant data leading to a conjecture about the components in the stable model of $X_0(p^n)$. In a future article, for example, we will employ these methods to explicitly describe the components in the stable model of $X_0(125)$.

2. Example 1, $X_0(49) \to X_0(7) \to X(1)$

Since the only supersingular value in characteristic 7 is $j = 1728$, the supersingular region of $X(1)$ is simply the disc described by $v_7(j - 1728) > 0$. Therefore we begin by choosing a parameter on $X_0(7)$ (which has genus 0) and determining an equation for π_1 centered around $j = 1728$. For a parameter we choose the following eta product (with the given cuspidal divisor).

$$t = \left(\frac{\eta_1}{\eta_7}\right)^4 \qquad (t) = (0) - (\infty).$$

Remark 2.1. *Whenever an eta product is used throughout the paper, we are implicitly defining a function via its canonical q-expansion at infinity and the formula*

$$\eta_d = q^{\frac{d}{24}} \prod_{n=1}^{\infty} (1 - q^{nd}).$$

In each case, the eta product has been verified to be a legitimate function on $X_0(N)$ by means of Ligozat's criteria as in [L, 3.2.1]. *Furthermore, the divisors of all eta products as well as the ramification indices of moduli-theoretic maps at the cusps have been calculated using families of Tate curves.*

Now we compare q-expansions to obtain Equation (1) for π_1 in terms of the parameters j and t. Then we use this equation to explicitly describe the pullback of the supersingular disc to $X_0(7)$ in Claim 2.2.

$$\pi_1^*(j - 1728) = \frac{(t^4 - 10 \cdot 7^2 t^3 - 9 \cdot 7^4 t^2 - 2 \cdot 7^6 t - 7^7)^2}{t^7}. \tag{1}$$

Claim 2.2. *The inverse image under π_1 of the supersingular disc $v_7(j - 1728) > 0$ is the annulus described by*

$$0 < v_7(t) < 2.$$

Proof. When $v_7(t) \leq 0$, Equation (1) implies that

$$v_7(\pi_1^*(j - 1728)) = 8v_7(t) - 7v_7(t) = v_7(t) \leq 0.$$

Similarly, when $v_7(t) \geq 2$, we see from Equation (1) that

$$v_7(\pi_1^*(j - 1728)) = 14 - 7v_7(t) \leq 0.$$

Finally we consider the case when $0 < v_7(t) < 2$. Over this entire region we have

$$v_7(\pi_1^*(j - 1728)) \geq \min\{8v_7(t), 14\} - 7v_7(t) > 0.$$ □

The picture for $X_0(7)$ is given in Figure 1 with the ordinary region shaded. Note that we have also included for comparison the picture of Deligne-Rapoport's model from [DR, VI.6.16], adding numbers to indicate the genera of the irreducible components. For the reader who wishes to further explore moduli-theoretic properties of our model we will also include now the formula for the Atkin-Lehner involution w_7. It is immediate that $w_7^*(t) = c/t$ for some constant c since the involution switches the cusps (0) and (∞). To compute c we first determine the formula for the moduli-theoretic map $\pi_7 : X_0(7) \to X(1)$ which has the effect of $q \to q^7$ on canonical q-expansions at ∞.

$$\pi_7^*(j) = \frac{(t^2 + 5t + 1)^3 (t^2 + 13t + 49)}{t}.$$

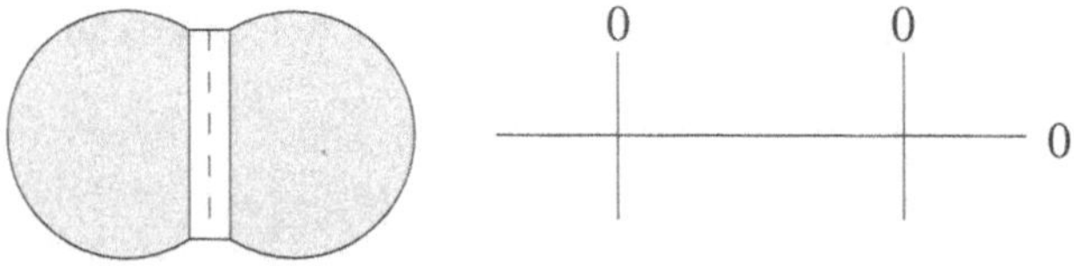

FIGURE 1. $X_0(7)/\mathbb{C}_7$ and Special Fiber of DR Model

Combining this formula with the compatibility relation $\pi_1 \circ w_7 = \pi_7$ we find that the constant $c = 49$.

Now we choose parameters for $X_0(49)$ (genus 1) and determine the equation for the forgetful map down to $X_0(7)$ in terms of these parameters.

$$x = \frac{\eta_1}{\eta_{49}} \qquad (x) = 2(0) - 2(\infty)$$

$$y = \left(\frac{\eta_7}{\eta_{49}}\right)^4 \qquad (y) = (0) + \sum_{i=1}^{6}(c_{7,i}) - 7(\infty)\,.$$

Note that by $\{c_{7,i}\}$ we refer to the 6 cusps of $X_0(49)$ representing the Néron 7-gon with its various level $\Gamma_0(49)$ structures. Comparison of q-expansions then yields the following (singular) equation for the curve.

$$\begin{aligned} f_{49}(x,y) &= y^2 - 7xy(x^2+5x+7) \\ &\quad - x(x^6+7x^5+21x^4+49x^3+147x^2+343x+343) = 0\,. \end{aligned} \tag{2}$$

Using the same parameter x it is of course possible to find a Weierstrass equation for this elliptic curve. Indeed, one such equation is the following.

$$z^2 = x(4x^2+21x+28) \tag{3}$$

The advantage of the model given by (2) is that it is easier to find an expression for the pullback of t in terms of these parameters. In fact, it follows directly from the definitions that

$$\pi_1^*(t) = x^4/y. \tag{4}$$

Using this formula for π_1 along with Equation (2) and Claim 2.2 it is now possible to completely describe the ordinary region of $X_0(49)$ in terms of the parameters x and y. Before we begin, however, we will need the following souped-up version of Hensel's Lemma in order to parameterize regions of $X_0(49)$ with open discs and annuli.

Lemma 2.3 (Hensel). *Let R^0 be the ring of analytic functions on an open disc or annulus over $\mathbb{C}_p$ with nonnegative valuation (everywhere). Let $f(y) \in R^0[y]$ and suppose $\exists y_0 \in R^0$ satisfying $v_p(f(y_0)) > 0$ and $v_p(f'(y_0)) = 0$ (everywhere). Then f has a root $y \in R^0$ satisfying $v_p(y - y_0) > 0$ (everywhere).*

Proof. As in the usual proof of Hensel's Lemma we define a sequence of functions by

$$y_{n+1} = y_n - \frac{f(y_n)}{f'(y_n)}.$$

On any affinoid disc the sequence converges to some y with respect to the supremum semi-norm. Now cover the open disc or annulus with affinoid discs. □

Claim 2.4. *The ordinary region of* $X_0(49)$ *consists of* 4 *genus* 0 *connected components, each bounded by a supersingular annulus. In terms of the parameters* x *and* y*, these components (and annuli) are described as follows:*

i) $v_7(x) \le 0$, *bounded by the annulus* $0 < v_7(x) < 1/2$

ii) $v_7(x) \ge 1$, *bounded by the annulus* $1/2 < v_7(x) < 1$

iii) $v_7(x + \sqrt{-7}) \ge 2/3$, *bounded by the annulus* $1/2 < v_7(x + \sqrt{-7}) < 2/3$

iv) $v_7(x - \sqrt{-7}) \ge 2/3$, *bounded by the annulus* $1/2 < v_7(x - \sqrt{-7}) < 2/3$.

Proof. (i) Let W_∞ be the region defined by $v_7(x) < 1/2$. On this entire region, y^2 and x^7 are the terms of Equation (2) with minimal valuation. In other words, over all of W_∞ we can rewrite Equation (2) as

$$y^2 = x^7 + \text{smaller terms}.$$

This suggests that we can parameterize W_∞ using the disc $D_s = \{s \mid v_7(s) > -1/4\}$ and a map close to $x = s^{-2}$, $y = s^{-7}$. Lemma 2.3 makes this totally precise. Just take x to be s^{-2} exactly and then let y be s^{-7} times the unique root of the polynomial $s^{14} f_{49}(s^{-2}, ys^{-7})$ close to $y_0 = 1$.

Now that W_∞ has been shown to be trivial, we turn our attention to the problem of determining which points of W_∞ are ordinary. The key fact is that on W_∞, as we have already noted, we know that $v_7(y) = \frac{7}{2} v_7(x)$. Therefore Equation (4) and Claim 2.2 imply that the ordinary points are those with $v_7(x) \le 0$.

(ii) The argument is similar to (i), but now y^2 and $343x$ are the terms of Equation (2) which have minimal valuation throughout the region

$$W_0 = \{(x, y) \mid v_7(x) > 1/2\}.$$

(iii) Equation (2) can be rewritten in the following way.

$$y^2 = 7xy((x^2 + 7) + 5x) + x(x^2 + 7)^3 + 7x^6 + 49x^4 + 343x^2.$$

This combined with the Equation (4) and Claim 2.2 shows that points satisfying $v_7(x) = 1/2$ are ordinary exactly when $v_7(x^2 + 7) \ge 7/6$. So to better understand the region described by $v_7(x) = 1/2$, we make the substitution $u = x + \sqrt{-7}$ and consider the following rigid subspaces of $X_0(49)$.

$$W^+ = \{(u, y) \mid v_7(u) > 1/2\} \qquad X^+ = \{(u, y) \mid v_7(u) \ge 2/3\}.$$

In the language of [C1], W^+ is a wide open space with underlying affinoid X^+. We have chosen X^+ so that it is precisely the ordinary region of W^+.

After the change of variables, the resulting equation is of the form

$$g(u, y) = y^2 - 8 \cdot 49u^3 + \text{smaller terms} = 0$$

where these two leading terms have minimal valuation whenever $1/2 < v_7(u) < 2/3$ (in other words on $W^+ - X^+$). This suggests that we can parameterize $W^+ - X^+$ using the annulus $A_s = \{s \mid 1/4 < v_7(s) < 1/3\}$ and a map close to $u = s^2/2$, $y = 7s^3$. As before this can be made precise by setting $u = s^2/2$ (exactly) and applying Hensel's Lemma to solve for y.

Because the affinoid X^+ is bounded by an annulus, we now know that it is connected, but we also now have a straightforward way to calculate its genus. First we extend A_s to the disc $D_s = \{s \mid v_7(s) < 1/3\}$, so that the parameterization map on A_s gives $D_s \cup X^+$ the structure of a proper curve. Now we calculate the genus of this curve by working out the divisor of the differential $du = sds$. It turns out to be $P - 3Q$, where Q is the infinite point with respect to s and P is given explicitly (by x and y coordinates) below.

$$P = \left(\frac{-21 - \sqrt{-7}}{8}, \frac{-49(31 + 3\sqrt{-7})}{256} \right).$$

Of course over all of $X_0(49)$ the differential $du = dx$ has three simple zeros corresponding to the three roots of the Weierstrass polynomial given in (3). However, only this one point P lies on X^+. Since we have computed $\deg(K_c) = -2$, Riemann-Roch then implies that X^+ has genus 0.

(iv) Identical to (iii) but we make the change of coordinates $u = x - \sqrt{-7}$. □

The resulting picture for $X_0(49)$ is given in Figure 2 with the ordinary region shaded and the special fiber of Edixhoven's model [E, 2.5.2] included for comparison. As before we now determine the formula for the Atkin-Lehner involution w_{49} so that further moduli-theoretic implications may be readily drawn from our model. Directly from the divisors of x and y we must have $w_{49}^*(x, y) = (c_1/x, c_2 y/x^4)$. We have chosen our parameters so that $y = \pi_7^*(t)$, where π_7 is the moduli-theoretic map defined as before (taking $q \to q^7$ at infinity). We have also already determined that $\pi_1^*(t) = x^4/y$ and $w_7^*(t) = 49/t$. Plugging these formulas into the compatibility relation $\pi_1 \circ w_{49} = w_7 \circ \pi_7$ we can conclude right away that $c_2 = c_1^4/49$. Acting on Equation (2) by w_{49} now shows that we must have $c_1 = 7$ and $c_2 = 49$. One immediate consequence is that w_{49} switches not only the ordinary affinoids of (i) and (ii), but also the ordinary affinoids in (iii) and (iv).

3. Example 2, $X_0(169) \to X_0(13) \to X(1)$

As in the first example, the supersingular region of $X(1)/\mathbb{C}_{13}$ is simply an open disc, this time the disc $v_{13}(j-5) > 0$. For a parameter on the genus 0 curve $X_0(13)$, we take $t = (\eta_1)^2/(\eta_{13})^2$ which again has the divisor $(0) - (\infty)$. The general form of the resulting formula for π_1 in terms of j and t is then given below in Equation (5). Since the valuations of the coefficients are sufficient for the proof of Claim 3.1, we list just the valuations in Table 1. Figure 3, which follows, is just a direct

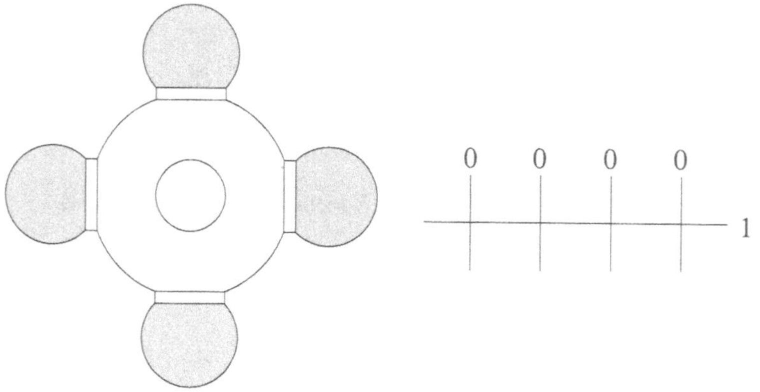

FIGURE 2. $X_0(49)/\mathbb{C}_7$ and Special Fiber of Edixhoven Model

visualization of Claim 3.1 along with a picture of the Deligne-Rapoport special fiber for comparison.

$$\pi_1^*(j-5) = \frac{t^{14} + a_{13}t^{13} + a_{12}t^{12} + \cdots + a_1 t + a_0}{t^{13}}. \qquad (5)$$

a_i	a_{13}	a_{12}	a_{11}	a_{10}	a_9	a_8	a_7	a_6	a_5	a_4	a_3	a_2	a_1	a_0
$v_{13}(a_i)$	1	2	3	4	5	7	7	8	9	10	11	12	13	13

TABLE 1. Valuations of Coefficients in Equation (5)

Claim 3.1. *The inverse image under π_1 of the supersingular disc $v_{13}(j-5) > 0$ is the annulus described by*

$$0 < v_{13}(t) < 1.$$

Proof. Identical to the proof of Claim 2.2. □

Remark 3.2. *As in the $p = 7$ example, it is straightforward to determine the formula for the moduli-theoretic map π_{13}. Then by the compatibility relation $\pi_1 \circ w_{13} = \pi_{13}$ we obtain the formula $w_{13}^*(t) = 13/t$ for the Atkin-Lehner involution.*

Now we generate a model for $X_0(169)$ (genus 8) by choosing the following parameters and comparing q-expansions to find Equation (6). Again the valuations of the coefficients in the equation for the curve are technically sufficient for our analysis (until more details are offered in the proof of Claim 3.3, part (iii)). So we

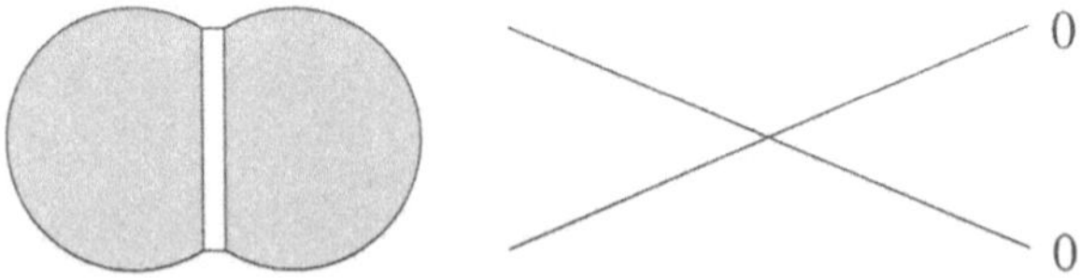

FIGURE 3. $X_0(13)/\mathbb{C}_{13}$ and Special Fiber of DR Model

offer this information in Table 2 below.

$$x = \frac{\eta_1}{\eta_{169}} \qquad (x) = 7(0) - 7(\infty)$$

$$y = \left(\frac{\eta_{13}}{\eta_{169}}\right)^2 \qquad (y) = (0) + \sum_{i=1}^{12}(c_{13,i}) - 13(\infty)$$

$$\begin{aligned} f_{169}(x,y) = y^7 + a_1 x y^6 + (b_3 x^3 + \cdots + b_1 x) y^5 + (c_5 x^5 + \cdots + c_1 x) y^4 \\ + (d_7 x^7 + \cdots + d_1 x) y^3 + (e_9 x^9 + \cdots + e_1 x) y^2 \\ + (f_{11} x^{11} + \cdots + f_1 x) y + (g_{13} x^{13} + \cdots + g_1 x) = 0 . \end{aligned} \quad (6)$$

	13	12	11	10	9	8	7	6	5	4	3	2	1
a													1
b											1	1	2
c									1	1	2	2	3
d							1	1	2	2	3	3	4
e					1	1	2	2	3	3	4	4	5
f			1	2	2	2	4	3	5	4	5	6	6
g	0	1	1	2	2	3	3	4	4	5	5	6	6

TABLE 2. Valuations of Coefficients in Equation (6)

In order to pull back the ordinary and supersingular regions of $X_0(13)$ we now determine the formula for π_1 in terms of the given parameters. Because of our choice of parameters it follows directly from the definitions that $\pi_1^*(t)$ is simply x^2/y. Now we do the same type of analysis that was used in the proof of Claim 2.4 to completely describe the ordinary region of $X_0(169)$.

Claim 3.3. *The ordinary region of $X_0(169)$ consists of 2 genus 0 connected components, and 2 genus 1 connected components. As with $X_0(49)$, each component is bounded by a supersingular annulus. In terms of the parameters x and y, these components (and annuli) are described as follows:*

i) $v_{13}(x) \leq 0$, *bounded by the annulus* $0 < v_{13}(x) < 1/2$
ii) $v_{13}(x) \geq 1$, *bounded by the annulus* $1/2 < v_{13}(x) < 1$
iii) $v_{13}(x+\sqrt{13}) \geq 7/12$, *bounded by the annulus* $1/2 < v_{13}(x+\sqrt{13}) < 7/12$
iv) $v_{13}(x-\sqrt{13}) \geq 7/12$, *bounded by the annulus* $1/2 < v_{13}(x-\sqrt{13}) < 7/12$.

Proof. (i–ii) Parameterize the wide open subspaces W_∞ and W_0 using open discs, exactly as in Claim 2.4.

(iii) Similiar to Claim 2.4 but with some subtle differences, such as the width of the bounding annuli and the field of definition. When $v_{13}(x) = 1/2$, we find that the valuation of y is initially determined completely by the valuation of $x^2 - 13$. More specifically, we have

$$y^7 = x(x^2-13)^6 + \text{smaller terms}$$

until $v_{13}(x^2-13) \geq 13/12$ and (consequently) $v(y) \geq 1$. Along with the formula $\pi_1^*(t) = x^2/y$ and Claim 3.1 this shows that the points satisfying $v_{13}(x) = 1/2$ are ordinary exactly when $v_{13}(x^2-13) \geq 13/12$. So to better understand the ordinary and supersingular regions when $v_{13}(x) = 1/2$, we make the substitution $u = x + \sqrt{13}$ and consider the rigid subspaces W^+ and X^+ defined below.

$$W^+ = \{(u,y) \mid v_{13}(u) > 1/2\} \qquad X^+ = \{(u,y) \mid v_{13}(u) \geq 7/12\}$$

Again we have chosen X^+ so that it is precisely the maximal ordinary underlying affinoid of the wide open space W^+.

Making the substitution $u = x + \sqrt{13}$ transforms Equation (6) into

$$g(u,y) = y^7 - 13^3\sqrt{13}u^6 + \text{smaller terms} = 0,$$

where these two terms have (strictly) minimal valuation on $W^+ - X^+$. Therefore we can parameterize $W^+ - X^+$ with the annulus $A_s = \{s \mid 1/14 < v_{13}(s) < 1/12\}$, by setting $u = s^7$ and extracting (via Hensel's Lemma) a root of $g(s^7, y)$ close to $y = \sqrt{13}s^6$. The parameterization map on A_s provides a way to glue the disc $D_s = \{s \mid v_{13}(s) < 1/12\}$ onto W^+ such that $X^+ \cup D_s$ inherits the structure of a proper curve. On this curve the differential $du = 7s^6 ds$ vanishes once at 8 distinct points of X^+ and has an order 8 pole at the infinite point of D_s. Therefore $\deg(K_c) = 0$ and we conclude that X^+ has genus 1.

(iv) Identical to (iii) but we make the change of coordinates $u = x - \sqrt{13}$. □

Remark 3.4. *The formula for* w_{169} *can be found in the same way that* w_{49} *was determined. From divisors we see immediately that* $w_{169}^*(x,y) = (c_1/x, c_2y/x^2)$. *Then compatibility of moduli-theoretic maps to* $X_0(13)$ *and substitution back into* (6) *shows that* $c_1 = c_2 = 13$. *So in constrast with the previous* $p = 7$ *example we see that* w_{169} *fixes the two inner ordinary affinoids.*

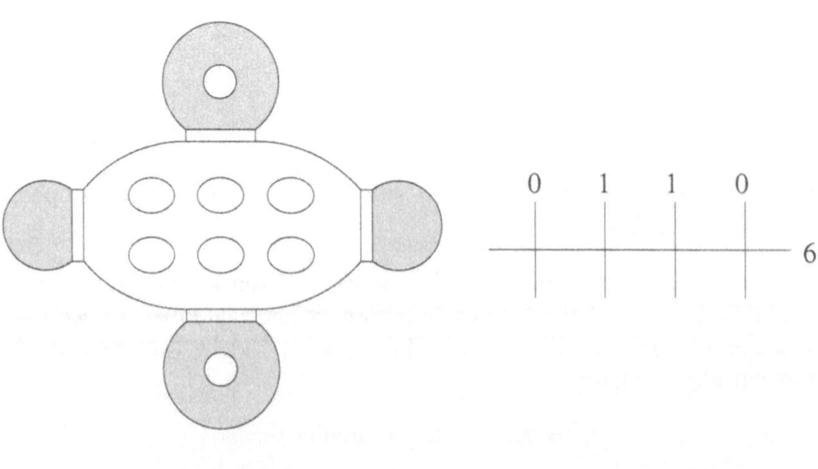

FIGURE 4. $X_0(169)/\mathbb{C}_{13}$ and Special Fiber of Edixhoven Model

Moduli-Theoretic Map	Equation
$\pi_1 : X_0(5) \to X(1)$	$\pi_1^*(j) = \frac{(t^2+2\cdot 5^3 t+5^5)^3}{t^5}$
$\pi_5 : X_0(5) \to X(1)$	$\pi_5^*(j) = \frac{(t^2+10t+5)^3}{t}$
$w_5 : X_0(5) \to X_0(5)$	$w_5^*(t) = \frac{125}{t}$
$\pi_1 : X_0(25) \to X_0(5)$	$\pi_1^*(t) = \frac{u^5}{u^4+5u^3+15u^2+25u+25}$
$\pi_5 : X_0(25) \to X_0(5)$	$\pi_5^*(t) = u(u^4 + 5u^3 + 15u^2 + 25u + 25)$
$w_{25} : X_0(25) \to X_0(25)$	$w_{25}^*(u) = 5/u$

TABLE 3. Moduli-Theoretic Maps for Analysis of $X_0(5)$ and $X_0(25)$

4. Example 3, $X_0(125) \to X_0(25) \to X_0(5) \to X(1)$

Both $X_0(5)$ and $X_0(25)$ have genus 0, so it is more straightforward to find formulas for the various moduli-theoretic maps than in the previous two examples. For this reason we simply summarize this data in Table 3. For parameters on $X_0(5)$ and $X_0(25)$ respectively we have chosen the eta products $t = (\eta_1)^6/(\eta_5)^6$ and $u = \eta_1/\eta_{25}$, both of which have the divisor $(0) - (\infty)$. We again have only one supersingular j-invariant in characteristic 5, namely $j = 0$. So we now determine

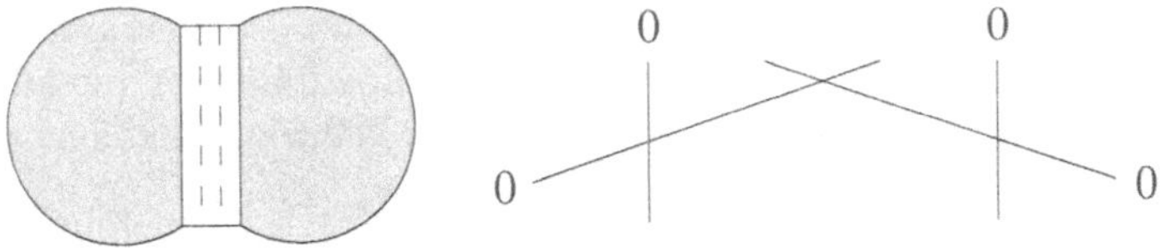

FIGURE 5. $X_0(5)/\mathbb{C}_5$ and Special Fiber of DR Model

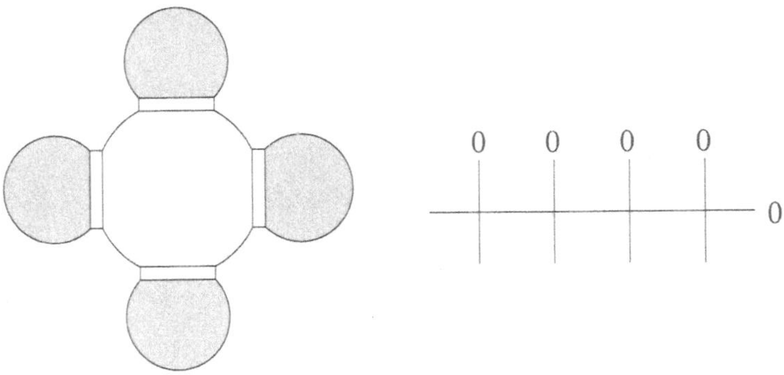

FIGURE 6. $X_0(25)/\mathbb{C}_5$ and Special Fiber of Edixhoven Model

the ordinary and supersingular regions of $X_0(5)$ and $X_0(25)$, in Claims 4.1 and 4.2, by pulling back the disc $v_5(j) > 0$ via the forgetful map. The resulting pictures of the two curves are then given as Figures 5 and 6.

Claim 4.1. *The inverse image under π_1 of the supersingular disc $v_5(j) > 0$ is the annulus described by*

$$0 < v_5(t) < 3.$$

Proof. Identical to Claim 2.2. □

Claim 4.2. *The ordinary region of $X_0(25)$ consists of 4 genus 0 connected components, each bounded by a supersingular annulus. In terms of the parameter u, these components (and annuli) are described as follows:*

i) $v_5(u) \leq 0$, *bounded by the annulus* $0 < v_5(u) < 1/2$
ii) $v_5(u) \geq 1$, *bounded by the annulus* $1/2 < v_5(u) < 1$
iii) $v_5(u + \sqrt{5}) \geq 3/4$, *bounded by the annulus* $1/2 < v_5(u + \sqrt{5}) < 3/4$
iv) $v_5(u - \sqrt{5}) \geq 3/4$, *bounded by the annulus* $1/2 < v_5(u - \sqrt{5}) < 3/4$.

Proof. Identical to Claim 2.4. □

At this point we would like to parameterize $X_0(125)$ (genus 8) and determine the equation for $\pi_1 : X_0(125) \to X_0(25)$. One way we can do this is to first find an equation for the smaller quotient,

$$X_0(125)^+ = X_0(125)/w_{125}.$$

To obtain a model for this smaller curve, we want to choose functions on $X_0(125)$ which are invariant under Atkin-Lehner. By compatibility of moduli-theoretic maps and the previously determined formulas, the following functions satisfy this condition of Atkin-Lehner invariance.

$$x = \frac{\pi_5^*(u)}{\pi_1^*(u)} \qquad (x) = \sum_{i=1}^{4}(c_{5,i}) + \sum_{i=1}^{4}(c_{25,i}) - 4(0) - 4(\infty)$$

$$y = \pi_5^*(u) + \frac{5}{\pi_1^*(u)} \qquad (y)_{\text{neg}} = -5(0) - 5(\infty)\,.$$

By comparing q-expansions we have the following equation relating x and y.

$$\begin{aligned} f_{125}^{+}(x,y) = x^5 - y^4 - 5xy^3 - 15x^2y^2 - 25x^3y - 25x^4 - 5y^3 \\ - 5xy^2 + 25x^3 - 15y^2 - 25x^2 - 25y + 25x - 25 = 0\,. \end{aligned} \tag{7}$$

Furthermore, straightforward algebra shows that $\pi_1^*(u)$ satisfies the following quadratic polynomial over x and y.

$$x(\pi_1^* u)^2 - y(\pi_1^* u) + 5 = 0\,. \tag{8}$$

One can easily check that the genus of the curve described by (7) is 2 and that the quadratic extension in (8) is ramified at 10 places. In particular this means that the larger curve must have genus 8, and therefore must be all of $X_0(125)$. This is convenient because we now have not only equations for $X_0(125)$, but also a priori an equation for the forgetful map $\pi_1 : X_0(125) \to X_0(25)$.

We now determine the ordinary and supersingular regions of $X_0(125)$ in two steps. First we show in Lemma 4.3 that on an affinoid neighborhood of the ordinary region the extension from $X_0(125)^+$ up to $X_0(125)$ is a trivial two-sheeted cover. Then in Claim 4.4 we show that the condition of being ordinary is independent of the choice of branch, and we identify three genus 0 affinoids of $X_0(125)^+$ which are the projection of the ordinary region. The result is that on $X_0(125)$ we must have six genus 0 connected components of the ordinary region.

Lemma 4.3. *The region of $X_0(125)^+$ described by $v_5(y) \leq 1/2$ splits trivially in the degree 2 extension up to $X_0(125)$. Furthermore, every point of $X_0(125)$ with $v_5(y) > 1/2$ is in the supersingular region.*

Proof. When $v_5(y) \leq 1/2$, Equation (7) implies that $v_5(x) \geq (4/5)v_5(y)$. This means that we can apply the quadratic formula to Equation (8) to solve for $\pi_1^*(u)$.

$$\pi_1^*(u) = \frac{y \pm y\sqrt{1 - \frac{20x}{y^2}}}{2x}\,. \tag{9}$$

The point is that we have a convergent power series for the square root in this region, which gives us two distinct embeddings into $X_0(125)$ (i.e., sections of the quotient map).

Now we consider points of $X_0(125)^+$ satisfying $v_5(y) > 1/2$. In this case, Equation (7) implies that $v_5(x) = 2/5$ exactly. Therefore the two roots of Equation (8) always satisfy the condition

$$\frac{1}{10} < v_5(\pi_1^*(u)) < \frac{1}{2}.$$

From Claim 4.2 these points must then be supersingular. □

Claim 4.4. *The ordinary region of $X_0(125)$ consists of six genus 0 connected components, two lying over each of the following genus 0 affinoids of $X_0(125)^+$.*

i) $v_5(y) \leq 0$, *bounded by the annulus* $0 < v_5(y) < 1/2$
ii) $v_5(y + \sqrt{5}) \geq 3/4$, *bounded by* $1/2 < v_5(y + \sqrt{5}) < 3/4$
iii) $v_5(y - \sqrt{5}) \geq 3/4$, *bounded by* $1/2 < v_5(y - \sqrt{5}) < 3/4$.

Proof. (i) Let $W_{\infty/0}$ be the region of $X_0(125)^+$ described by $v_5(y) < 1/2$. On this entire region, Equation (7) becomes

$$y^4 = x^5 + \text{smaller terms}.$$

Therefore we can parameterize $W_{\infty/0}$ using the disc $v_5(s) < 1/10$ and a map close to $x = s^4$, $y = s^5$. To determine the ordinary points of $X_0(125)$ lying over $W_{\infty/0}$ we consider the valuations of the two roots for $\pi_1^*(u)$ in Equation (8). When $v_5(y) < 1/2$ and $v_5(x) = (4/5)v_5(y)$ we must always have one root with valuation $1 - v_5(y)$ and one root with valuation $v_5(y) - v_5(x)$. Claim 4.2 then implies that (looking at either branch) the ordinary region lying over $W_{\infty/0}$ consists of those points with $v_5(y) \leq 0$.

(ii–iii) When $v_5(y) = 1/2$ we do some initial analysis to see that $v_5(x) \geq 2/5$. Now we look at (9) to determine when the two roots for $\pi_1^*(u)$ fall into the ordinary region as described in Claim 4.2. We find that one root, call it u_1, has valuation less than or equal to $1/10$ and satisfies the condition $v_5(u_1 - \frac{y}{x}) = 1/2$. The other root, say u_2, has valuation $1/2$ and satisfies the condition $v_5(u_2 - \frac{5}{y}) \geq 9/10$. By Claim 4.2, the point lying over (x, y) with $\pi_1^*(u) = u_1$ will be ordinary if and only if $v_5(x) \geq 1/2$. Similarly, the point lying over (x, y) with $\pi_1^*(u) = u_2$ will be ordinary if and only if $v_5(y^2 - 5) \geq 5/4$. It follows, however, from Equation (7) that when $v_5(y) = 1/2$ we have

$$v_5(x^5 - (y^2 - 5)^2) \geq 5/2.$$

So the two conditions, $v_5(x) \geq 1/2$ and $v_5(y^2 - 5) \geq 5/4$, are equivalent. This means that when $v_5(y) = 1/2$ the ordinary region of $X_0(125)$ can be described purely in terms of x and y and is independent of the branch of the $X_0(125)/X_0(125)^+$ cover.

To better understand the region $v_5(y^2 - 5) \geq 5/4$ we make the substitution $z = y + \sqrt{5}$ and consider the regions

$$W^+ = \{(x, z) \mid v_5(z) > 1/2\} \text{ and } X^+ = \{(x, z) \mid v_5(z) \geq 3/4\}.$$

This is of course now very similar to the previous examples. The goal is to show that $W^+ - X^+$ is in fact an annulus and then use Riemann-Roch to calculate the

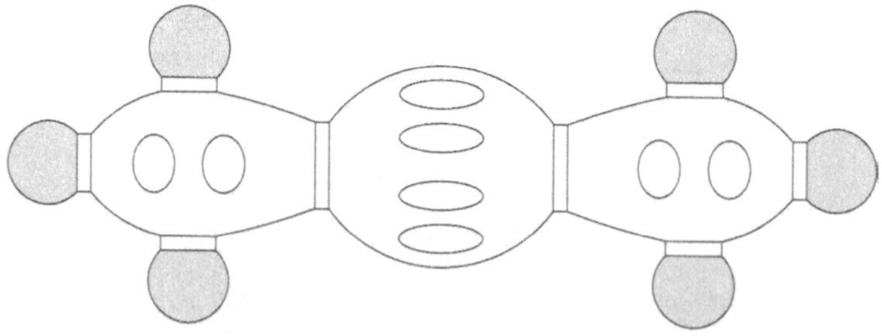

FIGURE 7. $X_0(125)/\mathbb{C}_5$

genus of X^+. On the entire region described by $1/2 < v_5(z) < 3/4$ we can rewrite Equation (7) as

$$g(x,z) = x^5 - 45z^2 + \text{smaller terms}.$$

Therefore $W^+ - X^+$ can be parameterized using the annulus $-3/10 < v_5(s) < -1/4$ and a map close to $x = 45s^2, z = 25s^5$. We simply set $x = 45s^2$ exactly and then solve for z using Hensel's Lemma. After gluing the disc $D_s = \{s \mid v_5(s) < -1/4\}$ onto W^+ we then calculate the divisor of the differential $dx = 90sds$ over $D_s \cup X^+$. As this differential has only one simple zero on X^+ and a triple pole at the infinite point of D_s, we have $\deg(K_c) = -2$ and the genus of X^+ is 0. Of course if we make the substitution $z = y - \sqrt{5}$ we obtain the analogous results for the region described in (iii). □

Remark 4.5. *The resulting picture for $X_0(125)$ is given in Figure 7 with the six ordinary components shaded. The picture is drawn to emphasize the 10 points fixed by w_{125} and the resulting genus 2 quotient $X_0(125)^+$. While the stable model of $X_0(125)$ is not yet known, much information can be drawn from the stable model for $X_0(125)^+$. We conclude the paper now by sketching a proof that $X_0(125)^+$ in fact has potential good reduction. Then the stable model of $X_0(125)$ should follow from understanding the distribution of the 10 ramified points over $X_0(125)^+$.*

Claim 4.6. *$X_0(125)^+$ has potential good reduction.*

Proof. We begin by choosing a degree two function f on $X_0(125)^+$ and determining the resulting hyperelliptic model for the curve.

$$f = \frac{y^2 + 5y + 20 - 10x}{xy + x^2 + 5x - 5}$$

$$g^2 = f^6 + 14f^5 + 85f^4 + 290f^3 + 590f^2 + 680f + 345\,.$$

The main idea of the proof is that the five roots of the right-hand side with $v_5(f) = 1/5$ are equidistant from each other. Letting α be one of these roots and making the substitution $f_0 = f - \alpha$ we obtain an equation of the form

$$g^2 = f_0^6 + a_5 f_0^5 + a_4 f_0^4 + a_3 f_0^3 + a_2 f_0^2 + a_1 f_0$$

where $v_5(a_5) = 0$, $v_5(a_1) = 6/5$, and the rest of the a_i's have valuation 1. Now scaling f_0 and g by appropriate constant factors we get an equation that reduces to

$$g_1^2 = f_1(f_1^4 + c), \qquad c \neq 0.$$

Therefore $X_0(125)^+$ has potential good reduction. □

Acknowledgements

We would like to thank first the organizers of MCAV02, which for this author was certainly a very educational and productive conference. Also the referee was extremely helpful in shaping the paper to make it both more focused and more readable. Finally, we would like to thank Robert Coleman for continuing support and guidance as well as a good deal of mathematical insight.

References

[B] K. Buzzard, *Analytic continuation of overconvergent eigenforms*, J. Amer. Math. Soc. **16** (2003), no. 1, 29–55.

[BGR] S. Bosch, U. Güntzer, R. Remmert, *Non-Archimedean Analysis*, Grundlehren der mathematischen Wissenschaften **261**, Springer-Verlag (1984).

[C1] R. Coleman, *Reciprocity laws on curves*, Compositio Mathematica **72** (1989), 205–235.

[C2] R. Coleman, *Comments on* $X_0(p^2)$ (in preparation)

[E] B. Edixhoven, *Minimal resolution and stable reduction of* $X_0(N)$, Ann. Inst. Fourier (Grenoble). **40** (1990), no. 1, 31–67.

[DR] P. Deligne, M. Rapoport, *Schemas de modules de courbes elliptiques*, Lecture Notes in Math. **349** (1973), 143–316.

[KM] N. Katz, B. Mazur, *Arithmetic moduli of elliptic curves*, Annals of Mathematics Studies **108**, Princeton University Press (1985).

[L] G. Ligozat, *Courbes modulaires de genre* 1, Bull. Soc. Math. France, Mém. **43** (1975).

Ken McMurdy
Department of Mathematics
University of Rochester
Rochester, NY 14627 USA
e-mail: kmcmurdy@math.rochester.edu

Progress in Mathematics, Vol. 224, 181–187

Elliptic $\mathbb{Q}$-Curves with Complex Multiplication

Tetsuo Nakamura

Abstract. Let H be the Hilbert class field of an imaginary quadratic field K. An elliptic curve E over H with complex multiplication by K is called a $\mathbb{Q}$-curve if E is isogenous over H to all its Galois conjugates. We classify $\mathbb{Q}$-curves over H, relating them with the cohomology group $H^2(H/\mathbb{Q}, \pm 1)$. The structures of the abelian varieties over $\mathbb{Q}$ obtained from $\mathbb{Q}$-curves by restriction of scalars are investigated.

1. Introduction

Let K be an imaginary quadratic field and H the Hilbert class field of K. Let E be an elliptic curve over H with complex multiplication by K. We say that E is a $\mathbb{Q}$-curve if E and E^σ are isogenous over H for all $\sigma \in \mathrm{Gal}(H/\mathbb{Q})$. Denote by ψ_E the Hecke character of H associated with E. Then E is a $\mathbb{Q}$-curve if and only if $\psi_E = \psi_E^\sigma$ for all $\sigma \in \mathrm{Gal}(H/\mathbb{Q})$.

As in no complex multiplication case (see [5]), we attach to a $\mathbb{Q}$-curve E a two-cocycle class $c(E) \in H^2(H/\mathbb{Q}, K^\times)$. For $\mathbb{Q}$-curves E, E', we see that $c(E) = c(E')$ if and only if $\psi_E = \psi_{E'} \cdot \chi \circ N_{H/\mathbb{Q}}$ with a quadratic Dirichlet character χ. Let Γ be the subset of $H^2(H/\mathbb{Q}, K^\times)$ consisting of $c(E)$ for all $\mathbb{Q}$-curves E over H. We show that there exists a bijection between Γ and a subspace Y of $H^2(H/\mathbb{Q}, \pm 1)$. Relating Y to an embedding problem associated with the exact sequence

$$1 \to \pm 1 \to G \to \mathrm{Gal}(H/\mathbb{Q}) \to 1,$$

we characterize the structure of Y and, as a consequence, we obtain that $\dim_{\mathbb{F}_2} Y = t(t-1)/2$, where t is the number of distinct prime factors of the discriminant of K. In some case (K is called exceptional), there are no $\mathbb{Q}$-curves over H. Replacing H by the ring class field of conductor 2, we obtain a similar classification of $\mathbb{Q}$-curves.

The abelian variety $B = R_{H/K}(E)$ obtained by restriction of scalars from a $\mathbb{Q}$-curve E can be defined over $\mathbb{Q}$. The structure of the endomorphism algebra $\mathrm{End}_{\mathbb{Q}} B \otimes \mathbb{Q}$ is studied according to this classification. In particular the cohomology class $c(E)$ determines $\mathrm{End}_{\mathbb{Q}} B \otimes \mathbb{Q}$.

This note is a summary of [4] and also includes a result of [3].

Notation. Throughout the note we fix the following notation.

K: an imaginary quadratic field of discriminant $D \neq -3, -4$.
t: the number of distinct primes dividing D.

H: the Hilbert class field of K.
Cl_K: the ideal class group of K.
ρ: the complex conjugation.
j_E: the j-invariant of an elliptic curve E.

Galois cohomology groups $H^i(\mathrm{Gal}(M/L), A)$ are denoted by $H^i(M/L, A)$. We call K exceptional if the discriminant D of K is of the form

$$D = -4p_1 \cdots p_{t-1} \quad (t \geq 2)$$

where $p_1, \ldots, p_{t-1}$ are primes satisfying $p_1 \equiv \cdots \equiv p_{t-1} \equiv 1 \bmod 4$.

2. Embedding problems associated with the Hilbert class field

An element γ of the Galois cohomology group $H^2(H/\mathbb{Q}, \pm 1)$ corresponds to an equivalence class of group extensions

$$1 \to \pm 1 \to G \to \mathrm{Gal}(H/\mathbb{Q}) \to 1. \tag{1}$$

If there exists a quadratic extension k of H such that $k/\mathbb{Q}$ is Galois and the natural map $\mathrm{Gal}(k/\mathbb{Q}) \to \mathrm{Gal}(H/\mathbb{Q})$ corresponds to the epimorphism in (1), we say that an embedding problem $(H/\mathbb{Q}, \pm 1, \gamma)$ has a solution k.

Let Y be the set of $\gamma \in H^2(H/\mathbb{Q}, \pm 1)$ such that $(H/\mathbb{Q}, \pm 1, \gamma)$ has a solution. We see that Y is a $\mathbb{F}_2$-subspace of $H^2(H/\mathbb{Q}, \pm 1)$. Write $\mathfrak{g} = \mathrm{Gal}(H/K) \cong Cl_K$ and denote by $\mathrm{Ext}(\mathfrak{g}, \pm 1)$ the elements of $H^2(\mathfrak{g}, \pm 1)$ corresponding to extensions of $\mathfrak{g}$ by $\{\pm 1\}$ that are abelian groups. The vector space over $\mathbb{F}_2$ of bilinear alternating form on $\mathfrak{g}/\mathfrak{g}^2$ is denoted by $\mathrm{Alt}(\mathfrak{g})$. Then we have an exact sequence

$$0 \to \mathrm{Ext}(\mathfrak{g}, \pm 1) \to H^2(\mathfrak{g}, \pm 1) \to \mathrm{Alt}(\mathfrak{g}) \to 0.$$

By [2, §1], $\dim \mathrm{Ext}(\mathfrak{g}, \pm 1) = t - 1$, $\dim H^2(\mathfrak{g}, \pm 1) = t(t-1)/2$, since $\dim \mathfrak{g}/\mathfrak{g}^2 = t - 1$ (t is the number of distinct primes dividing the discriminant of K).

Let res: $H^2(H/\mathbb{Q}, \pm 1) \to H^2(\mathfrak{g}, \pm 1)$ be the restriction map and put $Y_0 = \{\gamma \in Y \mid \mathrm{res}(\gamma) \in \mathrm{Ext}(\mathfrak{g}, \pm 1)\}$. Let k be a solution of $(H/\mathbb{Q}, \pm 1, \gamma)$ with $\gamma \in Y_0$. Then k is a quadratic extension of H such that $k/\mathbb{Q}$ is Galois and k/K is abelian.

Let p be a rational prime and $\mathfrak{p}$ a prime ideal of K dividing p. Denote by $U_\mathfrak{p}$ the group of local units for $\mathfrak{p}$ and put $U_p = \prod_{\mathfrak{p}|p} U_\mathfrak{p}$. We denote by

$$U_K = \prod_p U_p$$

the maximal compact subgroup of the idele group I_K of K and by $K_\infty^\times$ the archimedean part of I_K. Let $\chi = \chi_{k/H}$ be the character of I_H corresponding to k/H. Since k/K is abelian, there is a non-trivial character

$$\theta : U_K K^\times K_\infty^\times \quad \to \quad \pm 1$$

such that $\chi = \theta \circ N_{H/K}$ and $\theta(K^\times K_\infty^\times) = 1$; hence θ is determined by its restriction on U_K. Since $k/\mathbb{Q}$ is Galois, we have $\chi^\rho = \chi$ and this means that $\theta^\rho = \theta$. Conversely for any non-trivial character $\theta : U_K \to \pm 1$ such that

$$\theta^\rho = \theta \quad \text{and} \quad \theta(-1) = 1,$$

$\chi = \theta \circ N_{H/K}$ determines a solution k of $(H/\mathbb{Q}, \pm 1, \gamma)$ for some $\gamma \in Y_0$.

Proposition 2.1. *If K is exceptional, we have* $\dim Y_0 = t$. *Otherwise we have* $\dim Y_0 = t - 1$.

Proof. Let W be the set of characters $\theta : U_K \to \pm 1$ such that $\theta^\rho = \theta$ and $\theta(-1) = 1$. Denote by W_0 the set of $\theta \in W$ of the form $\theta = \kappa \circ N_{K/\mathbb{Q}}$ with a quadratic Dirichlet character κ. Noting that the characters in W_0 exactly correspond to the trivial class in $H^2(H/\mathbb{Q}, \pm 1)$, we obtain $Y_0 \cong W/W_0$. Using the prime divisors of the discriminant of K, we can construct elements of W which form a basis of W/W_0. For details, see [4, proof of Prop. 3]. □

Theorem 2.1. $\dim(Y/Y_0) = (t-1)(t-2)/2$.

Proof. If $t \leq 2$, then $\mathrm{Alt}(\mathfrak{g}) = (0)$, so that $Y = Y_0$ and our statement holds. Assume $t \geq 3$. Composing the natural map

$$H^2(\mathfrak{g}, \pm 1) \to H^2(\mathfrak{g}, \pm 1)/\mathrm{Ext}(\mathfrak{g}, \pm 1) \cong \mathrm{Alt}(\mathfrak{g})$$

with the restriction map $Y \subset H^2(H/\mathbb{Q}, \pm 1) \to H^2(\mathfrak{g}, \pm 1)$, we obtain a linear map $g : Y \to \mathrm{Alt}(\mathfrak{g})$. Since $\mathrm{Ker}\, g = Y_0$ and $\dim \mathrm{Alt}(\mathfrak{g}) = (t-1)(t-2)/2$, it suffices to show that g is surjective. This is proved as follows. First we construct an appropriate basis $\{f_{i,j} \mid 1 \leq i < j \leq t-1\}$ of $\mathrm{Alt}(\mathfrak{g})$. Then for each $f_{i,j}$, we can construct a quadratic extension k/H such that k is a solution of the embedding problem $(H/\mathbb{Q}, \pm 1, \gamma)$ with $g(\gamma) = f_{i,j}$. For details, see [4, proof of Theorem 1]. □

3. Classification of $\mathbb{Q}$-curves

Let L be a Galois extension over $\mathbb{Q}$ containing H. An elliptic curve E over L with complex multiplication by K is called a $\mathbb{Q}$-curve if E^σ and E are isogenous over L for all $\sigma \in \mathrm{Gal}(L/\mathbb{Q})$. Let ψ_E be the Hecke character of the idele group I_L of L associated with E. Then E is a $\mathbb{Q}$-curve if and only if $\psi_E = \psi_E^\sigma$ for all $\sigma \in \mathrm{Gal}(L/\mathbb{Q})$ (cf. [1, §11]).

For a $\mathbb{Q}$-curve E over L, choose isogenies $\varphi_\sigma : E^\sigma \to E$ for $\sigma \in \mathrm{Gal}(L/\mathbb{Q})$. Then

$$c(\sigma, \tau) = \varphi_\sigma \varphi_\tau^\sigma (\varphi_{\sigma\tau})^{-1} \in K^\times$$

defines a two-cocycle and the cohomology class of $\{c(\sigma, \tau)\}$ in $H^2(L/\mathbb{Q}, K^\times)$ depends only on the curve E, and not on the isogenies φ_σ chosen. We will denote by $c(E)$ this cohomology class. Let us denote by Γ_L the subset of $H^2(L/\mathbb{Q}, K^\times)$ consisting of elements of the form $c(E)$ for all $\mathbb{Q}$-curves E over L. Furthermore, we denote by Y_L the subspace of $H^2(L/\mathbb{Q}, \pm 1)$ consisting of all γ such that the embedding problem $(L/\mathbb{Q}, \pm 1, \gamma)$ is solvable.

Proposition 3.1. *If Γ_L is not empty, then Y_L operates on Γ_L simply transitively in an obvious manner. For $\mathbb{Q}$-curves E and E', we have $c(E) = c(E')$ if and only if $\psi_E = \psi_{E'} \cdot \kappa \circ N_{L/\mathbb{Q}}$, where κ is a quadratic Dirichlet character.*

Proof. We see that the natural map $H^2(L/\mathbb{Q}, \pm 1) \to H^2(L/\mathbb{Q}, K^\times)$ is injective. Hence $Y_L \subseteq H^2(L/\mathbb{Q}, \pm 1)$ operates on $H^2(L/\mathbb{Q}, K^\times)$ by multiplication. Let E and E' be $\mathbb{Q}$-curves over L. If E and E' are not isogenous over L, there exists a unique quadratic extension k over L that is Galois over $\mathbb{Q}$, such that the twist of E associated with k/L is isogenous to E' over L. The extension $k/\mathbb{Q}$ determines $\gamma \in Y_L$. Then we have $c(E) = c(E')\gamma$. □

In [6] a class of elliptic curves (more generally abelian varieties) with complex multiplication whose Hecke characters satisfy a certain condition are studied. We recall briefly what we need here.

For an integer $f \geq 1$, let $H^{(f)}$ denote the ring class field of K of conductor f. Let

$$U_{K,f} = \{u \in U_K | \ u(\mathbb{Z} + f\mathfrak{o}_K) = \mathbb{Z} + f\mathfrak{o}_K\}.$$

Then $P = U_{K,f}K^\times K_\infty^\times$ is the subgroup of I_K corresponding to $H^{(f)}$ by class field theory. Let E be an elliptic curve over $H^{(f)}$ with $\mathrm{End} E = \mathbb{Z} + f\mathfrak{o}_K$. Let us consider the following condition on the Hecke character ψ_E of E (see [6, Theorem 4]).

$$\text{There exists a Hecke character } \phi : U_{K,f}K^\times K_\infty^\times \to \mathbb{C}^\times \text{ such that } \psi_E = \phi \circ N_{H^{(f)}/K}. \tag{SH}$$

Here ϕ must satisfy the following conditions:

$$\phi(K^\times) = 1, \quad \phi(y) = y^{-1} \quad \text{for every } y \in K_\infty^\times, \tag{3}$$

$$\phi(U_{K,f}) = \pm 1 \text{ and } \phi(-1) = -1 \quad \text{for } -1 \in U_{K,f}. \tag{4}$$

If ψ_E satisfies (Sh), then clearly $\psi_E = \psi_E^\sigma$ for all $\sigma \in \mathrm{Gal}(H^{(f)}/K)$. Conversely from a character $\phi : U_{K,f} \to \pm 1$ with $\phi(-1) = -1$, extending it on $P = U_{K,f}K^\times K_\infty^\times$ by (3), we obtain $\psi = \phi \circ N_{H^{(f)}/K}$, which is a Hecke character of an elliptic curve E over $H^{(f)}$. Furthermore in this case E is a $\mathbb{Q}$-curve if and only if $\phi^\rho = \phi$ on $U_{K,f}$ (cf. [6, Proposition 9]).

If K is not exceptional, there exists a character $\phi : U_K \to \pm 1$ satisfying

$$\phi(-1) = -1, \quad \phi^\rho = \phi. \tag{5}$$

Therefore there exists a $\mathbb{Q}$-curve over H. If K is exceptional, there is no character $\phi : U_K \to \pm 1$ satisfying (5) (cf. [4, §4]).

Theorem 3.1. *There exist $\mathbb{Q}$-curves over the Hilbert class field H if and only if K is not exceptional. If K is exceptional, there exist $\mathbb{Q}$-curves over the ring class field $H^{(2)}$.*

Proof. In [1, §11] it is stated that if K is exceptional, there are no $\mathbb{Q}$-curves over H (see also [4, Prop. 5]). In case K is exceptional, we can find a character $\phi : U_{K,2} \to \pm 1$ such that $\phi(-1) = -1$ and $\phi^\rho = \phi$. Using this we get a $\mathbb{Q}$-curve over $H^{(2)}$ (see [4, proof of Theorem 2]). □

Applying Theorem 2.1, we obtain the following result concerning a classification of $\mathbb{Q}$-curves.

Theorem 3.2. *If K is not exceptional, the cohomology classes $c(E)$ classify isogeny classes of $\mathbb{Q}$-curves over H into $2^{t(t-1)/2}$ classes. Among them there are 2^{t-1} classes whose Hecke characters satisfy* (Sh). *If K is exceptional, take $H^{(2)}$, the ring class field of K of conductor 2, instead of H. Then exactly the same statements hold for isogeny classes of $\mathbb{Q}$-curves over $H^{(2)}$.*

4. Restriction of scalars of $\mathbb{Q}$-curves

First we briefly recall a result in [3]. An elliptic curve E over H with complex multiplication by K is called a K-curve if E^σ is isogenous to E over H for each $\sigma \in \mathrm{Gal}(H/K)$. For an elliptic curve E over H with complex multiplication by K, there exists an elliptic curve E_0 over H such that $j_E = j_{E_0}$ and ψ_{E_0} satisfies (Sh) by [6, Prop. 7]. Then E_0 is a K-curve. For a K-curve E over H, let us denote by $R_{H/K}(E)$ the abelian variety obtained from E by restriction of scalars from H to K. It is an abelian variety defined over K of dimension $h_K = [H : K]$.

Theorem 4.1. *Let E be a K-curve over H and put $B = R_{H/K}(E)$ and $R = \mathrm{End}_K B \otimes \mathbb{Q}$. If ψ_E satisfies* (Sh), *R is a field of degree h_K over K. If ψ_E does not satisfy* (Sh), *then the center Z of R is a field of degree h_0 over K with $h_K = 2^{2m}h_0 (m \geq 1)$ and one of the following two cases holds:*

(i) *$R \cong \mathrm{M}_{2^m}(Z)$. In this case, B is isogenous over K to a product of A with itself 2^m times, where A is K-simple, $2^m h_0$-dimensional and $Z = \mathrm{End}_K A \otimes \mathbb{Q}$.*

(ii) *$R \cong \mathrm{M}_{2^{m-1}}(D)$, where D is a division quaternion algebra over Z. In this case, B is isogenous over K to a product of A with itself 2^{m-1} times, where A is K-simple, $2^{m+1}h_0$-dimensional and $D = \mathrm{End}_K A \otimes \mathbb{Q}$.*

Proof. For a K-curve over H, choose E_0 and k as stated above. If ψ_E satisfies (Sh), we can show that R is a field of degree h over K by applying a similar argument as in [1, Chap. 4]. If ψ_E does not satisfy (Sh), then $G = \mathrm{Gal}(k/K)$ is non-commutative and G/C is an elementary abelian group of order 2^{2m} where C is the center of G. The commutator map $G \times G \in (x, y) \to [x, y] \in \{\pm 1\} = \mathrm{Gal}(k/H)$ induces a non-degenerate alternating form on $G/C \times G/C$. The proof is completed by using this form. For details, see [3, Theorem 3]. □

Henceforth we suppose that K is non-exceptional. Let E be a $\mathbb{Q}$-curve over H. Clearly E is a K-curve. Since E is defined over $\mathbb{Q}(j_E)$ (cf. [1, Theorem 10.1.3]), we have

$$B = R_{H/K}(E) \cong R_{\mathbb{Q}(j_E)/\mathbb{Q}}(E) \otimes K,$$

so that B is defined over $\mathbb{Q}$. Concerning the structure of the endomorphism algebra $R_0 = \mathrm{End}_{\mathbb{Q}}(B) \otimes \mathbb{Q}$ we obtain

Theorem 4.2. *The center Z_0 of $R_0 = \mathrm{End}_{\mathbb{Q}}(B) \otimes \mathbb{Q}$ is a field of degree h_0 over $\mathbb{Q}$ and $R_0 \cong M_{2^m}(Z_0)$ or $R_0 \cong M_{2^{m-1}}(D_0)$, where D_0 is a division quaternion algebra over Z_0 and $h_K = 2^{2m}h_0$. R_0 is commutative if and only if ψ_E satisfies* (Sh).

Since $R_0 = \{\alpha \in R \mid \rho(\alpha) = \alpha\}$, our assertions follow from the above theorem.

Proposition 4.1. *Let E, E' be $\mathbb{Q}$-curves over H and put:*

$$B = R_{H/K}(E),\ B' = R_{H/K}(E'),\ R_0 = \mathrm{End}_{\mathbb{Q}}(B) \otimes \mathbb{Q},\ R_0' = \mathrm{End}_{\mathbb{Q}}(B') \otimes \mathbb{Q}.$$

Then if $c(E) = c(E')$, we have $R_0 \cong R_0'$. Conversely if R_0 is commutative and $R_0 \cong R_0'$, we have $c(E) = c(E')$.

For a proof, see [4, Prop. 6].

Let us consider the case $K = \mathbb{Q}(\sqrt{-21})$. Let $\mathfrak{p}$ and $\mathfrak{p}'$ be the prime ideals of K such that $\mathfrak{p}^2 = (2 + \sqrt{-21})$ and $\mathfrak{p}'^2 = (10 + \sqrt{-21})$. We see that Cl_K is generated by $\mathfrak{p}$ and $\mathfrak{p}'$. Let $\mathfrak{q}$ be the prime ideal of K with $\mathfrak{q}^2 = (3)$. Let ϕ_0 be a character of I_K of conductor $\mathfrak{q}$ such that

$$\phi_0((\alpha)) = \left(\frac{\alpha}{\mathfrak{q}}\right)\alpha \quad \text{for every } \alpha \in K^\times,$$

where $\left(\frac{\alpha}{\mathfrak{q}}\right)$ denotes the norm residue symbol. Then ϕ_0 satisfies (5) and put $\psi_0 = \phi_0 \circ N_{H/K}$. For $p = 3$ and 7, we have a unique non-trivial character $\eta_p : U_p \to \pm 1$. Let $\eta_2 : U_2 \to \pm 1$ be such that Ker $\eta_2 = \{u \in U_2 \mid u \equiv 1 \bmod 2\}$. Here we view η_p as a character of U_K by composing with the projection $U_K \to U_p$. Then we define:

$$\omega_1 = \eta_3\,\eta_7 \circ N_{H/K}, \quad \omega_2 = \eta_2 \circ N_{H/K}.$$

For non-zero integers a, b, we denote by $(a,\ b) \in \mathrm{Br}_2(\mathbb{Q}) = H^2(\overline{\mathbb{Q}}/\mathbb{Q},\ \pm 1)$ the class of the quaternion algebra over $\mathbb{Q}$ generated by two elements I, J with $I^2 = a$, $J^2 = b$, $JI = -IJ$. Since $(21, -3)$ is trivial in $\mathrm{Br}_2(\mathbb{Q})$, there exists a D_4-extension k_0 over $\mathbb{Q}$ containing $\mathbb{Q}(\sqrt{-3}, \sqrt{21})$. Let χ be the character of I_H associated with k_0H/H. Then by Theorem 2, the equivalence classes of $\mathbb{Q}$-curves over H are exactly represented by the Hecke characters $\psi = \psi_0\omega$, $\omega \in \langle \omega_1, \omega_2, \chi \rangle$.

After some computations, we obtain the following table of R_0 (cf. [4, §6]).

ψ	R_0 (field)
ψ_0	$\mathbb{Q}(\sqrt{-2}, \sqrt{-14})$
$\psi_0\omega_1$	$\mathbb{Q}(\sqrt{-6}, \sqrt{2})$
$\psi_0\omega_2$	$\mathbb{Q}(\sqrt{-6}, \sqrt{-42})$
$\psi_0\omega_1\omega_2$	$\mathbb{Q}(\sqrt{-14}, \sqrt{-42})$

ψ	R_0 (quaternion alg.)
$\psi_0\chi$	$(-14, -42)$
$\psi_0\omega_1\chi$	$(-6, 42)$
$\psi_0\omega_2\chi$	$(-6, -2)$
$\psi_0\omega_1\omega_2\chi$	$(-14, 2)$

Remark. The division quaternion algebras $(-14, -42)$ and $(-6, -2)$ over $\mathbb{Q}$ are isomorphic because they ramify at the same primes 2 and ∞. The quaternion algebras $(-6, 42)$ and $(-14, 2)$ are isomorphic to $M_2(\mathbb{Q})$.

References

[1] B.H. Gross, *Arithmetic on elliptic curves with complex multiplication*, Lecture Notes in Math. **776**, Springer-Verlag, 1980.

[2] R. Massy, *Construction de p-extensions Galoisiennes d'un corps de caractéristique différente de p*, J. Algebra **109** (1987), 508–535.

[3] T. Nakamura, *Abelian varieties associated with elliptic curves with complex multiplication*, Acta Arith. **97** (2001), 379–385.

[4] T. Nakamura, *A classification of $\mathbb{Q}$-curves with complex multiplication*, to appear in J. Math. Soc. Japan (2004).

[5] J. Quer, $\mathbb{Q}$-curves and abelian varieties of GL_2-type, Proc. London Math. Soc. **81** (2000), 285–317.

[6] G. Shimura, *On the zeta function of an abelian variety with complex multiplication*, Ann. of Math. **94** (1971), 504–533.

Tetsuo Nakamura
Mathematical Institute
Tohoku University
Sendai 980-8578, Japan
e-mail: nakamura@math.tohoku.ac.jp

Progress in Mathematics, Vol. 224, 189–239

Abelian Varieties over $\mathbb{Q}$ with Large Endomorphism Algebras and Their Simple Components over $\overline{\mathbb{Q}}$

Elisabeth E. Pyle

Introduction

This dissertation expands upon the results of K. Ribet in [Ri 1] and [Ri 2] concerning abelian varieties "of $\mathbf{GL}_2$-type". An abelian variety $A/\mathbb{Q}$ is said to be of $\mathbf{GL}_2$-type if its algebra of $\mathbb{Q}$-endomorphisms $\mathbb{Q} \otimes \mathrm{End}_{\mathbb{Q}}(A)$ is a number field E of degree equal to the dimension of A.

The study of abelian varieties of $\mathbf{GL}_2$-type grew out of research surrounding the Taniyama-Shimura Conjecture. Recall that the usual statement of the Taniyama-Shimura Conjecture asserts that every elliptic curve over $\mathbb{Q}$ is a quotient of a modular curve of the form $X_0(M)$. Recent manuscripts circulated by Wiles and Taylor-Wiles prove the Taniyama-Shimura Conjecture for a large class of elliptic curves over $\mathbb{Q}$.

It is well known that an equivalent formulation of the Taniyama-Shimura Conjecture requires only that every elliptic curve over $\mathbb{Q}$ be a quotient of a standard modular curve of the form $X_1(N)$. Furthermore, in [Ma], Mazur proved the conjecture's equivalence to yet another seemingly weaker statement: an elliptic curve over $\mathbb{Q}$ is a quotient over $\mathbb{Q}$ of some $X_1(N)$ if and only if it is a quotient over $\overline{\mathbb{Q}}$ of some $X_1(N)_{/\overline{\mathbb{Q}}}$.

Now every map from $X_1(N)$ to an elliptic curve sending ∞ to O extends uniquely to a map from its Jacobian $J_1(N)$ to the elliptic curve. Hence, this line of inquiry leads naturally to the (closely related) questions: how may we characterize those simple abelian varieties over $\mathbb{Q}$ (respectively, over $\overline{\mathbb{Q}}$) that appear as quotients of $J_1(N)$ (respectively, $J_1(N)_{/\overline{\mathbb{Q}}}$) for some $N \geq 1$? We will loosely use the word "modular" to refer to such abelian varieties.

In [Ri 1], Ribet gives a conjectural answer to one of these questions.

Conjecture. *A simple abelian variety A over $\mathbb{Q}$ is modular if and only if A is of $\mathbf{GL}_2$-type.*

It is known that every simple quotient of $J_1(N)/\mathbb{Q}$ is of $\mathbf{GL}_2$-type. Indeed, the primary source of examples of abelian varieties of $\mathbf{GL}_2$-type is a construction resulting from extensive work by Shimura on the subject [Sh 2, Sh 4]. Let $f =$

$\sum a_n q^n$ be a normalized cuspidal eigenform of weight two on $\Gamma_1(N)$. Let E be the number field $\mathbb{Q}(\ldots, a_n, \ldots)$. Then Shimura constructs, as a quotient of $J_1(N)$, an abelian variety A_f over $\mathbb{Q}$ with the following properties: there is an action $E \xrightarrow{\sim} \mathbb{Q}\otimes \mathrm{End}_{\mathbb{Q}}(A_f)$, and the dimension of A_f is equal to $[E:\mathbb{Q}]$. Moreover, every simple quotient of $J_1(N)/\mathbb{Q}$ is isogenous to an A_f constructed in this way.

Thus the conjecture is reduced to: all abelian varieties over $\mathbb{Q}$ of $\mathbf{GL}_2$-type are isogenous to some A_f. In support of this assertion, Ribet has proved [Ri 1, Thm. 4.4] that if one assumes Serre's conjecture [Se 4, 3.2.4?] about representations of $\mathrm{Gal}(\overline{\mathbb{Q}}/\mathbb{Q})$, then it follows that all abelian varieties over $\mathbb{Q}$ of $\mathbf{GL}_2$-type are modular. (See also [Sh 1].)

Now in this dissertation, we look at the complete decomposition over $\overline{\mathbb{Q}}$ of abelian varieties of $\mathbf{GL}_2$-type. We determine necessary and sufficient conditions for an abelian variety over $\overline{\mathbb{Q}}$ to be a simple component of an abelian variety over $\mathbb{Q}$ of $\mathbf{GL}_2$-type; such abelian varieties will be called "building blocks". These results enable us to restate the above conjecture as: a simple abelian variety B over $\overline{\mathbb{Q}}$ is modular if and only if B is a building block.

We would like to emphasize that the work contained herein deals exclusively with the properties of abelian varieties of $\mathbf{GL}_2$-type, the properties of their component building blocks, and the relationships between these two kinds of abelian varieties, without any recourse to their possible "modular" structure.

In Chapter 1 we define our terminology and review the structure of abelian varieties over $\mathbb{Q}$ of $\mathbf{GL}_2$-type. Chapter 2 contains a discussion of the Brauer class of the endomorphism algebra of a building block; it culminates in a theorem we will apply in Chapter 4. Chapter 3 is a short preparation of a category-theoretic construction that we will also use in Chapter 4.

Chapter 4 contains the bulk of the technical work; it describes the correspondence between building blocks over $\overline{\mathbb{Q}}$ and abelian varieties over $\mathbb{Q}$ of $\mathbf{GL}_2$-type. Then Chapter 5 is a collection of further results, mostly connected to fields of definition for these abelian varieties and their endomorphisms. In Chapter 6 we illustrate the theory of Chapters 4 and 5 with explicit examples of a one-dimensional building block and an associated four-dimensional abelian variety of $\mathbf{GL}_2$-type. Chapter 7 is a brief consideration of some unanswered questions and avenues for future work.

Finally there are two appendices. In the first, we construct a counterexample to demonstrate the necessity of a certain condition we impose on building blocks. The second is a list of facts we quote from the theory of central simple algebras; we apply these theorems to the endomorphism algebras we encounter.

Notation

Throughout this dissertation we will work in the category of abelian varieties *up to isogeny.* That is the category whose objects are abelian varieties A, $B, \ldots$ with morphisms $\mathbb{Q}\otimes\mathrm{Hom}(A, B)$. In other words, we formally invert all isogenies $A \to B$. In this category, notation is simplified by writing $\mathrm{Hom}(A, B)$ for $\mathbb{Q}\otimes\mathrm{Hom}(A, B)$.

The endomorphism algebra $\mathrm{Hom}(A, A)$ will be denoted $\mathrm{End}(A)$, and the following variants of Hom will also apply to End.

If A and B are abelian varieties defined over a field K, then $\mathrm{Hom}(A, B)$ will refer to *all* morphisms $A_{/\overline{K}} \to B_{/\overline{K}}$. If we wish to refer to only those morphisms defined over K, we will write $\mathrm{Hom}_K(A, B)$.

In some cases, A and B may be equipped with an action of a ring $\mathcal{R}$; that is, there are fixed maps $\mathcal{R} \to \mathrm{End}(A)$ and $\mathcal{R} \to \mathrm{End}(B)$. A morphism $\varphi : A \to B$ will be said to "respect $\mathcal{R}$" (or "respect the actions of $\mathcal{R}$", or "be compatible with $\mathcal{R}$"...) if $\varphi \circ r = r \circ \varphi$ for all $r \in \mathcal{R}$. The subgroup of morphisms that respect $\mathcal{R}$ will be denoted $\mathrm{Hom}^{\mathcal{R}}(A, B)$.

We will write $Z(\mathcal{R})$ for the center of a ring $\mathcal{R}$.

The sign "$\sim$" will designate the equivalence relation of isogeny.

The symbol $G_{\mathbb{Q}}$ will denote the absolute Galois group $\mathrm{Gal}(\overline{\mathbb{Q}}/\mathbb{Q})$ with its usual profinite topology.

Acknowledgments

My warmest thanks go to my dissertation advisor, Ken Ribet, for all his helpful suggestions. Without his patient and continued support, this work would never have been realized.

This work was partially supported by NSF Grant #DMS 93-06898.

1. Abelian varieties over $\mathbb{Q}$ of $\mathbf{GL}_2$-type

In this first chapter we will recapitulate many of the results from [Ri 1], on which we rely heavily in the remainder of this dissertation. Our approach differs from Ribet's in that it focuses not on ℓ-adic representations, but rather on the structure of endomorphism algebras as semi-simple $\mathbb{Q}$-algebras. We remind the reader that for an abelian variety A, the notation $\mathrm{End}(A)$ will always refer to the $\mathbb{Q}$-*algebra* of endomorphisms of A.

Definition. An abelian variety $A/\mathbb{Q}$ is *of* $\mathbf{GL}_2$*-type* if $\mathrm{End}_{\mathbb{Q}}(A)$, the algebra of endomorphisms of A defined over $\mathbb{Q}$, is a number field of degree equal to the dimension of A.

Note that this definition applies only to abelian varieties defined over $\mathbb{Q}$, although we may not always say so explicitly; i.e., the phrase "A is of $\mathbf{GL}_2$-type" will imply that A is defined over $\mathbb{Q}$. Also note that abelian varieties of $\mathbf{GL}_2$-type are always simple over $\mathbb{Q}$, which is a modification of the terminology in [Ri 1].

Why do we say these abelian varieties are "of $\mathbf{GL}_2$-type"? If $A/\mathbb{Q}$ is an abelian variety such that $\mathrm{End}_{\mathbb{Q}}(A)$ is a number field E with $[E:\mathbb{Q}] = \dim A$, then it is well known that for any prime ℓ, the Tate module $V_\ell(A)$ is free of rank two over $E \otimes_{\mathbb{Q}} \mathbb{Q}_\ell$. Thus the action of the absolute Galois group $\mathrm{Gal}(\overline{\mathbb{Q}}/\mathbb{Q}) = G_{\mathbb{Q}}$ on $V_\ell(A)$ defines a representation $G_{\mathbb{Q}} \to \mathbf{GL}_2(E \otimes_{\mathbb{Q}} \mathbb{Q}_\ell)$.

Recall that in general, if a field E acts on an abelian variety $A/\overline{\mathbb{Q}}$, then $[E:\mathbb{Q}]$ divides $2\dim A$, and if $[E:\mathbb{Q}] = 2\dim A$, then A is said to have complex multiplication by E. [Mu]

Abelian varieties of $\mathbf{GL}_2$-type with complex multiplication over $\overline{\mathbb{Q}}$ have a particularly simple structure. Indeed, Shimura has proved the following in [Sh 1, Prop. 1.5]:

Proposition 1.1. *Let $A/\mathbb{Q}$ be an abelian variety of $\mathbf{GL}_2$-type. Suppose $A_{/\overline{\mathbb{Q}}}$ contains a simple abelian subvariety of CM-type. Then $A_{/\overline{\mathbb{Q}}}$ is isogenous to a power of a CM elliptic curve.*

The theory of elliptic curves with complex multiplication is well understood, and requires different techniques from the non-CM case. For these reasons, we will focus on abelian varieties with no complex multiplication. *Henceforth we will consider only abelian varieties of $\mathbf{GL}_2$-type with no complex multiplication over $\overline{\mathbb{Q}}$.*

We begin our analysis with a review of a main result from Ribet [Ri 1]. We will use the following terminology from the theory of central simple algebras. (See Appendix B for more details.) Let $\mathcal{A}$ be a simple $\mathbb{Q}$-algebra and let F be the center of $\mathcal{A}$. The *Schur index* of $\mathcal{A}$ is the integer $\sqrt{\dim_F \mathcal{D}}$, where $\mathcal{D}$ is a division algebra (determined up to isomorphism) such that $\mathcal{A} \cong \mathbf{M}_n(\mathcal{D})$ for some $n \geq 1$. A *subfield* of $\mathcal{A}$ is a $\mathbb{Q}$-subalgebra E of $\mathcal{A}$ such that E is a field. A *maximal subfield* of $\mathcal{A}$ is a subfield that is maximal with respect to inclusion.

Theorem 1.2. *Let $A/\mathbb{Q}$ be an abelian variety of $\mathbf{GL}_2$-type with no complex multipliation over $\overline{\mathbb{Q}}$. Let $E = \mathrm{End}_{\mathbb{Q}}(A)$. Then the full endomorphism algebra* $\mathrm{End}(A)$ *is a central simple algebra over a totally real field $F \subseteq E$; the Schur index of* $\mathrm{End}(A)$ *is* 1 *or* 2; *and* $\mathrm{End}(A)$ *contains E as a maximal subfield.*

Proof. Thanks to Jordi Quer of the Universitat Politècnica de Catalunya for pointing out an error in my original proof, and suggesting the following revised version.

First we show that $A/\overline{\mathbb{Q}}$ is isogenous to a power B^n, where $B/\overline{\mathbb{Q}}$ is a simple abelian variety. A priori we know that $A/\overline{\mathbb{Q}}$ is isogenous to a finite product $\prod B_i^{n_i}$, where the B_i's are pairwise non-isogenous simple subvarieties of $A/\overline{\mathbb{Q}}$. Since $E \subset \mathrm{End}(A)$ is a field, then E must act on each factor $B_i^{n_i}$. Thus $[E:\mathbb{Q}]$ divides $2n_i \dim B_i$ for each i. On the other hand, $[E:\mathbb{Q}] = \dim A = \sum n_i \dim B_i$, so $[E:\mathbb{Q}] \geq n_i \dim B_i$ for each i. Therefore $[E:\mathbb{Q}] = n_i \dim B_i$ or $2n_i \dim B_i$ for each i, but the latter case corresponds to $B_i^{n_i}$ having complex multiplication by E, which we have ruled out by hypothesis. Thus $[E:\mathbb{Q}] = n_i \dim B_i = \sum n_i \dim B_i$ for each i, and now it is clear that there is only such factor. That is, $A \sim B^n$.

We define $\mathcal{D} = \mathrm{End}(B)$ a division algebra, $F = Z(\mathcal{D})$ its center, and $t =$ the Schur index of $\mathcal{D}$ over F. At this point we have shown that $\mathrm{End}(A) \cong \mathbf{M}_n(\mathcal{D})$ is a central simple algebra over F of Schur index t.

The next step is to show that E is a maximal subfield of $\mathrm{End}(A)$. Here it suffices to prove that the centralizer $C(E)$ of E in $\mathrm{End}(A)$ is a field. Indeed, a field is always contained in its centralizer, so if E and $C(E)$ are both fields, then by the Double Centralizer Theorem [Pi, Prop. 12.7] they are equal, a condition which is equivalent by [Pi, Prop. 13.1] to E being a maximal subfield of $\text{End}(A)$. So let $\varphi \in C(E) \subset \text{End}(A)$ be a non-zero endomorphism. Then the image $\varphi(A)$ is isogenous to B^r for some $r \leq n$. Furthermore, E must act on B^r, because φ was chosen to commute with the endomorphisms in E. Therefore $[E:\mathbb{Q}]$ divides $2\dim B^r$, which implies that either $r = n$ or $r = n/2$. But the case $r = n/2$ corresponds to B^r having complex multiplication. Thus we find that $r = n$, and so φ is an isogeny, and so $C(E)$ is a field.

It follows immediately that the center F of $\text{End}(A)$ is contained in E, because all maximal subfields of a central simple algebra contain its center. Also, by [Pi, Prop. 13.1], we know $[E:F] = nt$, where t is the Schur index of $\text{End}(A)$. Therefore, using the relation $[E:\mathbb{Q}] = \dim A = n \dim B$, we discover that $\dim B = t[F:\mathbb{Q}]$.

We are now in a position to show that the Schur index t is 1 or 2. Recall Albert's classification (into Types I, II, III, and IV) of possible endomorphism algebras of simple abelian varieties, as found for example in [Mu, Chap. IV]. According to the table in [Mu, p. 202], either t is 1 or 2, or $\frac{1}{2}[F:\mathbb{Q}]t^2$ divides the dimension of B. But from above, we have $\dim B = t[F:\mathbb{Q}]$, and we conclude that t divides 2 in all cases. Let m be defined by $mt = 2$.

To complete the proof of the theorem, it remains to show that F is a totally real field. In other words, we need to rule out the possibility that $\text{End}(B)$ is a Type IV endomorphism algebra, in which case F would be a totally imaginary quadratic extension of a totally real field. The key point is that we have shown $F \subset E = \text{End}_{\mathbb{Q}} A$ consists of endomorphisms that are all defined over $\mathbb{Q}$. Hence we obtain a representation of F on the rational vector space $\text{Lie}(A/\mathbb{Q})$, and the trace of this representation takes values in $\mathbb{Q}$. (See [Mu, § 11] for a description of the space of tangent vectors $\text{Lie}(A/\mathbb{Q})$.)

On the other hand, when we extend coefficients to $\mathbb{C}$, the vector space $\text{Lie}(A/\mathbb{C})$ decomposes into n copies of $\text{Lie}(B/\mathbb{C})$. Following [Sh 3], let $\chi_1, \ldots, \chi_g$, $\overline{\chi}_1, \ldots, \overline{\chi}_g$ be the inequivalent absolutely irreducible complex representations of $\mathcal{D} = \text{End}(B)$. For each $\nu = 1, \ldots, g$, write r_ν and s_ν, respectively, for the multiplicities of χ_ν and $\overline{\chi}_\nu$ in the representation of $\mathcal{D}$ on $\text{Lie}(B/\mathbb{C})$. We know $r_\nu + s_\nu = mt = 2$ for each ν, but we cannot have $r_\nu = s_\nu = 1$ for all ν, by [Sh 3, Thm. 5(d)(e)]. Thus for some ν, we find that χ_ν, but not $\overline{\chi}_\nu$, appears in the representation of $\mathcal{D}$ on $\text{Lie}(B/\mathbb{C})$. Therefore the same is true for the representation of $\mathcal{D}$ on $\text{Lie}(A/\mathbb{C})$. Restricting to $F \subseteq \mathcal{D}$, we deduce that the trace of its representation on $\text{Lie}(A/\mathbb{C})$ cannot be fixed by complex conjugation, contradicting our first computation over $\mathbb{Q}$.

This completes the proof of the theorem. □

Thus we see that another way to characterize abelian varieties of $\mathbf{GL}_2$-type is as those abelian varieties that are simple over $\mathbb{Q}$ but have a "large" or "maximal" number of endomorphisms defined over $\mathbb{Q}$.

The simple components of $A_{/\overline{\mathbb{Q}}}$

Let A be an abelian variety of $\mathbf{GL}_2$-type. We now consider the properties of an abelian variety $B/\overline{\mathbb{Q}}$ appearing as a simple component of $A_{/\overline{\mathbb{Q}}}$. Note that the proof of the above theorem provides us with some information about $\operatorname{End}(B)$ as well as $\operatorname{End}(A)$. In fact, we have already proved the following proposition:

Proposition 1.3. *Let B be a simple component over $\overline{\mathbb{Q}}$ of an abelian variety of* $\mathbf{GL}_2$*-type. Then the endomorphism algebra* $\operatorname{End}(B)$ *is a central division algebra over a totally real field F; the Schur index t of* $\operatorname{End}(B)$ *is* 1 *or* 2*; and* $t[F:\mathbb{Q}] = \dim B$.

Thus B, like A, also has the property that its endomorphism algebra is maximal, if compared with other simple abelian varieties of the same dimension. We remark here that in the case $t = 2$, the condition $t[F:\mathbb{Q}] = \dim B$ means that $\operatorname{End} B$ must be an indefinite quaternion algebra over F. Indeed, the endomorphism algebra of a simple abelian variety is always totally definite or totally indefinite, and the equality $t[F:\mathbb{Q}] = \dim B$ can never hold if $\operatorname{End}(B)$ is a definite quaternion algebra [Sh 3, Prop. 15].

It turns out that B has another interesting property. Namely, it is isogenous to all of its Galois conjugates. More technically, we say that B is a $\mathbb{Q}$-abelian variety, which is defined as follows.

Definition. An abelian variety $X/\overline{\mathbb{Q}}$ is a $\mathbb{Q}$-*abelian variety* if for each $\sigma \in \operatorname{Gal}(\overline{\mathbb{Q}}/\mathbb{Q})$, there exists an isogeny $\mu_\sigma : {}^\sigma X \to X$ compatible with the endomorphisms of X; that is, $\mu_\sigma \circ {}^\sigma\varphi = \varphi \circ \mu_\sigma$ for all $\varphi \in \operatorname{End}(X)$.

It is easy to see that for each $\sigma \in G_{\mathbb{Q}}$, the conjugate ${}^\sigma B$ must be isogenous to B. Indeed, ${}^\sigma B^n \sim {}^\sigma A = A \sim B^n$, because A is defined over $\mathbb{Q}$. Then, by the uniqueness of decomposition of abelian varieties up to isogeny, we must have ${}^\sigma B \sim B$. It takes a little more work to show that there is actually an isogeny which is compatible with $\operatorname{End}(B)$.

Proposition 1.4. *Let B be a simple component over $\overline{\mathbb{Q}}$ of an abelian variety A of* $\mathbf{GL}_2$*-type. Then B is a $\mathbb{Q}$-abelian variety.*

Proof. Fix an isogeny $A_{/\overline{\mathbb{Q}}} \sim B^n$, or equivalently, an isomorphism $\operatorname{End}(A) \cong \mathbf{M}_n(\operatorname{End}(B))$. For each $\sigma \in G_{\mathbb{Q}}$, we form the σ-twist of this isogeny, ${}^\sigma B^n \sim {}^\sigma A$. Let $\varphi \in \operatorname{End}(B)$, and write $\widetilde{\varphi}$ for the endomorphism of A which is φ in each place. Then, because A is defined over $\mathbb{Q}$, there are commutative diagrams

$$\begin{array}{ccc} {}^\sigma B & \xrightarrow{\text{diag.}} {}^\sigma B^n \sim {}^\sigma A = & A \\ {\scriptstyle {}^\sigma\varphi}\downarrow & & \downarrow{\scriptstyle {}^\sigma\widetilde{\varphi}} \\ {}^\sigma B & \xrightarrow{\text{diag.}} {}^\sigma B^n \sim {}^\sigma A = & A \end{array} \qquad \begin{array}{ccc} A & \sim B^n \xrightarrow{\pi_j} & B \\ {\scriptstyle \widetilde{\varphi}}\downarrow & & \downarrow{\scriptstyle \varphi} \\ A & \sim B^n \xrightarrow{\pi_j} & B \end{array}$$

where π_j is projection onto the jth place of B^n.

Now $\operatorname{End}(A)$ is a central simple algebra, and for each $\sigma \in G_{\mathbb{Q}}$, there is an automorphism of $\operatorname{End}(A)$ given by $\psi \mapsto {}^\sigma\psi$. Hence by the Skolem-Noether Theorem, there exists an invertible $\alpha(\sigma) \in \operatorname{End}(A)$ such that ${}^\sigma\psi = \alpha(\sigma) \circ \psi \circ \alpha(\sigma)^{-1}$

for all $\psi \in \mathrm{End}(A)$. Therefore we may connect the above diagrams with

$$\begin{array}{ccc} A & \xrightarrow{\alpha(\sigma)^{-1}} & A \\ {\scriptstyle {}^\sigma\widetilde{\varphi}}\downarrow & & \downarrow{\scriptstyle \widetilde{\varphi}} \\ A & \xrightarrow{\alpha(\sigma)^{-1}} & A \end{array}$$

Then the composite map

$${}^\sigma B \to {}^\sigma B^n \sim {}^\sigma A = A \overset{\alpha(\sigma)^{-1}}{\longrightarrow} A \sim B^n \overset{\pi_j}{\longrightarrow} B$$

is a morphism from ${}^\sigma B$ to B that is compatible with $\mathrm{End}(B)$. For some j this morphism must be non-zero, hence an isogeny because B is simple. Thus, we have found an isogeny $\mu_\sigma : {}^\sigma B \to B$ such that $\mu_\sigma \circ {}^\sigma\varphi = \varphi \circ \mu_\sigma$ for all $\varphi \in \mathrm{End}(B)$, and so B is a $\mathbb{Q}$-abelian variety. □

We now make a definition based on these properties associated with simple components of abelian varieties of $\mathbf{GL}_2$-type.

Definition. A *building block* is an abelian variety $B/\overline{\mathbb{Q}}$ such that

(i) the endomorphism algebra $\mathrm{End}(B)$ is a central division algebra over a totally real field F with Schur index $t = 1$ or 2, and $t[F:\mathbb{Q}] = \dim B$.

(ii) B is a $\mathbb{Q}$-abelian variety.

We will show in Proposition 4.5 that every building block appears as a simple component of some abelian variety over $\mathbb{Q}$ of $\mathbf{GL}_2$-type.

The extension E/F

Let A be an abelian variety of $\mathbf{GL}_2$-type. Let $E = \mathrm{End}_{\mathbb{Q}}(A)$, and let $F = Z(\mathrm{End}(A))$ denote the center of $\mathrm{End}(A)$. We now look more closely at the field extension E/F. In the proof of the above proposition, we saw that for each $\sigma \in G_{\mathbb{Q}}$, the Skolem-Noether Theorem implies that there exists an invertible $\alpha(\sigma) \in \mathrm{End}(A)$ such that ${}^\sigma\varphi = \alpha(\sigma)\circ\varphi\circ\alpha(\sigma)^{-1}$ for all $\varphi \in \mathrm{End}(A)$. Notice that because $E = \mathrm{End}_{\mathbb{Q}}(A)$, the endomorphism $\alpha(\sigma)$ must be contained in the centralizer of E in $\mathrm{End}(A)$. But E is a maximal subfield of $\mathrm{End}(A)$, and hence its own centralizer, as F is a number field [Pi, Prop. 13.1]. Therefore $\alpha(\sigma) \in E$. Furthermore, each $\alpha(\sigma)$ is determined up to (non-zero) elements of the center F of $\mathrm{End}(A)$. We now fix a choice of $\alpha(\sigma) \in E$ for each $\sigma \in G_{\mathbb{Q}}$, choosing $\alpha(\sigma') = \alpha(\sigma)$ whenever ${}^{\sigma'}\varphi = {}^\sigma\varphi$ for all $\varphi \in \mathrm{End}(A)$. We may view α as a locally constant function $G_{\mathbb{Q}} \to E^*$ that is a lift of the evident homomorphism $G_{\mathbb{Q}} \to E^*/F^*$.

Proposition 1.5. *The field E is generated over F by the $\alpha(\sigma)$.*

Proof. Let $E' = F(\ldots,\alpha(\sigma),\ldots)$. Then $E' \subseteq E$, since each $\alpha(\sigma) \in E$. Now let φ be an endomorphism of A that lies in the centralizer of E'; that is, $\varphi\circ\alpha(\sigma) = \alpha(\sigma)\circ\varphi$ for all $\sigma \in G_{\mathbb{Q}}$. But then ${}^\sigma\varphi = \alpha(\sigma)\circ\varphi\circ\alpha(\sigma)^{-1} = \varphi$ for all σ, which means that φ is defined over $\mathbb{Q}$, and thus lies in E. Therefore the centralizer $C(E')$ of E' is also contained in E. Now we use the Double Centralizer Theorem [Pi, Thm. 12.7] to count dimensions, and we find $(\dim_F E')(\dim_F(C(E')) = \dim_F \mathrm{End}(A) = (\dim_F E)^2$. Thus we must have $E' = C(E') = E$, and so E is generated over F by the collection $\{\alpha(\sigma)\}$. □

Now recall that the Rosati involution induced by any polarization of $A/\mathbb{Q}$ is used to define an involution $e \mapsto \bar{e}$ on E which is independent of the polarization. This "complex conjugation" fixes F – indeed, any totally real subfield of E.

Lemma 1.6. *For all $\sigma \in G_{\mathbb{Q}}$, we have $\overline{\alpha(\sigma)} \circ \alpha(\sigma) \in F$.*

Proof. Fix a polarization (over $\mathbb{Q}$) of A and let $\varphi \mapsto \varphi'$ designate the Rosati involution on $\mathrm{End}(A)$ induced by the polarization. Fix $\sigma \in G_{\mathbb{Q}}$. Now for all $\varphi \in \mathrm{End}(A)$ we have ${}^\sigma\varphi = \alpha(\sigma) \circ \varphi \circ \alpha(\sigma)^{-1}$. Taking the Rosati involution of both sides, we find ${}^\sigma\varphi' = \overline{\alpha(\sigma)^{-1}} \circ \varphi' \circ \overline{\alpha(\sigma)}$ for all $\varphi' \in \mathrm{End}(A)$. Replacing φ' with φ, we find ${}^\sigma\varphi = \overline{\alpha(\sigma)^{-1}} \circ \varphi \circ \overline{\alpha(\sigma)}$ for all $\varphi \in \mathrm{End}(A)$. So we see that $\overline{\alpha(\sigma)}^{-1}$ acts via conjugation on $\mathrm{End}(A)$ exactly as $\alpha(\sigma)$ does. Therefore $\overline{\alpha(\sigma)} \circ \alpha(\sigma)$ lies in the center of $\mathrm{End}(A)$, which is F. □

Proposition 1.7. *The field E is an abelian Galois extension of F.*

Proof. Recall that E is generated over F by the $\alpha(\sigma)$, so it suffices to study the extensions $F(\alpha(\sigma))$. First we write $\alpha(\sigma)^2$ in a more complicated way:

$$\alpha(\sigma)^2 = \left(\alpha(\sigma)/\overline{\alpha(\sigma)}\right) \cdot \left(\overline{\alpha(\sigma)}\,\alpha(\sigma)\right).$$

We have seen that $\overline{\alpha(\sigma)}\,\alpha(\sigma)$ lies in F. Now we consider the first factor $\alpha(\sigma)/\overline{\alpha(\sigma)}$.

Note that by construction, the $\alpha(\sigma)$ satisfy $\alpha(\sigma) \circ \alpha(\tau) \circ \alpha(\sigma\tau)^{-1} \in F$ for all $\sigma, \tau \in G_{\mathbb{Q}}$. Therefore

$$\alpha(\sigma) \cdot \alpha(\tau) \cdot \alpha(\sigma\tau)^{-1} \quad = \quad \overline{\alpha(\sigma)} \cdot \overline{\alpha(\tau)} \cdot \overline{\alpha(\sigma\tau)}^{-1},$$

and rearranging terms, we find

$$\left(\alpha(\sigma)/\overline{\alpha(\sigma)}\right) \cdot \left(\alpha(\tau)/\overline{\alpha(\tau)}\right) \quad = \quad \alpha(\sigma\tau)/\overline{\alpha(\sigma\tau)} \qquad \text{for all } \sigma, \tau \in G_{\mathbb{Q}}.$$

That is, $\sigma \mapsto \alpha(\sigma)/\overline{\alpha(\sigma)}$ is a continuous character on $G_{\mathbb{Q}}$ with values in E^*. In particular, $\alpha(\sigma)/\overline{\alpha(\sigma)}$ is a root of unity.

Now we have expressed $\alpha(\sigma)^2$ as a product of a root of unity and an element of F. Clearly then, $\alpha(\sigma)$ is contained in an abelian Galois extension of F obtained by adjoining $\sqrt{\overline{\alpha(\sigma)}\,\alpha(\sigma)}$ and some number of roots of unity. Therefore $F(\alpha(\sigma))$ is an abelian Galois extension, and this is true for any σ. Finally, since E is the compositum of the $F(\alpha(\sigma))$, we see that the field E is also an abelian extension of F. □

Some cohomology

Let A be an abelian variety of $\mathbf{GL}_2$-type. Let $E = \mathrm{End}_{\mathbb{Q}}(A)$ and $F = Z(\mathrm{End}(A))$. As in the previous section, let $\alpha : G_{\mathbb{Q}} \to E^*$ be a locally constant function such that ${}^{\sigma}\varphi = \alpha(\sigma) \circ \varphi \circ \alpha(\sigma)^{-1}$ for all $\varphi \in \mathrm{End}(A)$. Recall that α is well defined modulo F^*.

We note that $c(\sigma, \tau) \stackrel{\mathrm{def}}{=} \alpha(\sigma)^{-1}\alpha(\tau)^{-1}\alpha(\sigma\tau)$ is a two-cocycle on $G_{\mathbb{Q}}$ with values in F^* (where $G_{\mathbb{Q}}$ acts trivially on F^*). Moreover, a different choice for α leads to a cocycle which is cohomologous to $c(\sigma, \tau)$. Thus the abelian variety A of $\mathbf{GL}_2$-type supplies us with a well-defined cohomology class $[c] \in \mathrm{H}^2(G_{\mathbb{Q}}, F^*)$.

Lemma 1.8. *The class of* $c(\sigma, \tau)$ *has order dividing two in* $\mathrm{H}^2(G_{\mathbb{Q}}, F^*)$.

Proof. Since $c(\sigma, \tau)$ lies in F^*, we have $c(\sigma, \tau) = \overline{c(\sigma, \tau)}$. Hence we may write

$$\begin{aligned} c(\sigma,\tau)^2 &= \overline{c(\sigma,\tau)}\, c(\sigma,\tau) \\ &= \left(\overline{\alpha(\sigma)}\,\alpha(\sigma)\right)^{-1} \left(\overline{\alpha(\tau)}\,\alpha(\tau)\right)^{-1} \left(\overline{\alpha(\sigma\tau)}\,\alpha(\sigma\tau)\right). \end{aligned}$$

Thus we have expressed $c(\sigma, \tau)^2$ as a coboundary, because we have already shown that $\overline{\alpha(\sigma)}\,\alpha(\sigma) \in F^*$ for all $\sigma \in G_{\mathbb{Q}}$. □

Now consider the $G_{\mathbb{Q}}$-module $\overline{\mathbb{Q}} \otimes_{\mathbb{Q}} F$, where $\sigma \in G_{\mathbb{Q}}$ acts via $\sigma(r \otimes f) = \sigma(r) \otimes f$. The natural inclusion $F \hookrightarrow \overline{\mathbb{Q}} \otimes_{\mathbb{Q}} F$ induces a map $\mathrm{H}^2(G_{\mathbb{Q}}, F^*) \to \mathrm{H}^2(G_{\mathbb{Q}}, (\overline{\mathbb{Q}} \otimes_{\mathbb{Q}} F)^*)$ and this latter group is canonically isomorphic to the Brauer group $\mathrm{Br}(F)$. (The isomorphism will be discussed in detail in Chapter 2.)

The endomorphism algebra $\mathrm{End}(A)$ is a central simple F-algebra, and as such, determines a Brauer class in $\mathrm{Br}(F)$. The class has order dividing two in $\mathrm{Br}(F)$, because the Schur index of $\mathrm{End}(A)$ is 1 or 2 [Pi, Thm. 18.6]. Furthermore, the field E splits $\mathrm{End}(A)$ because E is a maximal subfield of $\mathrm{End}(A)$ [Pi, Cor. 13.3], and E also splits $c(\sigma, \tau)$, by definition of the cocycle c. These observations motivate one to ask whether there is a connection between the class of $\mathrm{End}(A)$ in $\mathrm{Br}(F)$ and the class of c in $\mathrm{H}^2(G_{\mathbb{Q}}, (\overline{\mathbb{Q}} \otimes_{\mathbb{Q}} F)^*)$. This question is answered by the following proposition.

Proposition 1.9. *The image of* $[c]$ *in* $\mathrm{H}^2(G_{\mathbb{Q}}, (\overline{\mathbb{Q}} \otimes_{\mathbb{Q}} F)^*)$ *corresponds (under the natural isomorphism) to the class of* $\mathrm{End}(A)$ *in* $\mathrm{Br}(F)$.

Proof. [Ri 1, Thm. 5.6]. □

Alternatively, the results of Chapter 2 and Chapter 4 of this dissertation may be combined to give a different proof of Proposition 1.9.

2. The Brauer class of End(B)

In Chapter 1, we saw how an abelian variety of $\mathbf{GL}_2$-type decomposes over $\overline{\mathbb{Q}}$ into a product $B \times \cdots \times B$ of building blocks B with certain properties. Recall our definitions:

Definition. An abelian variety $B/\overline{\mathbb{Q}}$ is a $\mathbb{Q}$-*abelian variety* if for each $\sigma \in \mathrm{Gal}(\overline{\mathbb{Q}}/\mathbb{Q})$, there exists an isogeny $\mu_\sigma : {}^\sigma B \to B$ compatible with the endomorphisms of B; that is, $\mu_\sigma \circ {}^\sigma\varphi = \varphi \circ \mu_\sigma$ for all $\varphi \in \mathrm{End}(B)$.

Definition. A *building block* is an abelian variety $B/\overline{\mathbb{Q}}$ such that

(i) the endomorphism algebra $\mathrm{End}(B)$ is a central division algebra over a totally real field F with Schur index $t = 1$ or 2, and $t[F:\mathbb{Q}] = \dim B$.

(ii) B is a $\mathbb{Q}$-abelian variety.

Our ultimate goal is to show that every building block B appears as a simple quotient of some abelian variety $A/\mathbb{Q}$ of $\mathbf{GL}_2$-type. Towards that end, in this chapter we will study the class of the endomorphism algebra $\mathrm{End}(B)$ in the Brauer group $\mathrm{Br}(F)$.

Let $G_\mathbb{Q}$ be the absolute Galois group $\mathrm{Gal}(\overline{\mathbb{Q}}/\mathbb{Q})$ with its usual profinite topology. A building block $B/\overline{\mathbb{Q}}$ may be constructed from a model over a finite extension K of $\mathbb{Q}$; by abusing notation we will again call such a model B. Then ${}^\sigma B = {}^{\sigma'}B$ if σ, $\sigma' \in G_\mathbb{Q}$ restrict to the same map on K. Now because B is a $\mathbb{Q}$-abelian variety, we may fix a collection of isogenies $\{\, \mu_\sigma : {}^\sigma B \to B \,\}_{\sigma \in G_\mathbb{Q}}$ respecting $\mathrm{End}(B)$ such that $\mu_\sigma = \mu_{\sigma'}$ whenever ${}^\sigma B = {}^{\sigma'}B$. Define $c(\sigma,\tau) = \mu_\sigma \circ {}^\sigma\mu_\tau \circ \mu_{\sigma\tau}^{-1}$ for each pair $\sigma, \tau \in G_\mathbb{Q}$. Then $c(\sigma,\tau)$ is an element of $\mathrm{End}(B)$, as we may see from the diagram $B \xrightarrow{\mu_{\sigma\tau}^{-1}} {}^{\sigma\tau}B \xrightarrow{{}^\sigma\mu_\tau} {}^\sigma B \xrightarrow{\mu_\sigma} B$. Also, $c(\sigma,\tau) \circ \varphi = \varphi \circ c(\sigma,\tau)$ for all $\varphi \in \mathrm{End}(B)$; this follows from the compatibility conditions on the μ_σ. Thus $c(\sigma,\tau)$ lies in the center of $\mathrm{End}(B)$, and may be thought of as an non-zero element of the field F. Therefore, we may view c as a locally constant two-cocycle on $G_\mathbb{Q}$ with values in F^*, and we easily check that a different choice of isogenies ${}^\sigma B \to B$ changes c by a coboundary. Thus to the abelian variety B we associate a well-defined cohomology class $[c] \in \mathrm{H}^2(G_\mathbb{Q}, F^*)$, where $G_\mathbb{Q}$ acts trivially on F^*.

Next, consider the $G_\mathbb{Q}$-module $\overline{\mathbb{Q}} \otimes_\mathbb{Q} F$, where the action of $\sigma \in G_\mathbb{Q}$ is defined by $\sigma(r \otimes f) = \sigma(r) \otimes f$. The inclusion $F \hookrightarrow \overline{\mathbb{Q}} \otimes_\mathbb{Q} F$ induces a map $\mathrm{H}^2(G_\mathbb{Q}, F^*) \to \mathrm{H}^2(G_\mathbb{Q}, (\overline{\mathbb{Q}} \otimes_\mathbb{Q} F)^*)$, and this latter group is canonically isomorphic to the Brauer group of F, as we will see.

Theorem 2.1. *The image of* $[c]$ *in* $\mathrm{H}^2(G_\mathbb{Q}, (\overline{\mathbb{Q}} \otimes_\mathbb{Q} F)^*)$ *corresponds to the class of* $\mathrm{End}(B)$ *in the Brauer group* $\mathrm{Br}(F)$.

Theorem 2.1 is reminiscent of Proposition 1.9, which referred to another cohomology class $[c] \in \mathrm{H}^2(G_\mathbb{Q}, F^*)$ attached to an abelian variety of $\mathbf{GL}_2$-type. In Chapter 4 we will justify the use of this notation when we show that every building block B appears as a quotient of an abelian variety A over $\mathbb{Q}$ of $\mathbf{GL}_2$-type in such a way that the definitions of $[c]$ coincide. Since $\mathrm{End}(A)$ and $\mathrm{End}(B)$ determine the same class in $\mathrm{Br}(F)$ whenever $A \sim B \times \cdots \times B$, it is evident that Theorem 2.1 is a necessary condition for this assertion to be valid.

The proof of the Theorem 2.1 will be postponed until after we have reviewed the construction of the isomorphism $\mathrm{H}^2(G_\mathbb{Q}, (\overline{\mathbb{Q}} \otimes_\mathbb{Q} F)^*) \xrightarrow{\sim} \mathrm{Br}(F)$.

The identification of the Brauer group $\mathrm{Br}(F)$ **with** $\mathrm{H}^2(\mathrm{Gal}(\overline{F}/F),\, \overline{F}^*)$

Let G be a profinite group, and let X, X_1, X_2, X_3 be discrete (multiplicative) groups equipped with a left action of G. We recall the following definitions and facts from the theory of non-abelian cohomology. (See [Se 2, p. 131–134] for more details.)

- $\mathrm{H}^0(G,\, X) \stackrel{\text{def}}{=} \{x \in X : \sigma(x) = x \text{ for all } \sigma \in G\}$.
- A *one-cocycle* on G with values in X is a continuous map $G \to X$ sending $\sigma \mapsto x_\sigma$ such that $x_{\sigma\tau} = x_\sigma \cdot \sigma(x_\tau)$ for all $\sigma, \tau \in G$.
- The one-cocycles x_σ and y_σ are *cohomologous* if there exists an $x \in X$ such that $y_\sigma = x^{-1} \cdot x_\sigma \cdot \sigma(x)$ for all $\sigma \in G$. A one-cocycle is a *coboundary* if it is cohomologous to the trivial cocycle $\sigma \mapsto 1$.
- $\mathrm{H}^1(G,\, X) \stackrel{\text{def}}{=} \{\text{one-cocycles}\}/\{\text{coboundaries}\}$.
- If $1 \to X_1 \to X_2 \to X_3 \to 1$ is a short exact sequence of non-abelian G-modules then there exists an exact sequence of pointed sets

$$\begin{aligned} 1 \to \mathrm{H}^0(G,\, X_1) \to \mathrm{H}^0(G,\, X_2) \to \mathrm{H}^0(G,\, X_3) \\ \to \mathrm{H}^1(G,\, X_1) \to \mathrm{H}^1(G,\, X_2) \to \mathrm{H}^1(G,\, X_3). \end{aligned}$$

  Moreover, if X_1 is contained in the center of X_2, the sequence may be extended by one more term, with a map to $\mathrm{H}^2(G, X_1)$. (Recall that the "kernel" of a morphism of pointed sets is the inverse image of the distinguished element.)

In particular, let $G_F = \mathrm{Gal}(\overline{F}/F)$, where F is a number field. Then

$$1 \to \overline{F}^* \to \mathbf{GL}_n(\overline{F}) \to \mathbf{PGL}_n(\overline{F}) \to 1$$

is an exact sequence of G_F-modules, and the corresponding cohomology sequence is

$$\cdots \to \mathrm{H}^1(G_F,\, \mathbf{GL}_n(\overline{F})) \longrightarrow \mathrm{H}^1(G_F,\, \mathbf{PGL}_n(\overline{F})) \xrightarrow{\delta_n} \mathrm{H}^2(G_F,\, \overline{F}^*).$$

A generalization of Hilbert's Theorem 90 states that $\mathrm{H}^1(G_F,\, \mathbf{GL}_n(\overline{F}))$ is trivial [Se 2, Prop. X.3], so in fact the map δ_n is injective.

Now $\mathrm{H}^1(G_F,\, \mathbf{PGL}_n(\overline{F}))$ classifies the *forms* of $\mathbf{M}_n(\overline{F})$. That is, there is a canonical bijection

$$\mathrm{H}^1(G_F,\, \mathbf{PGL}_n(\overline{F})) \longrightarrow \mathrm{Br}_n(F), \tag{2.2}$$

where $\mathrm{Br}_n(F)$ denotes the set of Brauer classes of F-algebras $\mathcal{B}$ such that $\overline{F} \otimes_F \mathcal{B}$ is isomorphic to $\mathbf{M}_n(\overline{F})$ as an $\overline{F}$-algebra. This correspondence is described explicitly by associating to each one-cocycle $\sigma \mapsto P_\sigma \in \mathbf{PGL}_n(\overline{F})$ the Brauer class of the F-algebra

$$\mathcal{B} \stackrel{\text{def}}{=} \left\{ X \in \mathbf{M}_n(\overline{F}) \mid P_\sigma \cdot \sigma(X) \cdot P_\sigma^{-1} = X \text{ for all } \sigma \in G_F \right\}.$$

Now because $\mathrm{Br}(F)$ is the union (over n) of the $\mathrm{Br}_n(F)$, we may define a map from $\mathrm{Br}(F)$ to $\mathrm{H}^2(G_F, \overline{F}^*)$ by using the inverse of (2.2):

$$\begin{array}{ccc} \mathrm{H}^2(G_F, \overline{F}^*) & \dashleftarrow & \mathrm{Br}(F) \\ \uparrow \delta_n & & \uparrow \\ \mathrm{H}^1(G_F, \mathbf{PGL}_n(\overline{F})) & \overset{(2.2)}{\rightleftarrows} & \mathrm{Br}_n(F) \end{array}$$

This is the desired isomorphism. (See [Se 2, X.5] for the details.)

REMARK. In the above definition of $\mathcal{B}$ we have written $\sigma(X)$ when $\sigma \in G_F$ and $X \in \mathbf{M}_n(\overline{F})$. Here we are regarding $\mathbf{M}_n(\overline{F})$ as the ring of $\overline{F}$-endomorphisms of the vector space $\overline{F}^n$, with its standard G_F-module structure. Then $\sigma(X)$ is defined to be the endomorphism $\sigma \circ X \circ \sigma^{-1}$ on $\overline{F}^n$. It is useful to note, however, that we may also describe $\sigma(X)$ as the matrix obtained from $X \in \mathbf{M}_n(\overline{F})$ by applying σ to each entry.

Co-induced modules

Let H be a closed subgroup of a profinite group G, and let M be a discrete H-module (possibly non-abelian). The *co-induced* G-module $\mathrm{Coind}_H^G M$ is defined to be the set of continuous functions $\theta : G \to M$ such that $\theta(\eta x) = \eta(\theta(x))$ for all $x \in G$ and $\eta \in H$. The group structure on $\mathrm{Coind}_H^G M$ is defined by pointwise multiplication, and the action of $\sigma \in G$ is defined by $({}^\sigma\theta)(x) = \theta(x\sigma)$.

In the case $M = \overline{\mathbb{Q}}$, which we consider below, it will be helpful to put a $\overline{\mathbb{Q}}$-vector space structure on $\mathrm{Coind}_H^G \overline{\mathbb{Q}}$. This is achieved by defining $(r\theta)(x) = x(r) \cdot \theta(x)$ for $r \in \overline{\mathbb{Q}}$, $\theta \in \mathrm{Coind}_H^G \overline{\mathbb{Q}}$, and $x \in G_{\mathbb{Q}}$.

Now we return to the situation $G = G_{\mathbb{Q}}$ to study the $G_{\mathbb{Q}}$-module $\overline{\mathbb{Q}} \otimes_{\mathbb{Q}} F$. Consider the finite set of field embeddings $\{\iota : F \hookrightarrow \overline{\mathbb{Q}}\}$. The group $G_{\mathbb{Q}}$ acts (on the left) on this set in the usual manner; that is to say, if ι is an embedding of F into $\overline{\mathbb{Q}}$ and σ is an element of G, then $\sigma \circ \iota$ is another such embedding. We now fix an embedding ι_0, let $H = \mathrm{Gal}(\overline{\mathbb{Q}}/\iota_0(F))$, and view $\overline{\mathbb{Q}}$ as an H-module.

Proposition 2.3. *There is a $\overline{\mathbb{Q}}$-linear isomorphism of $G_{\mathbb{Q}}$-modules*

$$\overline{\mathbb{Q}} \otimes_{\mathbb{Q}} F \xrightarrow{\sim} \mathrm{Coind}_H^G \overline{\mathbb{Q}}, \tag{2.3}$$

in which $1 \otimes f$ is sent to the function $\theta_f : x \mapsto \iota_0(f)$.

Proof. Note that $\theta_f \in \mathrm{Coind}_H^G \overline{\mathbb{Q}}$ because H fixes $\iota_0(F)$ by definition. Also, one may check that ${}^\sigma(r\theta) = \sigma(r){}^\sigma\theta$, and use this formula to verify that (2.3) is a $G_{\mathbb{Q}}$-module homomorphism.

Now we want to show that this homomorphism is injective. Recall that we may write $\overline{\mathbb{Q}} \otimes_{\mathbb{Q}} F \xrightarrow{\sim} \prod_\iota \overline{\mathbb{Q}}$ via the map $r \otimes f \mapsto (r \cdot \iota(f))_\iota$. We use this isomorphism of $\overline{\mathbb{Q}}$-algebras to factor (2.3) through the product; the element $(r_\iota)_\iota \in \prod_\iota \overline{\mathbb{Q}}$ is then mapped to the function $x \mapsto x(r_{x^{-1}\iota_0})$ in $\mathrm{Coind}_H^G \overline{\mathbb{Q}}$. This correspondence is clearly injective.

Finally we note that a function $\theta \in \mathrm{Coind}_H^G \overline{\mathbb{Q}}$ is determined by its values on any set of right coset representatives of H in $G_\mathbb{Q}$. Thus the $\overline{\mathbb{Q}}$-dimension of $\mathrm{Coind}_H^G \overline{\mathbb{Q}}$ must be $[G_\mathbb{Q} : H] = [F : \mathbb{Q}]$, which is the same as the $\overline{\mathbb{Q}}$-dimension of $\overline{\mathbb{Q}} \otimes_\mathbb{Q} F$. But (2.3) is already injective, hence it must actually be an isomorphism. □

Proposition 2.4. *We obtain the following isomorphisms in an analogous manner.*

1. $(\overline{\mathbb{Q}} \otimes_\mathbb{Q} F)^* \cong \mathrm{Coind}_H^G \overline{\mathbb{Q}}^*$.
2. $\mathbf{M}_n(\overline{\mathbb{Q}} \otimes_\mathbb{Q} F) \cong \mathrm{Coind}_H^G \mathbf{M}_n(\overline{\mathbb{Q}})$.
3. $\mathbf{GL}_n(\overline{\mathbb{Q}} \otimes_\mathbb{Q} F) \cong \mathrm{Coind}_H^G \mathbf{GL}_n(\overline{\mathbb{Q}})$.
4. $\mathbf{PGL}_n(\overline{\mathbb{Q}} \otimes_\mathbb{Q} F) \cong \mathrm{Coind}_H^G \mathbf{PGL}_n(\overline{\mathbb{Q}})$.

Proof.

1. Follows directly from above because the $G_\mathbb{Q}$-action on $\overline{\mathbb{Q}} \otimes_\mathbb{Q} F$ respects the multiplicative structure as well as the additive structure.
2. We can see from the definition of "Coind_H^G" that $\mathbf{M}_n(\mathrm{Coind}_H^G \overline{\mathbb{Q}})$ is isomorphic to $\mathrm{Coind}_H^G \mathbf{M}_n(\overline{\mathbb{Q}})$.
3. The action of $G_\mathbb{Q}$ still respects the multiplicative structure of the matrix algebra.
4. Follows from (1) and (3), and the fact that the functor Coind_H^G commutes with the formation of quotient modules. □

REMARK. By using the isomorphism $\overline{\mathbb{Q}} \otimes_\mathbb{Q} F \xrightarrow{\sim} \prod_\iota \overline{\mathbb{Q}}$ described in the proposition, we may similarly write any of the above $G_\mathbb{Q}$-modules as a product over the set of embeddings $\{\iota\}$. The action of $\sigma \in G_\mathbb{Q}$ on such a product sends the ι-place to the $\sigma^{-1}\iota$-place via σ.

Cohomology of co-induced modules

Let G be a profinite group, H a closed subgroup of G, M a discrete H-module, and $\mathrm{Coind}_H^G M$ the associated co-induced G-module. In the usual case, where M is an abelian group, Shapiro's Lemma states that the G-cohomology of $\mathrm{Coind}_H^G M$ may be identified with the H-cohomology of M. See for example [Br, Prop. III.6.2]. This fact is also well known for non-abelian modules [Se 1, §5.8b], but for the convenience of the reader we include an explicit proof.

Proposition 2.5 ("Non-Abelian Shapiro's Lemma"). *There is a natural bijection of pointed sets* $\mathrm{H}^1(G, \mathrm{Coind}_H^G M) \to \mathrm{H}^1(H, M)$.

Proof. First note that the projection $\pi : \mathrm{Coind}_H^G M \to M$ defined by $\theta \mapsto \theta(1)$ is compatible with the inclusion $H \subset G$, in the sense that $\pi({}^\eta\theta) = \eta(\pi(\theta))$ for $\eta \in H$ and $\theta \in \mathrm{Coind}_H^G M$. Therefore there exists an induced map $\mathrm{H}^1(G, \mathrm{Coind}_H^G M) \to \mathrm{H}^1(H, M)$, given by sending the one-cocycle $\sigma \mapsto \theta_\sigma$ to the cocycle $\eta \mapsto \theta_\eta(1)$.

To show that this map is one-to-one, suppose that $\eta \mapsto \theta_\eta(1)$ is a coboundary, i.e., that there is a fixed $m \in M$ such that $\theta_\eta(1) = m^{-1} \cdot \eta(m)$ for all $\eta \in H$.

One defines $\psi(x) = m \cdot \theta_x(1)$ for $x \in G$, and checks that $\psi \in \operatorname{Coind}_H^G M$. Then $\theta_\sigma = \psi^{-1} \circ {}^\sigma\psi$ for all $\sigma \in G$, so θ_σ is also a coboundary. Thus we have an injective map $\mathrm{H}^1(G, \operatorname{Coind}_H^G M) \to \mathrm{H}^1(H, M)$.

In order to show this map is surjective, let $\eta \mapsto m_\eta$ be a one-cocycle on H with values in M. Fix a set of right coset representatives of H in G, with the representative for H being 1. Now for each $x \in G$, define h_x to be the element of H such that $h_x^{-1}x$ is the chosen coset representative of Hx. Clearly for $\eta \in H$ we have $h_\eta = \eta$, and more generally $h_{\eta x} = \eta h_x$ for all $x \in G$. Define $\theta_\sigma(x) = m_{h_x}^{-1} \cdot m_{h_{x\sigma}}$. One verifies

- $\theta_\sigma \in \operatorname{Coind}_H^G M$,
- $\sigma \mapsto \theta_\sigma$ is a one-cocycle on G, and
- $\theta_\eta(1) = m_\eta$ for $\eta \in H$

to finish the proof of surjectivity. □

The bijection between $\mathrm{H}^1(G_\mathbb{Q}, \mathbf{PGL}_n(\overline{\mathbb{Q}} \otimes_\mathbb{Q} F))$ and $\mathrm{Br}_n(F)$

Again we return to the specific case $G = G_\mathbb{Q}$ and $H = \operatorname{Gal}(\overline{\mathbb{Q}}/F)$, where we have identified F with $\iota_0(F) \subset \overline{\mathbb{Q}}$ for some fixed embedding ι_0. Shapiro's Lemma, together with our proof that $(\overline{\mathbb{Q}} \otimes_\mathbb{Q} F)^* \cong \operatorname{Coind}_H^G \overline{\mathbb{Q}}^*$, has already given us an isomorphism of the group $\mathrm{H}^2(G_\mathbb{Q}, (\overline{\mathbb{Q}} \otimes_\mathbb{Q} F)^*)$ with $\mathrm{H}^2(H, \overline{\mathbb{Q}}^*) \cong \mathrm{Br}(F)$. We will need a more precise description of the correspondence, however. We have the following commutative diagram for each n:

$$\begin{array}{ccccc}
\mathrm{H}^2(G_\mathbb{Q}, (\overline{\mathbb{Q}} \otimes_\mathbb{Q} F)^*) & \longrightarrow & \mathrm{H}^2(H, \overline{\mathbb{Q}}^*) & \longrightarrow & \mathrm{Br}(F) \\
\uparrow \delta_n & & \uparrow \delta_n & & \uparrow \\
\mathrm{H}^1(G_\mathbb{Q}, \mathbf{PGL}_n(\overline{\mathbb{Q}} \otimes_\mathbb{Q} F)) & \xrightarrow{(2.5)} & \mathrm{H}^1(H, \mathbf{PGL}_n(\overline{\mathbb{Q}})) & \xrightarrow{(2.2)} & \mathrm{Br}_n(F)
\end{array}$$

Here the δ_n are the connecting homomorphisms from the cohomology sequences derived from the two short exact sequences:

$$\begin{array}{lllllll}
1 \to (\overline{\mathbb{Q}} \otimes_\mathbb{Q} F)^* & \to & \mathbf{GL}_n(\overline{\mathbb{Q}} \otimes_\mathbb{Q} F) & \to & \mathbf{PGL}_n(\overline{\mathbb{Q}} \otimes_\mathbb{Q} F) & \to 1 & \text{(co-ind. } G_\mathbb{Q}\text{-modules)} \\
1 \to \quad \overline{\mathbb{Q}}^* & \to & \mathbf{GL}_n(\overline{\mathbb{Q}}) & \to & \mathbf{PGL}_n(\overline{\mathbb{Q}}) & \to 1 & (H\text{-modules}).
\end{array}$$

Note that Proposition 2.5 allows us to identify the two corresponding cohomology sequences. In particular, the map (2.5) is a bijection for which we may find an explicit formula. First write $\mathbf{PGL}_n(\overline{\mathbb{Q}} \otimes_\mathbb{Q} F)$ as the product $\prod_\iota \mathbf{PGL}_n(\overline{\mathbb{Q}})$. Let $\sigma \mapsto (P_{\iota,\sigma})_\iota$ represent a cohomology class in $\mathrm{H}^1(G_\mathbb{Q}, \mathbf{PGL}_n(\overline{\mathbb{Q}} \otimes_\mathbb{Q} F))$. Then the image under (2.5) of this class is the class of $\eta \mapsto P_{\iota_0,\eta}$ in $\mathrm{H}^1(H, \mathbf{PGL}_n(\overline{\mathbb{Q}}))$.

Now the image of the class of this latter cocycle under bijection (2.2) is the Brauer class of the algebra

$$\mathcal{B} = \left\{ X \in \mathbf{M}_n(\overline{\mathbb{Q}}) \mid P_{\iota_0,\eta} \cdot \eta(X) \cdot P_{\iota_0,\eta}^{-1} = X \text{ for all } \eta \in H \right\}.$$

But there is an alternate description of this algebra. Consider the parallel construction

$$\mathcal{B}' = \left\{ (X_\iota)_\iota \in \mathbf{M}_n(\overline{\mathbb{Q}} \otimes_{\mathbb{Q}} F) \;\middle|\; (P_{\iota,\sigma})_\iota \cdot \sigma(X_\iota)_\iota \cdot (P_{\iota,\sigma})_\iota^{-1} = (X_\iota)_\iota \text{ all } \sigma \in G_{\mathbb{Q}} \right\}. \quad \mathbf{(2.6)}$$

It is clear that $\mathcal{B}'$ is also an F-algebra, and that there is an F-algebra homomorphism $\mathcal{B}' \to \mathcal{B}$ given by $(X_\iota)_\iota \mapsto X_{\iota_0}$. In fact this map is an isomorphism. Indeed, if we fix left coset representatives $\{g_\iota\}$ of H in G, then $X \mapsto (g_\iota(X))_\iota$ is the inverse isomorphism (assuming that $(P_{\iota,\sigma})_\iota$ is constructed from $P_{\iota_0,\eta}$ as in Proposition 2.5). The verification, however, is tedious and we leave it to the reader.

Now we see that $\mathcal{B}$ and $\mathcal{B}'$ determine the same class in $\mathrm{Br}_n(F)$. Observe, however, that the construction of $\mathcal{B}'$ avoids the choice of the embedding $\iota_0 : F \hookrightarrow \overline{\mathbb{Q}}$. Therefore, even though bijections (2.5) and (2.2) depend on ι_0, their composition is independent of this choice, and thus the isomorphism $\mathrm{H}^2(G_{\mathbb{Q}}, (\overline{\mathbb{Q}} \otimes_{\mathbb{Q}} F)^*) \cong \mathrm{Br}(F)$ is completely canonical.

The proof of Theorem 2.1

Let B be a building block, and let $c(\sigma, \tau)$ be the associated two-cocycle on $G_{\mathbb{Q}}$ with values in F^*. Now that we have established the isomorphism $\mathrm{H}^2(G_{\mathbb{Q}}, (\overline{\mathbb{Q}} \otimes_{\mathbb{Q}} F)^*) \cong \mathrm{Br}(F)$, we aim to show that the image in $\mathrm{H}^2(G_{\mathbb{Q}}, (\overline{\mathbb{Q}} \otimes_{\mathbb{Q}} F)^*)$ of the cohomology class $[c]$ corresponds to the class of $\mathrm{End}(B)$ in $\mathrm{Br}(F)$. Note that the Brauer class of $\mathrm{End}(B)$ lies in $\mathrm{Br}_t(F)$ where $t = 1$ or 2, since by hypothesis, $\mathrm{End}(B)$ is a central division algebra over F with Schur index $t = 1$ or 2. Thus from the commutative diagram of the previous section we see it will suffice to find a one-cocycle with values in $\mathbf{PGL}_t(\overline{\mathbb{Q}} \otimes_{\mathbb{Q}} F)$ whose class maps to both the image of $[c]$ in $\mathrm{H}^2(G_{\mathbb{Q}}, (\overline{\mathbb{Q}} \otimes_{\mathbb{Q}} F)^*)$ and the Brauer class of $\mathrm{End}(B)$ in $\mathrm{Br}_t(F)$.

First, we consider the Lie algebra $\mathrm{Lie}\, B$. Recall that the space of tangent vectors $\mathrm{Lie}\, B$ is a $\overline{\mathbb{Q}}$-vector space canonically attached to the abelian variety B, whose dimension over $\overline{\mathbb{Q}}$ is equal to the dimension of B. (See, e.g., [Mu, §11].) Also, $\mathrm{Lie}\, B$ has the property that $\mathrm{Lie}\, {}^\sigma B \cong {}^\sigma(\mathrm{Lie}\, B) \overset{\text{def}}{=} \overline{\mathbb{Q}} \otimes_{\overline{\mathbb{Q}}} \mathrm{Lie}\, B$, where the map from the base $\overline{\mathbb{Q}}$ to the base extension $\overline{\mathbb{Q}}$ is given by σ. Furthermore, the action of F on B induces an F-vector space structure on $\mathrm{Lie}\, B$, and it is well known that since F is totally real, $\mathrm{Lie}\, B$ is free of rank t over $\overline{\mathbb{Q}} \otimes_{\mathbb{Q}} F$. (Here is where we use the condition that the dimension of B is exactly $t[F : \mathbb{Q}]$.)

Now each $\mu_\sigma : {}^\sigma B \to B$ induces an isomorphism of $\overline{\mathbb{Q}}$-vector spaces $\lambda_\sigma : \mathrm{Lie}\, {}^\sigma B \to \mathrm{Lie}\, B$, and the relation $\mu_\sigma \circ {}^\sigma\mu_\tau = c(\sigma,\tau)\, \mu_{\sigma\tau}$ implies that $\lambda_\sigma \circ {}^\sigma\lambda_\tau = c(\sigma,\tau)\, \lambda_{\sigma\tau}$ by functoriality. Here ${}^\sigma\lambda_\tau$ is the map $1 \otimes \lambda_\tau : {}^\sigma(\mathrm{Lie}\, {}^\tau B) \to {}^\sigma(\mathrm{Lie}\, B)$. Fix a basis $\{e_i\}_{i=1}^t$ for $\mathrm{Lie}\, B$ over $\overline{\mathbb{Q}} \otimes_{\mathbb{Q}} F$. Then for each $\sigma \in G_{\mathbb{Q}}$, $\{1 \otimes e_i\}_{i=1}^t$ is a basis for the σ-twist ${}^\sigma(\mathrm{Lie}\, B) \cong \mathrm{Lie}\, {}^\sigma B$. We may use this collection of bases to express the maps λ_σ as matrices $\Lambda_\sigma \in \mathbf{GL}_t(\overline{\mathbb{Q}} \otimes_{\mathbb{Q}} F)$, and the identity $\lambda_\sigma \circ {}^\sigma\lambda_\tau = c(\sigma,\tau)\, \lambda_{\sigma\tau}$ translates into $\Lambda_\sigma \cdot {}^\sigma\Lambda_\tau = c(\sigma,\tau)\, \Lambda_{\sigma\tau}$. At this point one easily verifies that the matrix ${}^\sigma\Lambda_\tau$ describing the map ${}^\sigma\lambda_\tau$ is the matrix Λ_τ with its entries twisted by σ, i.e., exactly $\sigma(\Lambda_\tau)$.

We are now provided with a one-cocycle with values in $\mathbf{PGL}_t(\overline{\mathbb{Q}}\otimes_{\mathbb{Q}}F)$, namely $\sigma \mapsto \Lambda_\sigma \bmod (\overline{\mathbb{Q}} \otimes_{\mathbb{Q}} F)^*$.

Proposition 2.7. *The class of* $\Lambda_\sigma \bmod (\overline{\mathbb{Q}} \otimes_{\mathbb{Q}} F)^*$ *in* $\mathrm{H}^1(G_{\mathbb{Q}}, \mathbf{PGL}_t(\overline{\mathbb{Q}} \otimes_{\mathbb{Q}} F))$ *is independent of the choice of basis for* $\operatorname{Lie} B$.

Proof. A different choice of basis for $\operatorname{Lie} B$ is described by a change-of-basis matrix $P \in \mathbf{GL}_t(\overline{\mathbb{Q}} \otimes_{\mathbb{Q}} F)$. The corresponding change of basis on $\operatorname{Lie} {}^\sigma B \cong {}^\sigma(\operatorname{Lie} B)$ will be given by ${}^\sigma P$. Thus the new matrix for the map $\lambda_\sigma : \operatorname{Lie} {}^\sigma B \to \operatorname{Lie} B$ will be $P \cdot \Lambda_\sigma \cdot {}^\sigma P^{-1}$, and the cocycle $\sigma \mapsto P \cdot \Lambda_\sigma \cdot {}^\sigma P^{-1}$ is cohomologous to $\sigma \mapsto \Lambda_\sigma$. □

It remains to show that the class of the cocycle $\sigma \mapsto \Lambda_\sigma \bmod (\overline{\mathbb{Q}} \otimes_{\mathbb{Q}} F)^*$ corresponds to both the image of $[c]$ and the Brauer class of $\operatorname{End}(B)$, as desired.

Proposition 2.8. *The class of* $\Lambda_\sigma \bmod (\overline{\mathbb{Q}}\otimes_{\mathbb{Q}} F)^*$ *in* $\mathrm{H}^1(G_{\mathbb{Q}}, \mathbf{PGL}_t(\overline{\mathbb{Q}}\otimes_{\mathbb{Q}} F))$ *maps to the image of* $[c]$ *in* $\mathrm{H}^2(G_{\mathbb{Q}}, (\overline{\mathbb{Q}} \otimes_{\mathbb{Q}} F)^*)$.

Proof. The connecting homomorphism which supplies the inclusion

$$\mathrm{H}^1(G_{\mathbb{Q}}, \mathbf{PGL}_t(\overline{\mathbb{Q}} \otimes_{\mathbb{Q}} F)) \hookrightarrow \mathrm{H}^2(G_{\mathbb{Q}}, (\overline{\mathbb{Q}} \otimes_{\mathbb{Q}} F)^*)$$

is defined by the following procedure. First, choose a one-cocycle to represent the given cohomology class. In our case we take $\Lambda_\sigma \bmod (\overline{\mathbb{Q}}\otimes_{\mathbb{Q}}F)^*$. Then lift the chosen cocycle to a map $G_{\mathbb{Q}} \to \mathbf{GL}_t(\overline{\mathbb{Q}}\otimes_{\mathbb{Q}}F)$; we choose $\sigma \mapsto \Lambda_\sigma$ for our application. Then $\Lambda_\sigma \cdot \sigma(\Lambda_\tau) \cdot \Lambda_{\sigma\tau}^{-1}$ is a two-cocycle with values in $(\overline{\mathbb{Q}} \otimes_{\mathbb{Q}} F)^*$, and the class of this two-cocycle in $\mathrm{H}^2(G_{\mathbb{Q}}, (\overline{\mathbb{Q}}\otimes_{\mathbb{Q}}F)^*)$ is independent of all the choices we made. In our situation, $\Lambda_\sigma \cdot \sigma(\Lambda_\tau) \cdot \Lambda_{\sigma\tau}^{-1} = c(\sigma, \tau)$, which is exactly what we wished to show. □

Proposition 2.9. *The class of* $\Lambda_\sigma \bmod (\overline{\mathbb{Q}} \otimes_{\mathbb{Q}} F)^*$ *in* $\mathrm{H}^1(G_{\mathbb{Q}}, \overline{\mathbb{Q}} \otimes_{\mathbb{Q}} F)$ *corresponds to the class of* $\operatorname{End}(B)$ *in* $\operatorname{Br}_t(F)$.

Proof. By applying Equation 2.6, we find the Brauer class of F-algebras corresponding to the cocycle $\Lambda_\sigma \bmod (\overline{\mathbb{Q}} \otimes_{\mathbb{Q}} F)^*$ is represented by the algebra

$$\mathcal{B} \stackrel{\text{def}}{=} \left\{ X \in \mathbf{M}_t(\overline{\mathbb{Q}} \otimes_{\mathbb{Q}} F) \mid \Lambda_\sigma \cdot \sigma(X) \cdot \Lambda_\sigma^{-1} = X \text{ for all } \sigma \in G_{\mathbb{Q}} \right\}.$$

On the other hand, by the functoriality of "Lie", there is a ring homomorphism $\operatorname{End}(B) \to \operatorname{End}(\operatorname{Lie} B)$. In fact this map is an injection, because $\operatorname{End}(B)$ is a division algebra. Then by using our fixed basis of $\operatorname{Lie} B$, we have

$$\operatorname{End}(B) \hookrightarrow \operatorname{End}(\operatorname{Lie} B) \cong \mathbf{M}_t(\overline{\mathbb{Q}} \otimes_{\mathbb{Q}} F).$$

We claim that under these maps, the image of $\operatorname{End}(B)$ lands in $\mathcal{B}$. Indeed, for $\varphi \in \operatorname{End}(B)$, let $\Phi \in \mathbf{M}_t(\overline{\mathbb{Q}} \otimes_{\mathbb{Q}} F)$ be the corresponding matrix. Recall that the original μ_σ satisfied $\varphi \circ \mu_\sigma = \mu_\sigma \circ {}^\sigma\varphi$. Hence by functoriality again, we have $\Phi \cdot \Lambda_\sigma = \Lambda_\sigma \cdot \sigma(\Phi)$ for all σ. Thus $\Lambda_\sigma \cdot \sigma(\Phi) \cdot \Lambda_\sigma^{-1} = \Phi$ for all $\sigma \in G_{\mathbb{Q}}$, which means $\Phi \in \mathcal{B}$.

Finally we see that both $\operatorname{End}(B)$ and $\mathcal{B}$ are t^2-dimensional over F, so the inclusion must in fact be an isomorphism $\operatorname{End}(B) \xrightarrow{\sim} \mathcal{B}$. □

This proves Theorem 2.1.

3. The tensor product of a module with an Abelian variety

Here we prepare some "abstract nonsense" we will need for our construction of the correspondence between building blocks and abelian varieties of $\mathbf{GL}_2$-type. The results in this chapter, however, are valid for arbitrary abelian varieties.

Let A be an abelian variety up to isogeny, let $\mathcal{R}$ be a ring with a map $\mathcal{R} \to \mathrm{End}(A)$, and let M be a finitely generated right $\mathcal{R}$-module. By abusing notation, we will identify $\mathcal{R}$ with its image in $\mathrm{End}(A)$, and say that A is an abelian variety with an action of $\mathcal{R}$.

Theorem 3.1. *Let $\mathfrak{C}$ be the category of abelian varieties up to isogeny, and let $\mathfrak{A}$ be the category of abelian groups. Let $T : \mathfrak{C} \to \mathfrak{A}$ be the covariant functor defined by $T(X) = \mathrm{Hom}_{\mathcal{R}}(M, \mathrm{Hom}(A, X))$. Then T is representable.*

Proof. [Mit, Thm. VI.3.1]. □

We will denote the object representing the functor T as $M \otimes_{\mathcal{R}} A$, and devote this chapter to summarizing some properties of this "tensor product". First, we may reformulate the above theorem as an assertion of the existence of an abelian variety $M \otimes_{\mathcal{R}} A$ characterized by the following universal property:

$M \otimes_{\mathcal{R}} A$ is an abelian variety together with a homomorphism of right $\mathcal{R}$-modules $\Delta : M \to \mathrm{Hom}(A, M \otimes_{\mathcal{R}} A)$, which, for an abelian variety X, induces an isomorphism of abelian groups $\mathrm{Hom}(M \otimes_{\mathcal{R}} A, X) \to \mathrm{Hom}_{\mathcal{R}}(M, \mathrm{Hom}(A, X))$ via $\psi \mapsto (m \mapsto \psi \circ \Delta_m)$. **(3.2)**

Next, standard procedures produce the following facts. In the list below, it is understood that each tensor product is equipped with the evident homomorphism Δ, and that the symbol "$\cong$" designates a natural equivalence that respects these homomorphisms.

Elementary Properties of $M \otimes_{\mathcal{R}} A$:

(i) If A is an abelian variety with an action of $\mathcal{R}$, then $\mathcal{R} \otimes_{\mathcal{R}} A \cong A$.

(ii) If M is a right $\mathcal{R}$-module, N is an $\mathcal{R}$-$\mathcal{S}$-bimodule, and A is an abelian variety with an action of $\mathcal{S}$, then $M \otimes_{\mathcal{R}} (N \otimes_{\mathcal{S}} A) \cong (M \otimes_{\mathcal{R}} N) \otimes_{\mathcal{S}} A$.

(iii) If $\{M_i\}$ is a finite collection of right $\mathcal{R}$-modules, and A is an abelian variety with an action of $\mathcal{R}$, then $(\oplus M_i) \otimes_{\mathcal{R}} A \cong \prod (M_i \otimes_{\mathcal{R}} A)$.

(iv) If M is a right $\mathcal{R}$-module and $\{A_i\}$ is a finite collection of abelian varieties, each with a map $\mathcal{R} \to \mathrm{End}(A_i)$, then $M \otimes_{\mathcal{R}} \prod A_i \cong \prod (M \otimes_{\mathcal{R}} A_i)$.

(v) $M \otimes_{\mathcal{R}} A$ is functorial in each variable separately. More explicitly, let M_1 and M_2 be right $\mathcal{R}$-modules and $f \in \mathrm{Hom}_{\mathcal{R}}(M_1, M_2)$. Let A_1 and A_2 be abelian varieties with actions of $\mathcal{R}$ and suppose $\varphi : A_1 \to A_2$ is a morphism of abelian varieties that respects $\mathcal{R}$; i.e., $r \circ \varphi = \varphi \circ r$ for all $r \in \mathcal{R}$. We write $\varphi \in \mathrm{Hom}^{\mathcal{R}}(A_1, A_2)$. Then $m \mapsto \Delta_{f(m)} \circ \varphi$ is a homomorphism of right $\mathcal{R}$-modules $M_1 \to \mathrm{Hom}(A_1, M_2 \otimes_{\mathcal{R}} A_2)$. Hence by (3.2), there exists a unique morphism $f \otimes \varphi \in \mathrm{Hom}(M_1 \otimes_{\mathcal{R}} A_1, M_2 \otimes_{\mathcal{R}} A_2)$ such that $(f \otimes \varphi) \circ \Delta_m = \Delta_{f(m)} \circ \varphi$ for all $m \in M$. Moreover, it is easy to check that if $g : M_2 \to M_3$ is another

$\mathcal{R}$-module homomorphism and $\psi : A_2 \to A_3$ another morphism of abelian varieties respecting $\mathcal{R}$, then $(g \otimes \psi) \circ (f \otimes \varphi) = (g \circ f) \otimes (\psi \circ \varphi)$.

If we apply property (v) in the particular case $A_1 = A_2 = A$ and $M_1 = M_2 = M$, we have ring homomorphisms

$$\begin{array}{rcl} \operatorname{End}_{\mathcal{R}} M & \to & \operatorname{End}(M \otimes_{\mathcal{R}} A) \\ f & \mapsto & f \otimes 1 \end{array} \qquad \begin{array}{rcl} \operatorname{End}^{\mathcal{R}}(A) & \to & \operatorname{End}(M \otimes_{\mathcal{R}} A) \\ \varphi & \mapsto & 1 \otimes \varphi \end{array}$$

whose images commute. Note that the center $Z(\mathcal{R}) \subseteq \operatorname{End}^{\mathcal{R}}(A)$, and that we may also map $Z(\mathcal{R}) \to \operatorname{End}_{\mathcal{R}} M$ via $r \mapsto \rho_r =$ right multiplication by r on M. (The condition $r \in Z(\mathcal{R})$ makes ρ_r a homomorphism of right $\mathcal{R}$-modules.)

Proposition 3.3. *The two maps $Z(\mathcal{R}) \to \operatorname{End}(M \otimes_{\mathcal{R}} A)$ coincide. Specifically, $1 \otimes r = \rho_r \otimes 1 \in \operatorname{End}(M \otimes_{\mathcal{R}} A)$.*

Proof. By (3.2), it suffices to show that $(\rho_r \otimes 1) \circ \Delta_m = (1 \otimes r) \circ \Delta_m$ for all $m \in M$. But this plainly follows from the definitions: $(\rho_r \otimes 1) \circ \Delta_m = \Delta_{\rho_r(m)} = \Delta_{mr} = \Delta_m \circ r = (1 \otimes r) \circ \Delta_m$. □

Thus we obtain a homomorphism of $Z(\mathcal{R})$-algebras

$$\operatorname{End}_{\mathcal{R}} M \otimes_{Z(\mathcal{R})} \operatorname{End}^{\mathcal{R}}(A) \longrightarrow \operatorname{End}^{Z(\mathcal{R})}(M \otimes_{\mathcal{R}} A). \tag{3.4}$$

Now suppose M is an $\mathcal{S}$-$\mathcal{R}$-bimodule. Then we may map $\mathcal{S} \to \operatorname{End}_{\mathcal{R}} M$ via $s \mapsto \lambda_s =$ left multiplication by s on M. By composing this map with $\operatorname{End}_{\mathcal{R}} M \to \operatorname{End}(M \otimes_{\mathcal{R}} A)$ from above, we find that we have an action of $\mathcal{S}$ on $M \otimes_{\mathcal{R}} A$. Again, we will usually identify $\mathcal{S}$ with its image in $\operatorname{End}(M \otimes_{\mathcal{R}} A)$. We now reconsider the defining property (3.2) of $M \otimes_{\mathcal{R}} A$.

Proposition 3.5. *Let M be an $\mathcal{S}$-$\mathcal{R}$-bimodule, and let A be an abelian variety with an action of $\mathcal{R}$. Suppose X is an abelian variety with an action of $\mathcal{S}$. Then Δ induces an isomorphism $\operatorname{Hom}^{\mathcal{S}}(M \otimes_{\mathcal{R}} A,\ X) \to \operatorname{Hom}_{\mathcal{S}\text{-}\mathcal{R}}(M,\ \operatorname{Hom}(A, X))$.*

Proof. It is clear that if $\psi : M \otimes_{\mathcal{R}} A \to X$ respects $\mathcal{S}$, then $m \mapsto \psi \circ \Delta_m$ is a homomorphism of $\mathcal{S}$-$\mathcal{R}$-bimodules. On the other hand, suppose $m \mapsto \varphi_m$ is an element of $\operatorname{Hom}_{\mathcal{S}\text{-}\mathcal{R}}(M,\ \operatorname{Hom}(A, X))$. Then there exists a morphism of abelian varieties $\widetilde{\varphi} \in \operatorname{Hom}(M \otimes_{\mathcal{R}} A,\ X)$ such that $\varphi_m = \widetilde{\varphi} \circ \Delta_m$ for all $m \in M$. We have to show that $\widetilde{\varphi}$ respects $\mathcal{S}$, i.e., that $s \circ \widetilde{\varphi} = \widetilde{\varphi} \circ s$ for all $s \in \mathcal{S}$. As usual, by (3.2) it suffices to show that $s \circ \widetilde{\varphi} \circ \Delta_m = \widetilde{\varphi} \circ s \circ \Delta_m$ for all $s \in \mathcal{S}$ and $m \in M$, and this follows directly from the relevant definitions: $s \circ \widetilde{\varphi} \circ \Delta_m = s \circ \varphi_m = \varphi_{sm} = \widetilde{\varphi} \circ \Delta_{sm} = \widetilde{\varphi} \circ s \circ \Delta_m$. □

Finally, if $M = \mathcal{S}$ is a ring, then it is an $\mathcal{S}$-$\mathcal{R}$-bimodule and the above results hold. Also, we have a ring homomorphism $\mathcal{R} \to \mathcal{S}$ via $r \mapsto 1 \cdot r$, so an abelian variety X with an action of $\mathcal{S}$ also has an action of $\mathcal{R}$. With the extra structure in this case, we find an isomorphism $\operatorname{Hom}_{\mathcal{S}\text{-}\mathcal{R}}(\mathcal{S},\ \operatorname{Hom}(A, X)) \xrightarrow{\sim} \operatorname{Hom}^{\mathcal{R}}(A, X)$, by taking the element $s \mapsto \varphi_s$ of $\operatorname{Hom}_{\mathcal{S}\text{-}\mathcal{R}}(\mathcal{S},\ \operatorname{Hom}(A, X))$ to φ_1. The reason that $\varphi_1 : A \to X$ respects $\mathcal{R}$ is because $\mathcal{R}$ commutes with $1 \in \mathcal{S}$. Thus we may now rephrase the defining property of $\mathcal{S} \otimes_{\mathcal{R}} A$ in the category of abelian varieties up to isogeny:

> Let A be an abelian variety with an action of $\mathcal{R}$, and let $\mathcal{S}$ be a ring which is also a right $\mathcal{R}$-module. Then $\mathcal{S} \otimes_{\mathcal{R}} A$ is an abelian variety with an action of $\mathcal{S}$, together with a morphism $\Delta_1 : A \to \mathcal{S} \otimes_{\mathcal{R}} A$ respecting $\mathcal{R}$, which satisfies the following universal property: given an abelian variety X with an action of $\mathcal{S}$, and a morphism $\varphi : A \to X$ respecting $\mathcal{R}$, there exists a unique morphism $\widetilde{\varphi} : \mathcal{S} \otimes_{\mathcal{R}} A \to X$ respecting $\mathcal{S}$ such that $\varphi = \widetilde{\varphi} \circ \Delta_1$.

Galois action

Now suppose that A is defined over $\overline{\mathbb{Q}}$; let us consider the effect of twisting by $\sigma \in \mathrm{Gal}(\overline{\mathbb{Q}}/\mathbb{Q}) = G_{\mathbb{Q}}$ on the tensor product construction. Note that if A is an abelian variety with an action of $\mathcal{R}$, then so is ${}^{\sigma}A$, for each $\sigma \in G_{\mathbb{Q}}$; we may map $\mathcal{R} \to \mathrm{End}(A) \to \mathrm{End}({}^{\sigma}A)$ via the usual "twist-by-σ".

Lemma 3.6. *Let A, A_1, A_2 be abelian varieties over $\overline{\mathbb{Q}}$ with actions of $\mathcal{R}$, and let M, M_1, M_2 be right $\mathcal{R}$-modules.*

1. *For each $\sigma \in G_{\mathbb{Q}}$, we have ${}^{\sigma}(M \otimes_{\mathcal{R}} A) \sim M \otimes_{\mathcal{R}} {}^{\sigma}A$.*
2. *Let $f \in \mathrm{Hom}_{\mathcal{R}}(M_1, M_2)$, let $\varphi \in \mathrm{Hom}^{\mathcal{R}}(A_1, A_2)$, and let $f \otimes \varphi$ be the induced map $M_1 \otimes_{\mathcal{R}} A_1 \to M_2 \otimes_{\mathcal{R}} A_2$ defined by property (v) above. Then for each $\sigma \in G_{\mathbb{Q}}$, we have ${}^{\sigma}(f \otimes \varphi) = f \otimes {}^{\sigma}\varphi$.*

Proof.

1. Fix $\sigma \in G_{\mathbb{Q}}$. We will demonstrate that ${}^{\sigma}(M \otimes_{\mathcal{R}} A)$ satisfies the defining property (3.2) of $M \otimes_{\mathcal{R}} {}^{\sigma}A$. First construct $M \otimes_{\mathcal{R}} A$ and $\Delta : M \to \mathrm{Hom}(A,\ M \otimes_{\mathcal{R}} A)$. Then for an abelian variety $X/\overline{\mathbb{Q}}$, the map Δ induces an isomorphism $\mathrm{Hom}(M \otimes_{\mathcal{R}} A,\ X) \to \mathrm{Hom}_{\mathcal{R}}(M,\ \mathrm{Hom}(A, X))$. Now if one defines ${}^{\sigma}\Delta : M \to \mathrm{Hom}_{\mathcal{R}}({}^{\sigma}A,\ {}^{\sigma}(M \otimes_{\mathcal{R}} A))$ as $({}^{\sigma}\Delta)_m \stackrel{\text{def}}{=} {}^{\sigma}(\Delta_m)$, then ${}^{\sigma}\Delta$ induces an isomorphism $\mathrm{Hom}({}^{\sigma}(M \otimes_{\mathcal{R}} A),\ {}^{\sigma}X) \to \mathrm{Hom}_{\mathcal{R}}(M,\ \mathrm{Hom}({}^{\sigma}A, {}^{\sigma}X))$. But of course any abelian variety may be written as ${}^{\sigma}X$ for some X, so ${}^{\sigma}(M \otimes_{\mathcal{R}} A)$ and ${}^{\sigma}\Delta$ do satisfy the characteristic property of $M \otimes_{\mathcal{R}} {}^{\sigma}A$.
2. The map $f \otimes \varphi$ is defined to be the unique morphism $M_1 \otimes_{\mathcal{R}} A_1 \to M_2 \otimes_{\mathcal{R}} A_2$ such that $(f \otimes \varphi) \circ \Delta_m = \Delta_{f(m)} \circ \varphi$ for all $m \in M_1$. Applying σ to both sides, we find ${}^{\sigma}(f \otimes \varphi) \circ {}^{\sigma}\Delta_m = {}^{\sigma}\Delta_{f(m)} \circ {}^{\sigma}\varphi = (f \otimes {}^{\sigma}\varphi) \circ {}^{\sigma}\Delta_m$ for all $m \in M_1$, which implies that ${}^{\sigma}(f \otimes \varphi) = f \otimes {}^{\sigma}\varphi$. □

4. Descent from $\overline{\mathbb{Q}}$ to $\mathbb{Q}$

In this chapter we will show how to construct, from a given building block B, an abelian variety $A/\mathbb{Q}$ of $\mathbf{GL}_2$-type with B as a simple quotient over $\overline{\mathbb{Q}}$. We will then describe a precise correspondence between these two types of abelian varieties. Recall the following definitions.

Definition. An abelian variety $B/\overline{\mathbb{Q}}$ is a $\mathbb{Q}$-*abelian variety* if for each $\sigma \in \mathrm{Gal}(\overline{\mathbb{Q}}/\mathbb{Q})$, there exists an isogeny $\mu_\sigma : {}^\sigma B \to B$ compatible with the endomorphisms of B; that is, $\mu_\sigma \circ {}^\sigma\varphi = \varphi \circ \mu_\sigma$ for all $\varphi \in \mathrm{End}(B)$.

Definition. A *building block* is an abelian variety $B/\overline{\mathbb{Q}}$ such that

(i) the endomorphism algebra $\mathrm{End}(B)$ is a central division algebra over a totally real field F with Schur index $t = 1$ or 2, and $t[F:\mathbb{Q}] = \dim B$.

(ii) B is a $\mathbb{Q}$-abelian variety.

Definition. An abelian variety $A/\mathbb{Q}$ is *of* $\mathbf{GL}_2$*-type* if $\mathrm{End}_{\mathbb{Q}}(A)$, the algebra of endomorphisms of A defined over $\mathbb{Q}$, is a number field of degree equal to the dimension of A.

Construction of an Abelian variety of $\mathbf{GL}_2$-type

We write $G_{\mathbb{Q}}$ for the absolute Galois group $\mathrm{Gal}(\overline{\mathbb{Q}}/\mathbb{Q})$ with its usual profinite topology. Let B be a building block. Then, by definition, B is a $\mathbb{Q}$-abelian variety. We have seen in Chapter 2 how to fix for each $\sigma \in G_{\mathbb{Q}}$ a choice of $\mu_\sigma : {}^\sigma B \to B$ respecting $\mathrm{End}(B)$ such that $c(\sigma,\tau) = \mu_\sigma \circ {}^\sigma\mu_\tau \circ \mu_{\sigma\tau}^{-1}$ is a locally constant two-cocycle on $G_{\mathbb{Q}}$ with values in F^* (a $G_{\mathbb{Q}}$-module with trivial action). The class $[c] \in \mathrm{H}^2(G_{\mathbb{Q}}, F^*)$ is independent of our choices for μ_σ.

Now consider the cohomology group $\mathrm{H}^2(G_{\mathbb{Q}}, \overline{F}^*)$, where $G_{\mathbb{Q}}$ acts trivially on $\overline{F}^*$. By a theorem of Tate [Se 3, Thm. 4], this group is trivial. Therefore the image of $[c]$ in $\mathrm{H}^2(G_{\mathbb{Q}}, \overline{F}^*)$ is a coboundary; that is, there exists a locally constant function $\beta : G_{\mathbb{Q}} \to \overline{F}^*$ such that $c(\sigma,\tau) = \beta(\sigma)\cdot\beta(\tau)\cdot\beta(\sigma\tau)^{-1}$ for all $\sigma,\tau \in G_{\mathbb{Q}}$. Fix such a function β. Let E be the field extension of F obtained by adjoining all the values $\beta(\sigma)$. Then E is a finite extension of F (as β is locally constant); set $d = [E:F]$.

Consider the abelian variety $E \otimes_F B$. From Equation 3.4 there is a map

$$\mathrm{End}_F E \otimes_F \mathrm{End}(B) \to \mathrm{End}(E \otimes_F B). \tag{4.1}$$

We claim this map is an isomorphism of F-algebras. Indeed, fix a basis of E over F. Then we see that $\mathrm{End}_F E$ is isomorphic to the matrix algebra $\mathbf{M}_d(F)$, and so the image of $\mathrm{End}_F E \otimes_F \mathrm{End}(B)$ in $\mathrm{End}(E \otimes_F B)$ is isomorphic to $\mathbf{M}_d(\mathrm{End}(B))$. On the other hand, the same choice of basis for E/F gives us $E \otimes_F B \sim (\bigoplus^d F) \otimes_F B \sim \prod^d B$, so the endomorphism algebra $\mathrm{End}(E \otimes_F B)$ must itself be isomorphic to $\mathbf{M}_d(\mathrm{End}(B))$. From now on we will frequently identify $\mathrm{End}(E \otimes_F B)$ with $\mathrm{End}_F E \otimes_F \mathrm{End}(B)$.

Now we claim that we may descend the abelian variety $E \otimes_F B$ to $\mathbb{Q}$.

For each $\sigma \in G_{\mathbb{Q}}$, define $\nu_\sigma \stackrel{\text{def}}{=} \beta(\sigma)^{-1} \otimes \mu_\sigma \in \mathrm{Hom}(E \otimes_F {}^\sigma B, E \otimes_F B)$, using our fixed $\beta(\sigma)$ and μ_σ. Recall that by Lemma 3.6, we have $E \otimes_F {}^\sigma B \sim {}^\sigma(E \otimes_F B)$; therefore ν_σ may be viewed as an isogeny ${}^\sigma(E \otimes_F B) \to E \otimes_F B$.

Proposition 4.2. *There exists an abelian variety A_0 over $\mathbb{Q}$ and an isogeny (over $\overline{\mathbb{Q}}$) $\kappa : E \otimes_F B \to A_0$ such that $\kappa^{-1} \circ {}^\sigma\kappa = \nu_\sigma$ for all $\sigma \in G_{\mathbb{Q}}$.*

Proof. First, we see that $\nu_\sigma \circ {}^\sigma\nu_\tau = \nu_{\sigma\tau}$ for all $\sigma, \tau \in G_{\mathbb{Q}}$. Indeed, by using the definitions of $c(\sigma,\tau)$ and $\beta(\sigma)$, and the properties of the tensor product from Chapter 3, we have

$$\begin{aligned}\nu_\sigma \circ {}^\sigma\nu_\tau &= \beta(\sigma)^{-1}\beta(\tau)^{-1} \otimes \mu_\sigma \circ {}^\sigma\mu_\tau \\ &= \beta(\sigma\tau)^{-1}c(\sigma,\tau)^{-1} \otimes c(\sigma,\tau)\mu_{\sigma\tau} \\ &= \beta(\sigma\tau)^{-1} \otimes \mu_{\sigma\tau} \\ &= \nu_{\sigma\tau}.\end{aligned}$$

Therefore, we may apply the method of [Ri 1, Thm. 8.2] to construct A_0. More explicitly, let K be a finite Galois extension of $\mathbb{Q}$ such that $E \otimes_F B$ is defined over K and all ν_σ are defined over K. Let X be the abelian variety $\mathrm{Res}_{K/\mathbb{Q}}(E \otimes_F B)$, where "Res" denotes Weil's "restriction of scalars" functor. (See [We, §1.3].) Then X is defined over $\mathbb{Q}$ and $X_{/K} \cong \prod_g {}^g(E \otimes_F B)$, where the product runs over $g \in \mathrm{Gal}(K/\mathbb{Q})$. We now define $\kappa = (\nu_g^{-1})_g : E \otimes_F B \to X_{/K}$. Lastly, write A_0 for the image of κ (or technically, the image of a suitable multiple of κ).

A direct (if tiresome) computation yields the formula $\kappa^{-1} \circ {}^\sigma\kappa = \nu_\sigma$. □

Next we ask which endomorphisms of A_0 are defined over $\mathbb{Q}$? Or more generally, how does $\sigma \in G_{\mathbb{Q}}$ act on $\mathrm{End}(A_0)$?

Proposition 4.3. *For each $\sigma \in G_{\mathbb{Q}}$, the automorphism $\varphi \mapsto {}^\sigma\varphi$ of $\mathrm{End}(A_0)$ is equal to conjugation by $\beta(\sigma)^{-1} \in E$. (Here the action of E on A_0 is induced by κ.)*

Proof. The isogeny κ induces an isomorphism $\mathrm{End}(A_0) \xrightarrow{\sim} \mathrm{End}(E \otimes_F B)$ via $\varphi \mapsto \kappa^{-1} \circ \varphi \circ \kappa$. Thus, the automorphism $\varphi \mapsto {}^\sigma\varphi$ on $\mathrm{End}(A_0)$ translates into the automorphism $\kappa^{-1} \circ \varphi \circ \kappa \mapsto \kappa^{-1} \circ {}^\sigma\varphi \circ \kappa$ on $\mathrm{End}(E \otimes_F B)$. Now we may write

$$\begin{aligned}\kappa^{-1} \circ {}^\sigma\varphi \circ \kappa &= \kappa^{-1} \circ ({}^\sigma\kappa \circ {}^\sigma\kappa^{-1}) \circ {}^\sigma\varphi \circ ({}^\sigma\kappa \circ {}^\sigma\kappa^{-1}) \circ \kappa \\ &= \nu_\sigma \circ {}^\sigma(\kappa^{-1} \circ \varphi \circ \kappa) \circ \nu_\sigma^{-1} \\ &= (\beta(\sigma)^{-1} \otimes 1) \circ (1 \otimes \mu_\sigma) \circ {}^\sigma(\kappa^{-1} \circ \varphi \circ \kappa) \circ (1 \otimes \mu_\sigma^{-1}) \circ (\beta(\sigma) \otimes 1) \\ &= (\beta(\sigma)^{-1} \otimes 1) \circ (\kappa^{-1} \circ \varphi \circ \kappa) \circ (\beta(\sigma) \otimes 1).\end{aligned}$$

The last equality follows because for all $\psi \in \mathrm{End}(E \otimes_F B)$, we have $(1 \otimes \mu_\sigma) \circ {}^\sigma\psi = \psi \circ (1 \otimes \mu_\sigma)$. Indeed, write ψ as an element of the tensor product $\mathrm{End}_F E \otimes_F \mathrm{End}(B)$ and use the fact that each μ_σ was originally chosen to be compatible with $\mathrm{End}(B)$.

Finally, translating back to $\mathrm{End}(A_0)$, we see ${}^\sigma\varphi = \beta(\sigma)^{-1} \circ \varphi \circ \beta(\sigma)$. □

Proposition 4.4. *Under the isomorphism $\mathrm{End}(A_0) \xrightarrow{\sim} \mathrm{End}_F E \otimes_F \mathrm{End}(B)$ induced by the isogeny κ, we have $\mathrm{End}_{\mathbb{Q}}(A_0) \cong E \otimes_F \mathrm{End}(B)$.*

Proof. By applying the above proposition we find that the endomorphisms of A_0 which are fixed by all $\sigma \in G_{\mathbb{Q}}$ are exactly those which commute with all $\beta(\sigma)$. Thus $\mathrm{End}_{\mathbb{Q}}(A_0)$ is an F-subalgebra of $\mathrm{End}(A_0)$; in fact, $\mathrm{End}_{\mathbb{Q}}(A_0)$ is the centralizer $C(E)$ of E in $\mathrm{End}(A_0)$, since E was defined to be the field generated over F by the $\beta(\sigma)$.

Now it is clear that $C(E)$ is contained in $E \otimes_F \mathrm{End}(B)$, so it suffices to show that these subalgebras have the same dimension over F. Recall that we have written $d = [E : F]$ and $t = 1$ or $2 =$ Schur index of $\mathrm{End}(B)$. Then

$\dim_F E \cdot \dim_F(E \otimes_F \mathrm{End}(B)) = d(dt^2) = \dim_F \mathrm{End}(A_0)$. Therefore, by the Double Centralizer Theorem [Pi, Thm. 12.7], $\dim_F C(E) = \dim_F(E \otimes_F \mathrm{End}(B))$. □

Proposition 4.5. *There exists an abelian variety $A/\mathbb{Q}$ of $\mathbf{GL}_2$-type containing B as a simple component.*

Proof. We have seen by Theorem 2.1 that if we send the class of the cocycle $c(\sigma, \tau)$, $[c] \in \mathrm{H}^2(G_{\mathbb{Q}}, F^*) \to \mathrm{H}^2(G_{\mathbb{Q}}, (\overline{\mathbb{Q}} \otimes_{\mathbb{Q}} F)^*) \xrightarrow{\sim} \mathrm{Br}(F)$ via the natural maps, then its image in the Brauer group is the class of $\mathrm{End}(B)$. Now consider what happens upon extending to the field E. There is a commutative diagram:

$$\begin{array}{ccccc} \mathrm{H}^2(G_{\mathbb{Q}}, F^*) & \longrightarrow & \mathrm{H}^2(G_{\mathbb{Q}}, (\overline{\mathbb{Q}} \otimes_{\mathbb{Q}} F)^*) & \cong & \mathrm{Br}(F) \\ \downarrow & & \downarrow & & \downarrow \\ \mathrm{H}^2(G_{\mathbb{Q}}, E^*) & \longrightarrow & \mathrm{H}^2(G_{\mathbb{Q}}, (\overline{\mathbb{Q}} \otimes_{\mathbb{Q}} E)^*) & \cong & \mathrm{Br}(E) \end{array}$$

On the one hand, $[c] \in \mathrm{H}^2(G_{\mathbb{Q}}, F^*)$ maps to the class of $\mathrm{End}(B)$ in $\mathrm{Br}(F)$, which maps to the class of $E \otimes_F \mathrm{End}(B)$ in $\mathrm{Br}(E)$. But on the other hand, c splits when taking values in E; in fact, this is how E was originally defined. In other words, the image of $[c]$ is trivial in $\mathrm{H}^2(G_{\mathbb{Q}}, E^*)$, and hence maps to the trivial class in $\mathrm{Br}(E)$. Therefore $E \otimes_F \mathrm{End}(B)$ must be isomorphic to the matrix algebra $\mathbf{M}_t(E)$, where $t = 1$ or 2 is the Schur index of $\mathrm{End}(B)$.

Fix an isomorphism $E \otimes_F \mathrm{End}(B) \xrightarrow{\sim} \mathbf{M}_t(E)$. Then by the previous proposition, we have $\mathrm{End}_{\mathbb{Q}}(A_0) \xrightarrow{\sim} \mathbf{M}_t(E)$. This isomorphism implies that A_0 must be isogenous over $\mathbb{Q}$ to A^t, where A is an abelian variety over $\mathbb{Q}$ with $\mathrm{End}_{\mathbb{Q}}(A) = E$. (For instance, we may define A to be the image of the endomorphism of A_0 corresponding to the matrix that has a 1 in the upper left corner and zeroes elsewhere.)

Now remember that $E \otimes_F B$ is isogenous to a product of d copies of B, where $d = [E : F]$. It follows that $A^t_{/\overline{\mathbb{Q}}} \sim B^d$, and therefore $A_{/\overline{\mathbb{Q}}} \sim B^{d/t}$. Thus B is a simple component of A. We have already seen that $\mathrm{End}_{\mathbb{Q}}(A) = E$, so to complete the proof we need only calculate the dimension of A in order to show that A is of $\mathbf{GL}_2$-type. Recall that because B is a building block, we have $\dim B = t[F : \mathbb{Q}]$. Thus the dimension of A is $\frac{d}{t} \dim B = d[F : \mathbb{Q}] = [E : \mathbb{Q}]$, and we see that A is of $\mathbf{GL}_2$-type, as desired. □

A more precise correspondence

Notice that it was necessary to make several choices in our construction of an abelian variety $A/\mathbb{Q}$ of $\mathbf{GL}_2$-type from a building block B. We now ask if there is any way to make the correspondence between these two types of abelian varieties more precise. If we start with an $A/\mathbb{Q}$ of $\mathbf{GL}_2$-type and decompose it into building blocks, what information must we record in order to recover the "same" $A/\mathbb{Q}$ via the above procedure? The main idea is to keep track of the "descent data" – the ν_σ – which provide the isogeny from $B \times \cdots \times B$ (defined over $\overline{\mathbb{Q}}$) to A (defined over $\mathbb{Q}$).

First, however, we must say what it means for two abelian varieties of $\mathbf{GL}_2$-type to be the "same".

Definition. A $\mathbf{GL}_2$*-type pair* is a pair (A, i) where A is an abelian variety over $\mathbb{Q}$ of $\mathbf{GL}_2$-type and i is an injection $\mathrm{End}_{\mathbb{Q}}(A) \hookrightarrow \overline{\mathbb{Q}}$.

Definition. Two $\mathbf{GL}_2$-type pairs (A, i) and (A', i') are *equivalent* if there exists an isogeny $a : A \to A'$ that is defined over $\mathbb{Q}$ and satisfies $i'(a \circ \varphi \circ a^{-1}) = i(\varphi)$ for all $\varphi \in \mathrm{End}_{\mathbb{Q}}(A)$.

We make similar definitions for building blocks.

Definition. A *building block triple* is a triple $(B, j, \{\nu_\sigma\}_{\sigma \in G_{\mathbb{Q}}})$ where

- B is a building block, and we write $\mathcal{B} = \mathrm{End}(B)$;
- j is an injection of the center $Z(\mathrm{End}(B)) \hookrightarrow \overline{\mathbb{Q}}$, and we write F for the image of j;
- for each $\sigma \in G_{\mathbb{Q}}$, ν_σ is a non-zero element of $\overline{\mathbb{Q}} \otimes_F \mathrm{Hom}^{\mathcal{B}}({}^\sigma B, B)$ such that the family $\{\nu_\sigma\}$ is locally constant and satisfies $\nu_\sigma \circ {}^\sigma\nu_\tau = \nu_{\sigma\tau}$ for all $\sigma, \tau \in G_{\mathbb{Q}}$.

Definition. Two building block triples $(B, j, \{\nu_\sigma\})$ and $(B', j', \{\nu'_\sigma\})$ are *equivalent* if there exists an isogeny $b : B \to B'$ such that $j'(b \circ \varphi \circ b^{-1}) = j(\varphi)$ for all $\varphi \in Z(\mathrm{End}(B))$, and $\nu'_\sigma = b \circ \nu_\sigma \circ {}^\sigma b^{-1}$ for all $\sigma \in G_{\mathbb{Q}}$.

Suppose we are given a building block triple $(B, j, \{\nu_\sigma\})$. Note that for each σ, $\mathrm{Hom}^{\mathcal{B}}({}^\sigma B, B)$ is a one-dimensional F-vector space; the choice of an isogeny $\mu_\sigma : {}^\sigma B \to B$ respecting $\mathcal{B}$ then determines $\beta(\sigma) \in \overline{\mathbb{Q}}^*$ such that ν_σ may be written as $\beta(\sigma)^{-1} \otimes \mu_\sigma$. Let $E \subset \overline{\mathbb{Q}}$ be the field that is generated over F by the values $\beta(\sigma)$; then because each $\beta(\sigma)$ is well defined modulo F^*, we see that E depends only on the ν_σ, and not on our choices for μ_σ. The field E is a finite extension of F because the collection $\{\nu_\sigma\}$ is locally constant. Now we may view each ν_σ as an element of $\mathrm{Hom}(E \otimes_F {}^\sigma B,\ E \otimes_F B)$ and construct an abelian variety of $\mathbf{GL}_2$-type using the procedure described in the previous section.

We claim there is the following relationship between building block triples and $\mathbf{GL}_2$-type pairs.

Theorem 4.6. *The above construction, which associates a* $\mathbf{GL}_2$*-type pair* (A, i) *to a building block triple* $(B, j, \{\nu_\sigma\})$*, induces a bijection on equivalence classes.*

We break the proof into several steps.

Step 1. *The correspondence* $(B, j, \{\nu_\sigma\}) \rightsquigarrow (A, i)$ *is well defined on equivalence classes.*

Suppose the triple $(B, j, \{\nu_\sigma\})$ is equivalent to $(B', j', \{\nu'_\sigma\})$. Fix an isogeny $b : B \to B'$ satisfying the stated conditions for equivalence. Our first observation is that the image of j' is equal to the image of j; we will call their common image F and identify $Z(\mathcal{B})$ and $Z(\mathcal{B}')$ with F. Also, the isogeny b induces an isomorphism $\mathrm{End}(B) \cong \mathrm{End}(B')$, and in particular, we note that $\mathrm{End}(B)$ and $\mathrm{End}(B')$ have the same Schur index $t = 1$ or 2 as central simple F-algebras.

Next, for each $\sigma \in G_{\mathbb{Q}}$, pick $\mu_\sigma \in \mathrm{Hom}^{\mathcal{B}}({}^\sigma B, B)$ and write $\nu_\sigma = \beta(\sigma)^{-1} \otimes \mu_\sigma$. Then $\nu'_\sigma = b \circ \nu_\sigma \circ {}^\sigma b^{-1}$ implies that we may write $\nu'_\sigma = \beta(\sigma)^{-1} \otimes (b \circ \mu_\sigma \circ {}^\sigma b^{-1})$. Thus the subfield of $\overline{\mathbb{Q}}$ generated over F by the coefficients of the ν'_σ is the same as the one generated by the coefficients of the ν_σ. Call this field E. Now our construction yields isogenies $\kappa : E \otimes_F B \to A^t$ and $\kappa' : E \otimes_F B' \to A'^t$ where A and A' are abelian varieties over $\mathbb{Q}$ of $\mathbf{GL}_2$-type with $E \xrightarrow{\sim} \mathrm{End}_{\mathbb{Q}}(A)$ and $E \xrightarrow{\sim} \mathrm{End}_{\mathbb{Q}}(A')$. These actions of E on A and A' define the pairs (A, i) and (A', i'), respectively.

Now $a = \kappa' \circ (1 \otimes b) \circ \kappa^{-1}$ is an isogeny $A^t \to A'^t$. We claim it is an isogeny defined over $\mathbb{Q}$. Indeed, for any $\sigma \in G_{\mathbb{Q}}$ we have:

$$\begin{aligned} {}^\sigma(\kappa' \circ (1 \otimes b) \circ \kappa^{-1}) &= (\kappa' \circ \kappa'^{-1}) \circ {}^\sigma\kappa' \circ (1 \otimes {}^\sigma b) \circ {}^\sigma\kappa^{-1} \circ (\kappa \circ \kappa^{-1}) \\ &= \kappa' \circ \nu'_\sigma \circ (1 \otimes {}^\sigma b) \circ \nu_\sigma^{-1} \circ \kappa^{-1} \\ &= \kappa' \circ (1 \otimes b) \circ \nu_\sigma \circ \nu_\sigma^{-1} \circ \kappa^{-1} \\ &= \kappa' \circ (1 \otimes b) \circ \kappa^{-1}. \end{aligned}$$

Moreover, κ, $1 \otimes b$ and κ' all respect the action of $E \otimes_F \mathcal{B}$. (Here we are using the isomorphism $\mathcal{B}' \cong \mathcal{B}$ induced by b to define the action of $E \otimes_F \mathcal{B}$ on $E \otimes_F B'$ and A'^t.) Therefore, a respects the action of $E \otimes_F \mathcal{B} \cong \mathbf{M}_t(E)$. But it is easy to see that the natural map $\mathrm{Hom}^E(A, A') \to \mathrm{Hom}^{\mathbf{M}_t(E)}(A^t, A'^t)$ is an isomorphism, so in fact we may regard a as a morphism $A \to A'$. Then a is an isogeny defined over $\mathbb{Q}$ that respects E; that is, a defines an equivalence of pairs (A, i) and (A', i'), as desired.

Step 2. *Define the inverse correspondence from pairs (A, i) to triples $(B, \jmath, \{\nu_\sigma\})$.*

Let (A, i) be a $\mathbf{GL}_2$-type pair. Write $\mathcal{A}$ for the full endomorphism algebra $\mathrm{End}(A)$, E for the image of i, and $F = i(Z(\mathcal{A}))$. By our usual abuse of notation, we will use i to identify E and F with the corresponding subalgebras of $\mathcal{A}$. Recall that $\mathcal{A}$ is a central simple algebra over F, by Theorem 1.2. Thus $\mathcal{A}$ has a unique (up to isomorphism) simple right $\mathcal{A}$-module. Fix such a simple module M.

Let $B = M \otimes_{\mathcal{A}} A$. We claim that B is a building block. By Propositions 1.3 and 1.4, it suffices to show that B is a simple component of A. Now because $\mathcal{A}$ is a simple algebra, for some n we have $\mathcal{A} \cong \bigoplus^n M$ as right $\mathcal{A}$-modules. Fix such an isomorphism. Then

$$A \;=\; \mathcal{A} \otimes_{\mathcal{A}} A \;\sim\; (\textstyle\bigoplus^n M) \otimes_{\mathcal{A}} A \;\sim\; \prod^n (M \otimes_{\mathcal{A}} A) \;\sim\; \prod^n B.$$

Thus B is a subvariety of A. It remains to show that B is simple.

By [Pi, Cor. 3.5a], $\mathcal{A}$ is isomorphic to $\mathbf{M}_n(\mathcal{B})$, where $\mathcal{B} \stackrel{\mathrm{def}}{=} \mathrm{End}_{\mathcal{A}} M$ is a central division algebra over F. But the above isogeny $A \sim \prod^n B$ implies that $\mathcal{A} \cong \mathbf{M}_n(\mathrm{End}(B))$. Therefore we must have that $\mathrm{End}(B) \cong \mathcal{B}$ is a division algebra, and so B is simple.

We remark that in fact we have an explicit isomorphism $\mathcal{B} \to \mathrm{End}(B)$, namely the one arising from the general map (3.4):

$$\begin{array}{ccc} \mathrm{End}_{\mathcal{A}} M \otimes_F \mathrm{End}^{\mathcal{A}}(A) & \longrightarrow & \mathrm{End}(B) \\ \mathcal{B} \otimes_F F & \xrightarrow{\sim} & \mathrm{End}(B). \end{array} \tag{4.7}$$

This map induces an isomorphism $F \xrightarrow{\sim} Z(\mathrm{End}(B))$, whose inverse provides us with the definition of j in the building block triple we will associate with B.

For future reference, we write t for the common Schur index of $\mathrm{End}(A)$ and $\mathrm{End}(B)$ (remember $t = 1$ or 2), and note that $[E:F] = nt$, because E is a maximal subfield of $\mathcal{A}$ by Theorem 1.2.

Now we need to define ν_σ for each $\sigma \in G_{\mathbb{Q}}$. We will do so by constructing an isogeny $\kappa : E \otimes_F B \to A^t$. Because $B = M \otimes_{\mathcal{A}} A$, we will view $E \otimes_F B$ as $(E \otimes_F M) \otimes_{\mathcal{A}} A$.

Note that $E \otimes_F M$ is an $(E \otimes_F \mathcal{B})$-$\mathcal{A}$-bimodule. Furthermore, we know that E splits $\mathcal{B}$ (because E is a maximal subfield of $\mathcal{A}$), so we may fix an isomorphism

$$E \otimes_F \mathcal{B} \xrightarrow{\sim} \mathbf{M}_t(E) \cong \mathrm{End}_{\mathbb{Q}}(A^t).$$

In this way we obtain an action of $E \otimes_F \mathcal{B}$ on A^t, and so, a left action of $E \otimes_F \mathcal{B}$ on $\mathrm{Hom}(A, A^t)$. Now choose an isomorphism of $(E \otimes_F \mathcal{B})$-$\mathcal{A}$-bimodules $E \otimes_F M \to \mathrm{Hom}(A, A^t)$. (They are abstractly isomorphic because they have the same dimension – n^2t^3 – over F [Pi, Cor. 3.3b].) Then by the defining property of the tensor product (3.2), there is a morphism $\kappa : (E \otimes_F M) \otimes_{\mathcal{A}} A \to A^t$ that respects the actions of $E \otimes_F \mathcal{B}$ (Proposition 3.5).

We claim this morphism is an isogeny. Indeed, the image of κ is a subvariety of A^t with an action of $E \otimes_F \mathcal{B} \cong \mathbf{M}_t(E)$. But the only subvarieties of A^t mapped into themselves by $\mathbf{M}_t(E)$ are the zero subvariety and all of A^t. Now κ cannot be the zero morphism, because the homomorphism of modules used to define κ was non-zero. Therefore the image of κ must be A^t. Now we note that as right $\mathcal{A}$-modules, $E \otimes_F M \cong \bigoplus^{nt} M \cong \bigoplus^{t} \mathcal{A}$, so the dimension of $(E \otimes_F M) \otimes_{\mathcal{A}} A$ is equal to the dimension of A^t. Thus κ is an isogeny.

We are now able to define the ν_σ. Consider $\kappa^{-1} \circ {}^\sigma\kappa : E \otimes_F {}^\sigma B \to E \otimes_F B$, which is an isogeny respecting $E \otimes_F \mathcal{B}$. The fact that

$$\mathrm{Hom}^{E\otimes\mathcal{B}}(E \otimes_F {}^\sigma B,\ E \otimes_F B) \cong E \otimes_F \mathrm{Hom}^{\mathcal{B}}({}^\sigma B, B)$$

follows from $\mathrm{End}^{E\otimes\mathcal{B}}(E \otimes_F B) \cong E$ (which we know from the proof of Proposition 4.4). Hence we may write $\nu_\sigma = \kappa^{-1} \circ {}^\sigma\kappa \in E \otimes_F \mathrm{Hom}^{\mathcal{B}}({}^\sigma B, B)$. Plainly, $\nu_\sigma \circ {}^\sigma\nu_\tau = \nu_{\sigma\tau}$ as desired.

Lastly we must check that the ν_σ do not depend on the various module isomorphisms chosen above. Suppose we make different choices and construct $\kappa' : E \otimes_F B \to A^t$ respecting $E \otimes_F \mathcal{B}$. Then $\kappa' \circ \kappa^{-1} : A^t \to A^t$ is an isogeny respecting $E \otimes_F \mathcal{B} \cong \mathbf{M}_t(E)$. But the only endomorphisms of A^t that respect $\mathbf{M}_t(E)$ lie in E, and are therefore defined over $\mathbb{Q}$. Hence ${}^\sigma(\kappa' \circ \kappa^{-1}) = \kappa' \circ \kappa^{-1}$ for all $\sigma \in G_{\mathbb{Q}}$, and we see $\nu'_\sigma = \kappa'^{-1} \circ {}^\sigma\kappa' = \kappa^{-1} \circ {}^\sigma\kappa = \nu_\sigma$ is independent of our choices.

Thus to the $\mathbf{GL}_2$-type pair (A, i) we associate the triple $(B, j, \{\nu_\sigma\})$.

Step 3. *The correspondence defined in Step 2 is well defined on equivalence classes.*

Suppose (A', i') is a pair that is equivalent to (A, i). Write $\mathcal{A} = \mathrm{End}(A)$ and $\mathcal{A}' = \mathrm{End}(A')$. Let $a : A \to A'$ be an isogeny which satisfies the stated conditions for the equivalence of pairs. Then a induces an isomorphism $\mathcal{a} \xrightarrow{\sim} \mathcal{a}'$ via $\varphi \mapsto a \circ \varphi \circ a^{-1}$.

First notice that the image of i' is the same as the image of i; we will call their common image E. Also $i'(Z(\mathcal{a}')) = i(Z(\mathcal{a}))$, which we will call F. Henceforth we will identify E and F with these subalgebras of the endomorphism algebras.

Now let M be a simple right $\mathcal{a}$-module, and M' a simple right $\mathcal{a}'$-module. Form $B = M \otimes_{\mathcal{a}} A$ and $B' = M' \otimes_{\mathcal{a}'} A'$, and the accompanying data j, j', ν_σ, and ν'_σ as above. Now we use the isomorphism $\mathcal{a} \xrightarrow{\sim} \mathcal{a}'$ to view A' as an abelian variety with an action of $\mathcal{a}$, and M' as an $\mathcal{a}$-module. Then we may write $B' = M' \otimes_{\mathcal{a}} A'$.

Because $\mathcal{a}$ has a unique simple right module, there must be an isomorphism of $\mathcal{a}$-modules $f : M \to M'$. Define $b : B \to B'$ to be the morphism $f \otimes a : M \otimes_{\mathcal{a}} A \to M' \otimes_{\mathcal{a}} A'$. Then b is an isogeny because both f and a are invertible, and b respects the action of F, because F commutes with everything in sight.

It remains to show that $\nu'_\sigma = b \circ \nu_\sigma \circ {}^\sigma b^{-1}$ for each $\sigma \in G_{\mathbb{Q}}$. Remember that ν_σ is defined as $\kappa^{-1} \circ {}^\sigma\kappa$ where κ is an isogeny $E \otimes_F B \to A^t$ respecting $E \otimes_F \mathcal{B}$, and similarly ν'_σ is defined in terms of $\kappa' : E \otimes_F B' \to A'^t$. Now notice that $a^{-1} \circ \kappa' \circ (1 \otimes b) \circ \kappa^{-1}$ is an endomorphism of A^t respecting $E \otimes_F \mathcal{B} \cong \mathbf{M}_t(E)$. But the only such endomorphisms are elements of E, which are endomorphisms defined over $\mathbb{Q}$. Therefore $\kappa' \circ (1 \otimes b) \circ \kappa^{-1}$ is defined over $\mathbb{Q}$, so ${}^\sigma(\kappa' \circ (1 \otimes b) \circ \kappa^{-1}) = \kappa' \circ (1 \otimes b) \circ \kappa^{-1}$ for all $\sigma \in G_{\mathbb{Q}}$, and it follows that $\nu'_\sigma = b \circ \nu_\sigma \circ {}^\sigma b^{-1}$.

Step 4. *The above correspondences are inverse to each other.*

Suppose we start with a $\mathbf{GL}_2$-type pair (A, i), and use it to construct a building block triple $(B, j, \{\nu_\sigma\})$. Then from the map (4.7) we have $Z(\mathrm{End}(A)) \xrightarrow{\sim} Z(\mathrm{End}(B))$, so the field F is the same from both viewpoints. Next, we need to show the field E = image of i is the same as the field E generated over F by the coefficients of the ν_σ. Write $\nu_\sigma = \beta(\sigma)^{-1} \otimes \mu_\sigma$, and remember these ν_σ are defined by an isogeny $\kappa : E \otimes_F B \to A^t$, where $t = 1$ or 2. Then by Proposition 4.3 we find that for $\varphi \in \mathrm{End}(A^t)$ and $\sigma \in G_{\mathbb{Q}}$, we have ${}^\sigma\varphi = \beta(\sigma)^{-1} \circ \varphi \circ \beta(\sigma)$. Hence the same relationship holds for $\varphi \in \mathrm{End}(A)$. In other words, these $\beta(\sigma)^{-1}$ are the same (up to F^*) as the $\alpha(\sigma)$ appearing in Chapter 1, and we have seen there (Proposition 1.5) that these elements do indeed generate $\mathrm{End}_{\mathbb{Q}}(A)$ over F.

Now clearly when we construct $E \otimes_F B$ and descend it to $\mathbb{Q}$ via the map κ defined by the ν_σ, we are just recovering the map $\kappa : E \otimes_F (M \otimes_{\mathcal{a}} A) \to A^t$ we used to define the ν_σ to begin with. Therefore the map from pairs to triples and back to pairs is the identity on equivalence classes.

On the other hand, what happens when we start with a triple $(B, j, \{\nu_\sigma\})$ and construct a pair (A, i)? From (4.1) we have $Z(\mathrm{End}(B)) \xrightarrow{\sim} Z(\mathrm{End}(A^t)) \cong Z(\mathrm{End}(A))$, and so again we may write F with no confusion. Now we want to show that the building block triple constructed from (A, i) is equivalent to the triple $(B, j, \{\nu_\sigma\})$.

Let M be a simple right $\mathcal{a}$-module; we may view it as an F-$\mathcal{a}$-bimodule. Choose an isomorphism of F-$\mathcal{a}$-bimodules $M \to \mathrm{Hom}(A, B)$; they are isomorphic because they have the same F-dimension. Then by the defining property of the tensor product (3.2), there is a morphism $M \otimes_a A \to B$ respecting F. This morphism is an isogeny because it is non-zero by construction and because $M \otimes_a A$ and B are simple abelian varieties. Call the inverse of this map b.

Finally, we have to show that the ν'_σ defined by (A, i) above are equal to $b \circ \nu_\sigma \circ {}^\sigma b^{-1}$ for the original ν_σ associated with B. Let $\kappa : E \otimes_F B \to A^t$ be defined by the original ν_σ. Then $\kappa' = \kappa \circ (1 \otimes b)^{-1}$ is an isogeny $E \otimes_F (M \otimes_a A) \to A^t$ respecting $E \otimes_F \mathrm{End}(B)$, where the action of $\mathrm{End}(B)$ on $M \otimes_a A$ is induced by b. We have already seen that any such isogeny is valid for defining the ν'_σ, and so $\nu'_\sigma = \kappa'^{-1} \circ {}^\sigma\kappa' = (1 \otimes b) \circ \kappa^{-1} \circ {}^\sigma\kappa \circ (1 \otimes {}^\sigma b^{-1}) = b \circ \nu_\sigma \circ {}^\sigma b^{-1}$.

This proves Theorem 4.6. □

Summary

Let $(B, j, \{\nu_\sigma\})$ be a building block triple; we will write $\nu_\sigma = \beta(\sigma)^{-1} \otimes \mu_\sigma$ with the understanding that the following formulas are independent of the transcriptions chosen. Let (A, i) be a $\mathbf{GL}_2$-type pair corresponding to $(B, j, \{\nu_\sigma\})$ under Theorem 4.6. The chart below summarizes how to define, from either perspective, the most frequently used constructs associated to this situation.

Construct	**$\mathbf{GL}_2$-type pair** (A, i)	**Building block triple** $(B, j, \{\nu_\sigma\}_{\sigma \in G_\mathbb{Q}})$
number field F	$i(Z(\mathrm{End}(A)))$	$j(Z(\mathrm{End}(B)))$
number field E	$i(\mathrm{End}_\mathbb{Q}(A))$	extension of F gen. by $\{\beta(\sigma) : \sigma \in G_\mathbb{Q}\}$
$\alpha : G_\mathbb{Q} \to E^*$ (well-defined mod F^*)	${}^\sigma\varphi = \alpha(\sigma) \circ \varphi \circ \alpha(\sigma)^{-1}$ for all $\varphi \in \mathrm{End}(A)$	β^{-1}
$[c] \in \mathrm{H}^2(G_\mathbb{Q}, F^*)$	class of $c(\sigma, \tau) =$ $\alpha(\sigma\tau)\,\alpha(\sigma)^{-1}\alpha(\tau)^{-1}$	class of $c(\sigma, \tau) =$ $\mu_\sigma \circ {}^\sigma\mu_\tau \circ \mu_{\sigma\tau}^{-1}$

5. Fields of definition for building blocks

In this chapter we generalize results of Shimura and Ribet concerning various properties of building blocks, in particular, regarding their fields of definition. Recall that a (finite) compositum of quadratic extensions of a field is said to be a $(2, \ldots, 2)$-extension, and that such an extension is Galois with group $(\mathbb{Z}/2\mathbb{Z})^r$ for some r.

Theorem 5.1. *Let $B/\overline{\mathbb{Q}}$ be a building block. Then B is isogenous over $\overline{\mathbb{Q}}$ to an abelian variety defined over a $(2, \ldots, 2)$-extension of $\mathbb{Q}$.*

Elkies first proved this in the case when B is an elliptic curve [El]. Ribet generalized the result to the case of a building block with $\mathrm{End}(B)$ a totally real field [Ri 2, Thm. 1.2]. In fact, similar arguments apply to the general case, and we follow his approach as we outline the proof below.

Remember that to a building block B we have associated a well-defined cohomology class $[c] \in \mathrm{H}^2(G_{\mathbb{Q}}, F^*)$ (where $G_{\mathbb{Q}}$ acts trivially on F^*) by choosing isogenies $\mu_\sigma : {}^\sigma B \to B$ which respect $\mathrm{End}(B)$ and defining a two-cocycle $c(\sigma,\tau) = \mu_\sigma \circ {}^\sigma\mu_\tau \circ \mu_{\sigma\tau}^{-1}$. We will see that the class $[c]$ provides certain information about a field of definition for B.

Definition. Let K be a number field. A *building block over* K is an abelian variety B over K such that $B_{/\overline{\mathbb{Q}}}$ is a building block and all endomorphisms of B are defined over K.

Proposition 5.2. *Let K be a number field. Let B be a building block and $[c]$ its associated cohomology class. Then B is isogenous over $\overline{\mathbb{Q}}$ to a building block B_0 over K if and only if $[c]$ lies in the kernel of the restriction map* $\mathrm{H}^2(G_{\mathbb{Q}}, F^*) \to \mathrm{H}^2(\mathrm{Gal}(\overline{\mathbb{Q}}/K), F^*)$.

Proof. Suppose B_0 is an abelian variety over K that is isogenous to B. Fix an isogeny $\kappa : B \to B_0$. Then for each $\sigma \in \mathrm{Gal}(\overline{\mathbb{Q}}/K)$, define $\mu_\sigma = \kappa^{-1} \circ {}^\sigma\kappa : {}^\sigma B \to B$; these isogenies respect $\mathrm{End}(B)$ because by hypothesis all endomorphisms of B_0 are defined over K. Clearly now $c(\sigma,\tau) = 1$ for all $\sigma,\tau \in \mathrm{Gal}(\overline{\mathbb{Q}}/K)$, and so $[c] = 1 \in \mathrm{H}^2(\mathrm{Gal}(\overline{\mathbb{Q}}/K), F^*)$.

Conversely, suppose $[c]$ is trivial in $\mathrm{H}^2(\mathrm{Gal}(\overline{\mathbb{Q}}/K), F^*)$. Choose a collection $\{\mu_\sigma\}$ as usual to define the two-cocycle $c(\sigma,\tau)$. Then for $\sigma,\tau \in \mathrm{Gal}(\overline{\mathbb{Q}}/K)$, we may write $c(\sigma,\tau)$ as a coboundary; i.e., $c(\sigma,\tau) = \beta(\sigma)\cdot\beta(\tau)\cdot\beta(\sigma\tau)^{-1}$ where $\beta : \mathrm{Gal}(\overline{\mathbb{Q}}/K) \to F^*$ is a locally constant function. Now replace each μ_σ with $\beta(\sigma)^{-1}\circ\mu_\sigma$, which is again an isogeny ${}^\sigma B \to B$ respecting $\mathrm{End}(B)$. The new family $\{\mu_\sigma\}$ satisfies $\mu_\sigma \circ {}^\sigma\mu_\tau = \mu_{\sigma\tau}$ for all $\sigma,\tau \in \mathrm{Gal}(\overline{\mathbb{Q}}/K)$. Therefore we may apply [Ri 1, Thm. 8.2] to construct B_0/K and an isogeny $\kappa : B \to B_{0/\overline{\mathbb{Q}}}$.

Finally, we need to check that all endomorphisms of B_0 are defined over K, but this follows easily from the facts that the μ_σ are compatible with $\mathrm{End}(B)$ and that $\mu_\sigma = \kappa^{-1}\circ{}^\sigma\kappa$ by construction of κ. Indeed, for $\varphi \in \mathrm{End}(B_0)$, we have $\kappa^{-1}\circ\varphi\circ\kappa \in \mathrm{End}(B)$, so $\mu_\sigma \circ {}^\sigma(\kappa^{-1}\circ\varphi\circ\kappa) = (\kappa^{-1}\circ\varphi\circ\kappa)\circ\mu_\sigma$ for all $\sigma \in \mathrm{Gal}(\overline{\mathbb{Q}}/K)$. We may rewrite this as $\kappa^{-1}\circ{}^\sigma\varphi\circ{}^\sigma\kappa = \kappa^{-1}\circ\varphi\circ{}^\sigma\kappa$, and therefore ${}^\sigma\varphi = \varphi$. Since this holds for all $\sigma \in \mathrm{Gal}(\overline{\mathbb{Q}}/K)$, we see that $\varphi \in \mathrm{End}(B_0)$ is defined over K. □

Note that $[c]$ has order dividing two in $\mathrm{H}^2(G_{\mathbb{Q}}, F^*)$. Indeed, we know from Chapter 4 that there exists an abelian variety A of $\mathbf{GL}_2$-type containing the building block B as a simple component, and that the class of the cocycle $c(\sigma,\tau)$ associated to B is exactly that of the cocycle we associated to A in Chapter 1. We saw there (Lemma 1.8) that the class of this cocycle has order one or two.

Write $\mathrm{H}^2(G_{\mathbb{Q}}, F^*)[2]$ for the group of elements of the cohomology group $\mathrm{H}^2(G_{\mathbb{Q}}, F^*)$ of order dividing two.

Proposition 5.3. *Let F be a totally real field and define $P = F^*/\{\pm 1\}$. Then $\mathrm{H}^2(G_{\mathbb{Q}}, F^*)[2]$ is abstractly isomorphic to $\mathrm{H}^2(G_{\mathbb{Q}}, \{\pm 1\}) \times \mathrm{Hom}(G_{\mathbb{Q}}, P/P^2)$.*

Proof. (Ribet) We follow the procedure described in [Ri 2, Thm. 3.3].

Consider the exact sequence of trivial $G_{\mathbb{Q}}$-modules

$$1 \to \{\pm 1\} \to F^* \to P \to 1.$$

This sequence is split because P is a free group [Ri 2, Lemma 3.5]. Thus we may write $F^* \cong \{\pm 1\} \times P$. We note that this isomorphism is in general non-canonical (although when $[F:\mathbb{Q}]$ is odd, we may embed $P \hookrightarrow F^*$ as the group of elements of F^* of positive norm). In any case, there is an isomorphism

$$\mathrm{H}^2(G_{\mathbb{Q}}, F^*)[2] \cong \mathrm{H}^2(G_{\mathbb{Q}}, \{\pm 1\})[2] \times \mathrm{H}^2(G_{\mathbb{Q}}, P)[2].$$

Clearly all elements of $\mathrm{H}^2(G_{\mathbb{Q}}, \{\pm 1\})$ are of order dividing two, so $\mathrm{H}^2(G_{\mathbb{Q}}, \{\pm 1\})[2] = \mathrm{H}^2(G_{\mathbb{Q}}, \{\pm 1\})$.

We claim that there is an isomorphism $\mathrm{Hom}(G_{\mathbb{Q}}, P/P^2) \xrightarrow{\sim} \mathrm{H}^2(G_{\mathbb{Q}}, P)[2]$. Indeed, note that the squaring map $x \mapsto x^2$ is an injective map $P \to P$, because P is torsion-free. Now look at the exact sequence

$$1 \longrightarrow P \xrightarrow{\text{sq.}} P \longrightarrow P/P^2 \longrightarrow 1.$$

The corresponding long exact cohomology sequence is

$$\cdots \longrightarrow \mathrm{H}^1(G_{\mathbb{Q}}, P) \longrightarrow \mathrm{H}^1(G_{\mathbb{Q}}, P/P^2) \longrightarrow \mathrm{H}^2(G_{\mathbb{Q}}, P) \xrightarrow{\text{sq.}} \mathrm{H}^2(G_{\mathbb{Q}}, P).$$

Now $G_{\mathbb{Q}}$ is acting trivially, so we may regard each H^1 as a Hom. Furthermore, P is torsion-free, so $\mathrm{H}^1(G_{\mathbb{Q}}, P) = \mathrm{Hom}(G_{\mathbb{Q}}, P) = 1$. Also, the image of $\mathrm{H}^1(G_{\mathbb{Q}}, P/P^2)$ in $\mathrm{H}^2(G_{\mathbb{Q}}, P)$ is the kernel of the squaring map $\mathrm{H}^2(G_{\mathbb{Q}}, P) \to \mathrm{H}^2(G_{\mathbb{Q}}, P)$, which consists exactly of the subgroup of elements of order dividing two. Therefore we may rewrite the cohomology sequence as

$$1 \to \mathrm{Hom}(G_{\mathbb{Q}}, P/P^2) \twoheadrightarrow \mathrm{H}^2(G_{\mathbb{Q}}, P)[2].$$

Hence $\mathrm{Hom}(G_{\mathbb{Q}}, P/P^2)$ is isomorphic to $\mathrm{H}^2(G_{\mathbb{Q}}, P)[2]$, proving the proposition. □

We are now ready to prove Theorem 5.1.

Proof. By the previous propositions, it suffices to show that the images of the cohomology class $[c]$ in $\mathrm{H}^2(G_{\mathbb{Q}}, \{\pm 1\})$ and $\mathrm{Hom}(G_{\mathbb{Q}}, P/P^2)$ under the isomorphism of Proposition 5.3 are trivialized by restriction to $\mathrm{Gal}(\overline{\mathbb{Q}}/K)$ where K is a $(2, \ldots, 2)$-extension of $\mathbb{Q}$.

Recall that $\mathrm{H}^2(G_{\mathbb{Q}}, \{\pm 1\})$ may be identified with $\mathrm{Br}(\mathbb{Q})[2] = \{$Brauer classes of quaternion algebras over $\mathbb{Q}\}$. A quaternion algebra over $\mathbb{Q}$ may always be split by a quadratic extension; therefore the image of $[c]$ in $\mathrm{H}^2(G_{\mathbb{Q}}, \{\pm 1\})$ is trivial under restriction to some quadratic extension $K_{\pm}$ of $\mathbb{Q}$.

The other "piece" of $[c]$ lies in $\mathrm{Hom}(G_{\mathbb{Q}}, P/P^2)$. Since the image of any continuous homomorphism $G_{\mathbb{Q}} \to P/P^2$ is a finite elementary 2-group, the kernel of the homomorphism corresponding to the image of $[c]$ is $\mathrm{Gal}(\overline{\mathbb{Q}}/K_P)$, where K_P is a $(2,\ldots,2)$-extension of $\mathbb{Q}$.

Now let K be the composite of the fields $K_\pm$ and K_P. Then K is a $(2,\ldots,2)$-extension such that $[c]$ is in the kernel of the restriction map $\mathrm{H}^2(G_{\mathbb{Q}},\ F^*) \to \mathrm{H}^2(\mathrm{Gal}(\overline{\mathbb{Q}}/K),\ F^*)$. □

Elkies's proof of the theorem for elliptic curves shows that the field K_P already trivializes $[c]$ in that case. Ribet's method demonstrates this interesting fact to be true whenever $[F:\mathbb{Q}]$ is odd [Ri 2, Cor. 4.5]; it is unknown, however, whether it is true in general. In any case, the field K_P is a significant part of the $(2,\ldots,2)$-extension K, and in order to determine K_P, we will now discuss an explicit description of the image of $[c]$ in $\mathrm{Hom}(G_{\mathbb{Q}},\ P/P^2)$.

The "degree" of μ_σ

Let B_1 and B_2 be isogenous building blocks, and fix polarizations $\theta_1 : B_1 \to \widehat{B}_1$ and $\theta_2 : B_2 \to \widehat{B}_2$. We will say an isogeny $\mu : B_1 \to B_2$ is *Rosati compatible* if the isomorphism $\mathrm{End}(B_1) \cong \mathrm{End}(B_2)$ induced by μ is compatible with the Rosati involutions defined by θ_1 and θ_2. For such an isogeny, we define the "degree" $\delta(\mu) = \mu \circ \theta_1^{-1} \circ \hat{\mu} \circ \theta_2 \in Z(\mathrm{End}(B_2))$. Note that $\delta(\mu)$ depends on the chosen polarizations θ_1, θ_2 (much as the Rosati involution of an endomorphism algebra depends on a choice of polarization).

The following properties of δ are elementary. If $\varphi : B_2 \to B_2$, then $\delta(\varphi) = \varphi \circ \varphi'$, where φ' is the Rosati involution of φ with respect to the polarization θ_2. Also, δ is multiplicative in the following sense: if $\lambda : B_2 \to B_3$ is another isogeny that is Rosati compatible, then $\delta(\lambda \circ \mu) = (\lambda \circ \delta(\mu) \circ \lambda^{-1}) \circ \delta(\lambda)$ (assuming we use a fixed set of polarizations to calculate the degrees). In particular, if each building block is equipped with an action $F \xrightarrow{\sim} Z(\mathrm{End}(B_i))$, and λ and μ respect F, then we may write $\delta(\lambda \circ \mu) = \delta(\lambda) \cdot \delta(\mu)$ as elements of F.

Proposition 5.4. *Let B_1, B_2 be building blocks with polarizations θ_1, θ_2, respectively, and let $\mu : B_1 \to B_2$ be an isogeny that is Rosati compatible. Then the degree $\delta(\mu)$ is a totally positive element of the totally real field F.*

Proof. We will regard the isogeny $\mu : B_1 \to B_2$ as an element of $\mathrm{End}(B_1 \times B_2)$, and use the properties of the Rosati involution on this endomorphism algebra to prove the proposition.

First define an isomorphism

$$\begin{array}{ccc} \mathbf{M}_2(\mathrm{End}(B_2)) & \longrightarrow & \mathrm{End}(B_1 \times B_2) \\ \begin{pmatrix} \varphi_{11} & \varphi_{12} \\ \varphi_{21} & \varphi_{22} \end{pmatrix} & \longmapsto & \begin{pmatrix} \mu^{-1} \circ \varphi_{11} \circ \mu & \mu^{-1} \circ \varphi_{12} \\ \varphi_{21} \circ \mu & \varphi_{22} \end{pmatrix}. \end{array}$$

Now the polarizations θ_1, θ_2 define a polarization $\theta_1 \times \theta_2$ on $\mathrm{End}(B_1 \times B_2)$ which, in turn, induces a Rosati involution on $\mathrm{End}(B_1 \times B_2)$. Pulling back this involution to $\mathbf{M}_2(\mathrm{End}(B_2))$, we find it is given by the formula

$$\Phi = \begin{pmatrix} \varphi_{11} & \varphi_{12} \\ \varphi_{21} & \varphi_{22} \end{pmatrix} \longmapsto \begin{pmatrix} \varphi_{11}' & \delta(\mu)\circ\varphi_{21}' \\ \varphi_{12}'\circ\delta(\mu)^{-1} & \varphi_{22}' \end{pmatrix} = \Phi'.$$

where $\varphi \mapsto \varphi'$ is the Rosati involution on $\mathrm{End}(B_2)$ coming from θ_2.

Now the Rosati involution is a positive involution; that is, for a non-zero endomorphism Φ, the trace $\mathrm{Tr}(\Phi\Phi') \in \mathbb{Q}$ is positive [Mu, §21 Thm. 1]. In particular, consider $\Phi \in \mathrm{End}(B_1 \times B_2)$ of the form $\left(\begin{smallmatrix} 0 & 0 \\ \varphi & 0 \end{smallmatrix}\right)$. Then

$$\Phi' = \begin{pmatrix} 0 & \delta(\mu)\circ\varphi' \\ 0 & 0 \end{pmatrix} \quad \text{and} \quad \Phi\Phi' = \begin{pmatrix} 0 & 0 \\ 0 & \delta(\mu)\circ\varphi\circ\varphi' \end{pmatrix}.$$

Therefore $\mathrm{Tr}(\Phi\Phi') = \mathrm{Tr}(\delta(\mu)\circ\varphi\circ\varphi') > 0$ for all $\varphi \in \mathrm{End}(B_2)$.

We now claim the fact that this inequality holds for all φ implies that $\delta(\mu)$ must be totally positive. Indeed, let I be the finite set of embeddings $\{\iota : F \hookrightarrow \mathbb{R}\}$. Suppose some embedding, call it ι_o, is such that $\iota_o(\delta(\mu)) < 0$. Now by the Strong Approximation Theorem, we may choose $\varphi \in F$ such that $|\iota_o(\varphi))|^2$ is large, say greater than $\#I/|\iota_o(\delta(\mu))|$, but $|\iota(\varphi)|^2$ is small, say less than $1/\max_{\iota\neq\iota_0}\{|\iota(\delta(\mu))|\}$ for all $\iota \neq \iota_o$.

Now by [Mu, p. 182], $\mathrm{Tr}(\delta(\mu)\circ\varphi\circ\varphi')$ is a positive multiple of $\mathrm{Tr}_{F/\mathbb{Q}}(\delta(\mu)\cdot\varphi^2)$, which we may calculate as

$$\begin{aligned} \mathrm{Tr}_{F/\mathbb{Q}}(\delta(\mu)\cdot\varphi^2) &= \textstyle\sum_{\iota\in I}\iota(\delta(\mu))\cdot\iota(\varphi)^2 \\ &= \iota_o(\delta(\mu))\cdot\iota_o(\varphi)^2 \;+\; \textstyle\sum_{\iota\neq\iota_0}\iota(\delta(\mu))\cdot\iota(\varphi)^2 \\ &< -\#I \;+\; \textstyle\sum_{\iota\neq\iota_0} 1 \\ &< 0 \end{aligned}$$

because we chose φ carefully.

This contradicts the fact that $\mathrm{Tr}(\delta(\mu)\circ\varphi\circ\varphi')$ is positive for all φ. Therefore $\iota(\delta(\mu))$ must be positive for all embeddings $\{\iota : F \hookrightarrow \mathbb{R}\}$. □

Next, we look at the relationship between $\delta(\mu)$ and the standard degree $\deg\mu \in \mathbb{Q}$.

Proposition 5.5. *Let $\delta(\mu)$ be the "degree" of an isogeny of building blocks $\mu : B_1 \to B_2$ defined in terms of polarizations θ_1 and θ_2 with $\deg\theta_1 = \deg\theta_2$. Then $\deg\mu = \mathrm{N}_{\mathcal{B}/\mathbb{Q}}(\delta(\mu))$, where $\mathrm{N}_{\mathcal{B}/\mathbb{Q}}$ is the canonical reduced norm of $\mathcal{B} = \mathrm{End}(B_2)$ over $\mathbb{Q}$.*

Proof. Write $t = \sqrt{\dim_F \mathcal{B}}$ (= 1 or 2) as usual. First, because B_2 is simple, we know $\deg\delta(\mu) = (\mathrm{N}_{\mathcal{B}/\mathbb{Q}}(\delta(\mu)))^n$ for some $n \geq 1$ [Mu, p. 182]. But "deg" is a polynomial function of degree $2\dim B_2$, and $N_{\mathcal{B}/\mathbb{Q}}$ is a polynomial function of degree $t[F:\mathbb{Q}] = \dim B_2$ (since B_2 is a building block). Thus $n = 2$ and $\deg\delta(\mu) = (\mathrm{N}_{\mathcal{B}/\mathbb{Q}}(\delta(\mu)))^2$.

On the other hand, $\delta(\mu) = \mu\circ\theta_1^{-1}\circ\hat{\mu}\circ\theta_2$ means that

$$\deg\delta(\mu) = (\deg\mu)(\deg\theta_1)^{-1}(\deg\hat{\mu})(\deg\theta_2) = (\deg\mu)^2,$$

because $\deg\hat{\mu} = \deg\mu$, and $\deg\theta_1 = \deg\theta_2$ by hypothesis.

Therefore, $(\deg\mu)^2 = (\mathrm{N}_{\mathcal{B}/\mathbb{Q}}(\delta(\mu))^2$, and it follows that $\deg\mu = \pm\,\mathrm{N}_{\mathcal{B}/\mathbb{Q}}(\delta(\mu)) = \pm(\mathrm{N}_{F/\mathbb{Q}}(\delta(\mu)))^t$, because $\delta(\mu)$ lies in F. Now we have seen that $\delta(\mu)$ is totally positive; thus $\mathrm{N}_{F/\mathbb{Q}}(\delta(\mu)) > 0$. Of course, $\deg\mu$ is also positive, and so we must have $\deg\mu = +\,\mathrm{N}_{\mathcal{B}/\mathbb{Q}}(\delta(\mu))$. □

Now let B be a building block and fix a collection $\{\mu_\sigma : {}^\sigma B \to B\}$ respecting $\mathrm{End}(B)$, as usual. Let $\theta : B \to \widehat{B}$ be a polarization of B. Then for each $\sigma \in G_\mathbb{Q}$, the isogeny ${}^\sigma\theta : {}^\sigma B \to {}^\sigma\widehat{B}$ is a polarization of ${}^\sigma B$ and we define, as above, the degree $\delta(\mu_\sigma) = \mu_\sigma \circ {}^\sigma\theta^{-1} \circ \hat{\mu}_\sigma \circ \theta$. It is easy to check that the isogenies $\{\mu_\sigma\}$ are Rosati compatible, because each μ_σ was chosen to be compatible with $\mathrm{End}(B)$. Note, however, that by requiring the polarization of ${}^\sigma B$ to be ${}^\sigma\theta$, now $\delta(\mu_\sigma) \in F$ is independent of the chosen polarization θ.

Moreover, the μ_σ are well defined modulo F^*, and therefore $\delta(\mu_\sigma) \bmod F^{*2}$ depends only on the building block B and not on the choices for μ_σ or the polarization defining δ. Also, $\mu_\sigma \circ {}^\sigma\mu_\tau \circ \mu_{\sigma\tau}^{-1} \in F^*$ implies $\delta(\mu_\sigma)\cdot\delta(\mu_\tau)\cdot\delta(\mu_{\sigma\tau})^{-1} \in F^{*2}$, and hence $\sigma \mapsto \delta(\mu_\sigma) \bmod F^{*2}$ is a locally constant homomorphism $G_\mathbb{Q} \to F^*/F^{*2}$.

Recall from the previous section the two-cocycle $c(\sigma,\tau) = \mu_\sigma \circ {}^\sigma\mu_\tau \circ \mu_{\sigma\tau}^{-1}$ associated to the building block B. We have seen there is an isomorphism $\mathrm{H}^2(G_\mathbb{Q}, F^*)[2] \cong \mathrm{H}^2(G_\mathbb{Q}, \{\pm1\}) \times \mathrm{Hom}(G_\mathbb{Q}, P/P^2)$, where P is defined to be $F^*/\{\pm1\}$, and we are interested in the image of $[c]$ in $\mathrm{Hom}(G_\mathbb{Q}, P/P^2)$.

Proposition 5.6. *The image of* $[c]$ *in* $\mathrm{Hom}(G_\mathbb{Q}, P/P^2)$ *is exactly the homomorphism* $\sigma \mapsto \delta(\mu_\sigma) \bmod \{\pm1\}F^{*2}$.

Proof. The projection map $\mathrm{H}^2(G_\mathbb{Q}, F^*)[2] \to \mathrm{Hom}(G_\mathbb{Q}, P/P^2)$ is broken into two parts: first, the map $\mathrm{H}^2(G_\mathbb{Q}, F^*)[2] \to \mathrm{H}^2(G_\mathbb{Q}, P)[2]$ induced by $F^* \to F^*/\{\pm1\} = P$, followed by the inverse of the isomorphism $\mathrm{Hom}(G_\mathbb{Q}, P/P^2) \xrightarrow{\sim} \mathrm{H}^2(G_\mathbb{Q}, P)[2]$ coming from the exact sequence $1 \to P \xrightarrow{\text{sq.}} P \longrightarrow P/P^2 \to 1$. Thus we have to show that the image of the homomorphism $\sigma \mapsto \delta(\mu_\sigma) \bmod \{\pm1\}F^{*2}$ in $\mathrm{H}^2(G_\mathbb{Q}, P)[2]$ is the same as the image of $[c]$ in this group.

The isomorphism $\mathrm{Hom}(G_\mathbb{Q}, P/P^2) \xrightarrow{\sim} \mathrm{H}^2(G_\mathbb{Q}, P)[2]$ maps the homomorphism $\sigma \mapsto \delta(\mu_\sigma) \bmod \{\pm1\}F^{*2}$ to a cohomology class as follows: Lift the homomorphism to a map $G_\mathbb{Q} \to P$, for example, $\sigma \mapsto \delta(\mu_\sigma) \bmod \{\pm1\}$. Then the element $\delta(\mu_\sigma)\cdot\delta(\mu_\tau)\cdot\delta(\mu_{\sigma\tau})^{-1} \bmod \{\pm1\}$ is a square in P, and $\sigma,\tau \mapsto \sqrt{\delta(\mu_\sigma)\cdot\delta(\mu_\tau)\cdot\delta(\mu_{\sigma\tau})^{-1}} \bmod \{\pm1\}$ is the two-cocycle that defines the image in $\mathrm{H}^2(G_\mathbb{Q}, P)[2]$.

On the other hand, $c(\sigma,\tau) = \mu_\sigma \circ {}^\sigma\mu_\tau \circ \mu_{\sigma\tau}^{-1}$, and taking the degree δ of both sides we find $c(\sigma,\tau)^2 = \delta(\mu_\sigma)\cdot\delta(\mu_\tau)\cdot\delta(\mu_{\sigma\tau})^{-1}$. Hence $c(\sigma,\tau) \bmod \{\pm1\}$ gives the same class in $\mathrm{H}^2(G_\mathbb{Q}, P)[2]$ that the homomorphism $\sigma \mapsto \delta(\mu_\sigma) \bmod \{\pm1\}F^{*2}$ does. □

Some characters

Let B be a building block. As we have seen in Chapter 4, we form a building block triple $(B, j, \{\nu_\sigma\})$ by attaching some extra data to B: namely, $j : Z(\mathrm{End}(B)) \hookrightarrow \overline{\mathbb{Q}}$, whose image we call F, and for all $\sigma \in G_{\mathbb{Q}}$, a non-zero $\nu_\sigma \in \overline{\mathbb{Q}} \otimes_F \mathrm{Hom}^{\mathcal{B}}({}^\sigma B, B)$ such that $\nu_\sigma \circ {}^\sigma\nu_\tau = \nu_{\sigma\tau}$. Because $\mathrm{Hom}^{\mathcal{B}}({}^\sigma B, B)$ is a one-dimensional F-vector space, we can, and do, write $\nu_\sigma = \beta(\sigma)^{-1} \otimes \mu_\sigma$ for some $\mu_\sigma : {}^\sigma B \to B$ respecting $\mathrm{End}(B)$ and some $\beta(\sigma) \in \overline{\mathbb{Q}}^*$. Let E be the field extension of F generated by the $\beta(\sigma)$.

Now let (A, i) be a pair of $\mathbf{GL}_2$-type corresponding to $(B, j, \{\nu_\sigma\})$ under Theorem 4.6. Then i is an isomorphism of $\mathrm{End}_{\mathbb{Q}}(A)$ onto the field E that maps $Z(\mathrm{End}(A))$ onto the field F. Recall that E is a Galois extension of F. In [Ri 1], Ribet associates a character $\varepsilon : G_{\mathbb{Q}} \to E^*$ to the abelian variety A. Our goal in this section is to recover ε directly from the triple $(B, j, \{\nu_\sigma\})$.

The condition $\nu_\sigma \circ {}^\sigma\nu_\tau = \nu_{\sigma\tau}$ tells us that

$$\mu_\sigma \circ {}^\sigma\mu_\tau \circ \mu_{\sigma\tau}^{-1} = c(\sigma, \tau) = \beta(\sigma) \cdot \beta(\tau) \cdot \beta(\sigma\tau)^{-1} \in F^*,$$

and therefore, if δ is the degree map defined in the previous section,

$$\delta(\mu_\sigma) \cdot \delta(\mu_\tau) \cdot \delta(\mu_{\sigma\tau})^{-1} = \delta(c(\sigma, \tau)) = c(\sigma, \tau)^2 = \beta(\sigma)^2 \cdot \beta(\tau)^2 \cdot \beta(\sigma\tau)^{-2}.$$

Hence we see that $\sigma \mapsto \delta(\mu_\sigma)/\beta(\sigma)^2$ is a continuous character $G_{\mathbb{Q}} \to E^*$. Note that this character depends only on the triple $(B, j, \{\nu_\sigma\})$ and not on our choices for μ_σ; we claim that it is in fact equal to the character ε.

In order to define ε and to prove our claim, we need to study the "λ-adic" representations of $G_{\mathbb{Q}}$ provided by the abelian variety A of $\mathbf{GL}_2$-type.

Let ℓ be a prime number and let $V_\ell = V_\ell(A)$ be the Tate module associated to A. Because A is of $\mathbf{GL}_2$-type, $V_\ell(A)$ is free of rank two over $E \otimes_{\mathbb{Q}} \mathbb{Q}_\ell$. Let $\{\lambda\}$ be the set of primes of E lying over ℓ, and for each λ, write E_λ for the completion of E at λ; then $E \otimes_{\mathbb{Q}} \mathbb{Q}_\ell \cong \bigoplus_{\lambda | \ell} E_\lambda$. Now define $V_\lambda = E_\lambda \otimes_{E \otimes \mathbb{Q}_\ell} V_\ell$. We see immediately that each V_λ is a two-dimensional E_λ-vector space, and that $V_\ell \cong \bigoplus_{\lambda | \ell} V_\lambda$. Each V_λ is still also a $G_{\mathbb{Q}}$-module, since the action of $G_{\mathbb{Q}}$ on V_ℓ commutes with $E \otimes_{\mathbb{Q}} \mathbb{Q}_\ell$.

Our starting point for studying the V_λ will be the following result of Faltings.

Theorem 5.7. *Let K be a finite Galois extension of $\mathbb{Q}$ such that all endomorphisms of A are defined over K, and let $H = \mathrm{Gal}(\overline{\mathbb{Q}}/K)$. Then*

$$\mathrm{End}(A) \otimes_{\mathbb{Q}} \mathbb{Q}_\ell \to \mathrm{End}_{\mathbb{Q}_\ell[H]} V_\ell$$

is an isomorphism of $\mathbb{Q}_\ell[G_{\mathbb{Q}}]$-modules.

The $G_{\mathbb{Q}}$-module structure on V_ℓ induces an action of $G_{\mathbb{Q}}$ on $\mathrm{End}_{\mathbb{Q}_\ell} V_\ell$ via ${}^\sigma f = \sigma \circ f \circ \sigma^{-1}$. Now $\mathrm{End}_{\mathbb{Q}_\ell[H]} V_\ell$ is the submodule of $\mathrm{End}_{\mathbb{Q}_\ell} V_\ell$ fixed by H; it is still a $G_{\mathbb{Q}}$-submodule because H is a normal subgroup. This defines the action of $G_{\mathbb{Q}}$ on $\mathrm{End}_{\mathbb{Q}_\ell[H]} V_\ell$ in the theorem. Of course, the action of $G_{\mathbb{Q}}$ on $\mathrm{End}(A) \otimes_{\mathbb{Q}} \mathbb{Q}_\ell$ comes from the natural action of $G_{\mathbb{Q}}$ on $\mathrm{End}(A)$.

Corollary 5.8. *Let $H \subset G_{\mathbb{Q}}$ be a normal subgroup of finite index with the properties described in the above theorem. Then $\mathrm{End}_{E_\lambda[H]} V_\lambda \cong E_\lambda$ for all $\lambda \mid \ell$.*

Proof. Recall that E is its own centralizer in $\mathrm{End}(A)$. Therefore the centralizer of $E\otimes_{\mathbb{Q}}\mathbb{Q}_\ell$ in $\mathrm{End}(A)\otimes_{\mathbb{Q}}\mathbb{Q}_\ell$ is $E\otimes_{\mathbb{Q}}\mathbb{Q}_\ell$. Applying the isomorphism in the theorem, we find that the centralizer of $E\otimes_{\mathbb{Q}}\mathbb{Q}_\ell$ in $\mathrm{End}_{\mathbb{Q}_\ell[H]} V_\ell$ is again $E\otimes_{\mathbb{Q}}\mathbb{Q}_\ell$. Therefore,

$$\begin{aligned} \mathrm{End}_{(E\otimes\mathbb{Q})[H]} V_\ell &\cong E\otimes_{\mathbb{Q}}\mathbb{Q}_\ell \\ \bigoplus\nolimits_{\lambda|\ell} \mathrm{End}_{E_\lambda[H]} V_\lambda &\cong \bigoplus\nolimits_{\lambda|\ell} E_\lambda \\ \mathrm{End}_{E_\lambda[H]} V_\lambda &\cong E_\lambda. \end{aligned}$$ □

Corollary 5.9. *Let ℓ be a prime that splits completely in E. Then $\mathrm{Hom}_{\mathbb{Q}_\ell[H]}(V_\lambda, V_{\lambda'})$ is non-zero if and only if $\lambda'=\gamma(\lambda)$ for some $\gamma\in\mathrm{Gal}(E/F)$.*

Proof. Regard the isomorphisms

$$\mathrm{End}(A)\otimes_{\mathbb{Q}}\mathbb{Q}_\ell \cong \mathrm{End}_{\mathbb{Q}_\ell[H]} V_\ell \cong \bigoplus\nolimits_{\lambda,\lambda'} \mathrm{Hom}_{\mathbb{Q}_\ell[H]}(V_\lambda, V_{\lambda'})$$

as isomorphisms of $(F\otimes_{\mathbb{Q}}\mathbb{Q}_\ell)$-modules. Because $F\otimes_{\mathbb{Q}}\mathbb{Q}_\ell$ is the center of the algebra $\mathrm{End}(A)\otimes_{\mathbb{Q}}\mathbb{Q}_\ell$, every non-zero $f\in\mathrm{Hom}_{\mathbb{Q}_\ell[H]}(V_\lambda,V_{\lambda'})$ must be an $(F\otimes_{\mathbb{Q}}\mathbb{Q}_\ell)$-isomorphism. But it is clear that $V_\lambda\cong V_{\lambda'}$ as $(F\otimes_{\mathbb{Q}}\mathbb{Q}_\ell)$-modules if and only if $E_\lambda\cong E_{\lambda'}$ as F-modules, which is equivalent to the condition $\lambda'=\gamma(\lambda)$ for some $\gamma\in\mathrm{Gal}(E/F)$. In other words, if $\lambda'\neq\gamma(\lambda)$, then $\mathrm{Hom}_{\mathbb{Q}_\ell[H]}(V_\lambda,V_{\lambda'})=0$.

Now because ℓ splits completely in E, there are $[E:\mathbb{Q}]$ primes λ lying over ℓ, and we have $E_\lambda\cong\mathbb{Q}_\ell$ for each λ. By the previous corollary, $\mathrm{End}_{\mathbb{Q}_\ell[H]} V_\lambda\cong\mathbb{Q}_\ell$. Therefore, the $\mathbb{Q}_\ell$-dimension of $\mathrm{Hom}_{\mathbb{Q}_\ell[H]}(V_\lambda,V_{\gamma(\lambda)})$ must be zero or one. Hence the $\mathbb{Q}_\ell$-dimension of $\bigoplus_\gamma\bigoplus_\lambda \mathrm{Hom}_{\mathbb{Q}_\ell[H]}(V_\lambda,V_{\gamma(\lambda)})$ is less than or equal to $[E:F][E:\mathbb{Q}]$, with equality holding if and only if each $\mathrm{Hom}_{\mathbb{Q}_\ell[H]}(V_\lambda,V_{\gamma(\lambda)})$ is one-dimensional. On the other hand, the $\mathbb{Q}_\ell$-dimension of $\mathrm{End}(A)\otimes_{\mathbb{Q}}\mathbb{Q}_\ell$ is $[E:F]^2\cdot[F:\mathbb{Q}]$, which forces equality above. □

Now consider the action of $G_{\mathbb{Q}}$ on $\mathrm{End}(A)\otimes_{\mathbb{Q}}\mathbb{Q}_\ell$. Recall for each $\sigma\in G_{\mathbb{Q}}$ we have defined $\alpha(\sigma)\in E^*$ to be an element of $\mathrm{End}(A)$ such that ${}^\sigma\varphi=\alpha(\sigma)\circ\varphi\circ\alpha(\sigma)^{-1}$. That is to say, $\sigma\in G_{\mathbb{Q}}$ acts on $\mathrm{End}(A)$ via conjugation by $\alpha(\sigma)$, where $\alpha(\sigma)\in E^*$ is well defined modulo F^*. We fix a choice of $\alpha(\sigma)$ for each σ; these $\alpha(\sigma)$ satisfy $\alpha(\sigma)\cdot\alpha(\tau)\cdot\alpha(\sigma\tau)^{-1}\in F^*$, as always. Therefore, for each $\gamma\in\mathrm{Gal}(E/F)$, we may define a character $\chi_\gamma: G_{\mathbb{Q}}\to E^*$ by $\chi_\gamma(\sigma)=\gamma(\alpha(\sigma))/\alpha(\sigma)$. Note that these characters depend only on the abelian variety A.

NOTATION: Let ℓ be a prime which splits completely in E, and let λ be a prime of E dividing ℓ. Then E_λ will denote $\mathbb{Q}_\ell$ with the map $E\hookrightarrow\mathbb{Q}_\ell$ induced by λ, and the image in $\mathbb{Q}_\ell$ of an element $e\in E$ will be written e_λ.

Lemma 5.10. *Let ℓ be a prime which splits completely in E. Let $\gamma\in\mathrm{Gal}(E/F)$. Then the $\mathbb{Q}_\ell[G_{\mathbb{Q}}]$-module V_λ is isomorphic to $V_{\gamma(\lambda)}\otimes\chi_{\gamma,\lambda}^{-1}$.*

Proof. We know that V_λ and $V_{\gamma(\lambda)}$ are isomorphic $\mathbb{Q}_\ell[H]$-modules for some normal subgroup H of $G_{\mathbb{Q}}$ of finite index. Let $f\in\mathrm{Hom}_{\mathbb{Q}_\ell[H]}(V_\lambda,V_{\gamma(\lambda)})$ be an isomorphism. Now $\mathrm{End}(A)\otimes_{\mathbb{Q}}\mathbb{Q}_\ell$ is $G_{\mathbb{Q}}$-isomorphic to $\mathrm{End}_{\mathbb{Q}_\ell[H]} V_\ell$, and we know σ acts on $\mathrm{End}(A)\otimes_{\mathbb{Q}}\mathbb{Q}_\ell$ via conjugation by $\alpha(\sigma)$. Also, $\alpha(\sigma)\in E$ acts on V_λ as $\alpha(\sigma)_\lambda$ for each λ. Thus we may write ${}^\sigma f = \alpha(\sigma)_{\gamma(\lambda)} \cdot f \cdot \alpha(\sigma)_\lambda^{-1}$ in $\mathrm{Hom}_{\mathbb{Q}_\ell[H]}(V_\lambda, V_{\gamma(\lambda)})$. Now because $E_\lambda \cong \mathbb{Q}_\ell$, we have

$$ {}^\sigma f = \frac{\alpha(\sigma)_{\gamma(\lambda)}}{\alpha(\sigma)_\lambda} \cdot f = \left(\frac{\gamma(\alpha(\sigma))}{\alpha(\sigma)}\right)_\lambda \cdot f = \chi_{\gamma,\lambda}(\sigma) \cdot f. $$

Hence for all $\sigma \in G_{\mathbb{Q}}$ and $v \in V_\lambda$, we have $\sigma(f(v)) = \chi_{\gamma,\lambda}(\sigma) \cdot f(\sigma(v))$, and we see that f is an isomorphism of $\mathbb{Q}_\ell[G_{\mathbb{Q}}]$-modules $V_\lambda \xrightarrow{\sim} V_{\gamma(\lambda)} \otimes \chi_{\gamma,\lambda}^{-1}$. □

Now what is the character ε? The action of $G_{\mathbb{Q}}$ on V_λ defines a representation $\rho_\lambda : G_{\mathbb{Q}} \to \mathrm{Aut}\, V_\lambda$. By [Ri 1, Lemma 3.1] there exists a character of finite order $\varepsilon : G_{\mathbb{Q}} \to E^*$ such that for each prime $\lambda \mid \ell$ of E, we have $\det \rho_\lambda = \varepsilon \cdot \chi_\ell$, where $\chi_\ell : G_{\mathbb{Q}} \to \mathbb{Q}_\ell^*$ is the ℓ-adic cyclotomic character. As we will soon see, ε is equal to the character $\chi_{\mathrm{conj.}}^{-1}$, where "conj." $\in \mathrm{Gal}(E/F)$ is the canonical complex conjugation $e \mapsto \bar{e}$ on E.

Lemma 5.11. *Let ℓ be a prime that splits completely in E. Then the $G_{\mathbb{Q}}$-module V_λ is isomorphic to $V_{\bar{\lambda}} \otimes \varepsilon_\lambda$.*

Proof. (Ribet) Let $\langle\ ,\ \rangle : V_\ell \times V_\ell \to \mathbb{Q}_\ell(1)$ be the Weil pairing on $V_\ell(A)$ induced by a fixed polarization of $A/\mathbb{Q}$. Here $\mathbb{Q}_\ell(1)$ denotes the ℓ-adic Tate module $\mathbb{Q}_\ell \otimes_{\mathbb{Z}_\ell} \varprojlim \boldsymbol{\mu}_{\ell^n}$ of ℓ-power roots of unity in $\overline{\mathbb{Q}}$. Recall that the Weil pairing is non-degenerate and Galois invariant; that is, for $\sigma \in G_{\mathbb{Q}}$ and $v, w \in V_\ell$ we have ${}^\sigma\langle v, w\rangle = \langle \sigma(v), \sigma(w)\rangle$. Thus we obtain an isomorphism of $G_{\mathbb{Q}}$-modules

$$ V_\ell \cong \mathrm{Hom}_{\mathbb{Q}_\ell}(V_\ell, \mathbb{Q}_\ell(1)). $$

By writing V_ℓ as the direct sum of the V_λ, we then have

$$ \bigoplus_{\lambda \mid \ell} V_\lambda \cong \bigoplus_{\lambda \mid \ell} \mathrm{Hom}_{\mathbb{Q}_\ell}(V_\lambda, \mathbb{Q}_\ell(1)). \qquad (*) $$

We claim that under this isomorphism, the $\bar{\lambda}$-component on the left-hand side corresponds to the λ-component on the right-hand side. Indeed, recall that if $\varphi \mapsto \varphi'$ is the Rosati involution on $\mathrm{End}(A)$ coming from our fixed polarization, then φ and φ' are adjoint with respect to the Weil pairing; in particular, $\langle ev, w\rangle = \langle v, \bar{e}w\rangle$ for $e \in E$ and $v, w \in V_\ell$. Now we may extend the automorphism "conj." to $E \otimes_{\mathbb{Q}} \mathbb{Q}_\ell$; it is simply the map which permutes the components of $\bigoplus_{\lambda \mid \ell} E_\lambda$ by taking the λ-component to the $\bar{\lambda}$-component. The new "conj." still has the same property with respect to $\langle\ ,\ \rangle$.

Now let $\{\pi_\lambda\}$ be a set of orthogonal idempotents for $E \otimes_{\mathbb{Q}} \mathbb{Q}_\ell$, so that $V_\lambda = \pi_\lambda V_\ell$. Note that $\overline{\pi_\lambda} = \pi_{\bar{\lambda}}$. By using $\langle \pi_\lambda v, w\rangle = \langle v, \overline{\pi_\lambda} w\rangle$, it is easy to verify that if $v = \pi_\lambda v \in V_\lambda$, then under $(*)$ the corresponding homomorphism f_v is $f_v \circ \pi_{\bar{\lambda}} \in \mathrm{Hom}_{\mathbb{Q}_\ell}(V_{\bar{\lambda}}, \mathbb{Q}_\ell(1))$. This proves the claim, and we have

$$ V_{\bar{\lambda}} \cong \mathrm{Hom}_{\mathbb{Q}_\ell}(V_\lambda, \mathbb{Q}_\ell(1)). $$

On the other hand, V_λ is a two-dimensional E_λ-vector space. Therefore the determinant gives a pairing $V_\lambda \times V_\lambda \to E_\lambda$ which, by definition of ε, defines an isomorphism of $\mathbb{Q}_\ell[G_\mathbb{Q}]$-modules

$$V_\lambda \cong \operatorname{Hom}_{\mathbb{Q}_\ell}(V_\lambda,\, \mathbb{Q}_\ell(1) \otimes \varepsilon_\lambda) \cong \operatorname{Hom}_{\mathbb{Q}_\ell}(V_\lambda, \mathbb{Q}_\ell(1)) \otimes \varepsilon_\lambda.$$

Combining this with the previous isomorphism, we see $V_\lambda \cong V_{\overline{\lambda}} \otimes \varepsilon_\lambda$. □

Theorem 5.12. *Let $(B, j, \{\nu_\sigma\})$ be a building block triple, and let (A, i) be a pair of $\mathbf{GL}_2$-type corresponding to the triple under Theorem 4.6. Then the character $\sigma \mapsto \delta(\mu_\sigma)/\beta(\sigma)^2$ defined by $(B, j, \{\nu_\sigma\})$ is equal to the character ε defined by (A, i).*

Proof. Let ℓ be a prime that splits completely in E. Together, the previous two lemmas imply that $\varepsilon_\lambda = \chi^{-1}_{\text{conj.},\lambda}$ for all λ dividing ℓ, and so $\varepsilon = \chi^{-1}_{\text{conj.}} : G_\mathbb{Q} \to E^*$. That is, we may write $\varepsilon(\sigma) = \alpha(\sigma)/\overline{\alpha(\sigma)}$ for the usual collection $\{\alpha(\sigma)\}_{\sigma \in G_\mathbb{Q}}$ defined by A.

On the other hand, we know that under the correspondence $(B, j, \{\nu_\sigma\}) \rightsquigarrow (A, i)$, we may take $\alpha(\sigma) = \beta(\sigma)^{-1}$. Thus $\varepsilon(\sigma) = \overline{\beta(\sigma)}/\beta(\sigma)$.

Now we compare our two characters by taking their quotient; we find a new character $\sigma \mapsto \varepsilon(\sigma)^{-1} \cdot \delta(\mu_\sigma)/\beta(\sigma)^2 = \delta(\mu_\sigma)/\beta(\sigma)\overline{\beta(\sigma)}$, which takes values in F^*. Therefore $\delta(\mu_\sigma)/\beta(\sigma)\overline{\beta(\sigma)} \in \{\pm 1\}$ for all $\sigma \in G_\mathbb{Q}$, because F is a totally real field. But $\delta(\mu_\sigma)$ and $\beta(\sigma)\overline{\beta(\sigma)}$ are both totally positive elements of F^*, so their quotient must be $+1$. Hence we have $\delta(\mu_\sigma) = \beta(\sigma)\overline{\beta(\sigma)}$ for all σ, and $\varepsilon(\sigma) = \delta(\mu_\sigma)/\beta(\sigma)^2$. □

This theorem has some interesting consequences. First, recall our discussion of fields of definition for building blocks. We found that a building block B is always isogenous to a building block B_0 over a $(2, \ldots, 2)$-extension K of $\mathbb{Q}$. Furthermore, that extension may frequently be taken to be the fixed field K_P of the subgroup $\{\sigma : \delta(\mu_\sigma) \in \{\pm 1\}F^{*2}\} \subset G_\mathbb{Q}$, and in all cases, a quadratic extension of K_P suffices. Now if A is an abelian variety of $\mathbf{GL}_2$-type corresponding to B, this information translates into the knowledge of a field K where A breaks up into its simple components, i.e., $A_{/K} \sim B_0 \times \cdots \times B_0$. The theorem lets us compute the field K_P from data associated with A only.

Corollary 5.13. *Let $\boldsymbol{\mu}_E$ denote the group of roots of unity in E^*. Then the field K_P is the fixed field of $\{\sigma : \alpha(\sigma) \in \boldsymbol{\mu}_E F^*\} \subset G_\mathbb{Q}$.*

Proof. Fix $\sigma \in G_\mathbb{Q}$. It suffices to show that $\delta(\mu_\sigma) \in \{\pm 1\}F^{*2}$ if and only if $\alpha(\sigma) \in \boldsymbol{\mu}_E F^*$. Suppose that $\delta(\mu_\sigma) \in \{\pm 1\}F^{*2}$. Then by the theorem, $\alpha(\sigma)^2 = \varepsilon(\sigma)/\delta(\mu_\sigma) \in \boldsymbol{\mu}_E F^{*2}$, which implies that $\alpha(\sigma) \in \boldsymbol{\mu}_E F^*$. Conversely, suppose that $\alpha(\sigma) \in \boldsymbol{\mu}_E F^*$. Then $\delta(\mu_\sigma) = \varepsilon(\sigma)/\alpha(\sigma)^2 \in (\boldsymbol{\mu}_E \cap F^*)F^{*2} = \{\pm 1\}F^{*2}$. □

Finally, we may formulate a criterion to determine when, given a building block, we may construct from it an abelian variety A of $\mathbf{GL}_2$-type with $\varepsilon = 1$. This question is motivated by the discussion in the introduction of a possible "modular" structure on A. Recall that A is "modular" if it appears as a quotient of the Jacobian $J_1(N)$ for some N. In this event, A is isogenous to an abelian variety A_f, where f is a normalized cuspidal eigenform of weight two on $\Gamma_1(N)$. (Conjecturally, all abelian varieties of $\mathbf{GL}_2$-type may be constructed in this way.) Now when A is modular, the character ε is exactly the character of the modular form f on $\Gamma_0(N)$ [Ri 1, Lemma 4.3]. Thus the case $\varepsilon = 1$ is especially interesting because the morphism $J_1(N) \to A$ then factors through $J_0(N)$.

The previous theorem tells us ε depends on our choice of $\nu_\sigma = \beta(\sigma)^{-1} \otimes \mu_\sigma$.

Corollary 5.14. *Let B be a building block, and $[c]$ its associated cohomology class. Then the following are equivalent:*

(i) *There exists a choice of $\{\nu_\sigma\}$ such that the character ε is trivial.*
(ii) *The class $[c]$ lies in the kernel of the natural map*

$$\mathrm{H}^2(G_\mathbb{Q},\, F^*) \to \mathrm{H}^2(G_\mathbb{Q},\, F(\ldots, \sqrt{\delta(\mu_\sigma)}, \ldots)).$$

Proof. First note that the field extension $F_1 \stackrel{\text{def}}{=} F(\ldots, \sqrt{\delta(\mu_\sigma)}, \ldots)$ is independent of our choices for μ_σ, because $\delta(\mu_\sigma)$ is well defined modulo F^{*2}. Furthermore, since all $\delta(\mu)$ are totally positive elements of F^*, we see that the field F_1 is totally real.

Now suppose $\varepsilon = 1$. Then by the theorem, $\beta(\sigma)^2 = \delta(\mu_\sigma)$ for all $\sigma \in G_\mathbb{Q}$. Thus $\beta(\sigma) \in F(\sqrt{\delta(\mu_\sigma)})$, and so $c(\sigma, \tau)$ splits in F_1.

On the other hand, if c splits in F_1, then we may choose all $\beta(\sigma)$ in this totally real field. Therefore $\overline{\beta(\sigma)} = \beta(\sigma)$ for all σ, and so $\varepsilon = 1$. □

6. An example of a $\mathbb{Q}$-curve

In this chapter we give a concrete example of a building block of dimension one, and we calculate many of the associated functions and characters that were discussed in the previous chapter.

We see from its definition that a building block of dimension one is simply an elliptic curve over $\overline{\mathbb{Q}}$ without complex multiplication that is isogenous to all of its Galois conjugates. The following example of such a $\mathbb{Q}$-curve was suggested by N. Elkies of Harvard University.

Let B be the elliptic curve over $\mathbb{Q}(\sqrt{-3})$ defined by the equation

$$B:\ y^2 = x^3 + 2x^2 + bx, \text{ where } b \in \mathbb{Q}(\sqrt{-3}) \text{ is an algebraic integer of trace } 1.$$

Proposition 6.1. *The curve B does not have complex multiplication.*

Proof. Let ω be a primitive cube root of unity in $\mathbb{Q}(\sqrt{-3})$, and let $\mathfrak{p} = (1-\omega)$ be the unique prime of $\mathbb{Q}(\sqrt{-3})$ lying over 3. We claim that $b \equiv -1 \bmod \mathfrak{p}$, because b has trace 1. Indeed, write $b = r + s\omega$ with $r, s \in \mathbb{Z}$. Then $\mathrm{Tr}(b) = 1$ implies that $s = 2r - 1$, and so $b + 1 = (1-\omega)(1 + r\omega) \in \mathfrak{p}$.

Now the discriminant of B is $64b^2(1-b) \not\equiv 0 \bmod \mathfrak{p}$, so $\mathfrak{p}$ is a prime of good reduction for B. Let $B_\mathfrak{p}$ be the reduction of B modulo $\mathfrak{p}$; the equation for $B_\mathfrak{p}$ over the field of three elements $\mathbb{F}_3$ is then $y^2 = x^3 + 2x^2 - x$. Recall that $\mathrm{End}(B)$ is a subfield of $\mathrm{End}(B_\mathfrak{p})$, and that it is possible to determine $\mathrm{End}(B_\mathfrak{p})$ by calculating the characteristic polynomial of the Frobenius endomorphism at $\mathfrak{p}$. We can write down this polynomial after counting the number of $\mathbb{F}_3$-rational points on $B_\mathfrak{p}$. It is easy to see there are only two such points: ∞ and $(0,0)$. Therefore the trace of Frobenius is $1+3-2=2$, and the characteristic polynomial is X^2-2X+3, by [Si, Chap. V]. Hence $\mathrm{End}(B_\mathfrak{p}) \cong \mathbb{Q}(\sqrt{-2})$, which is a field with class number one.

Now $\mathrm{End}(B) \hookrightarrow \mathbb{Q}(\sqrt{-2})$, but B cannot have complex multiplication by this field. Indeed, an elliptic curve over $\overline{\mathbb{Q}}$ with complex multiplication by a quadratic imaginary field $\mathcal{K}$ of class number one must have a rational j-invariant [Si, Thm. C.11.2], and the j-invariant of B is $(4-3b)^3/b^2(1-b) \notin \mathbb{Q}$. Thus $\mathrm{End}(B) \cong \mathbb{Q}$. □

Let τ be the non-trivial automorphism of $\mathbb{Q}(\sqrt{-3})$ over $\mathbb{Q}$. Then $\tau(b)=1-b$, because b has trace 1, and we may define an isogeny $\mu : {}^\tau B \to B$ of degree two by

$$\begin{array}{ccc} {}^\tau B:\ y^2 = x^3+2x^2+(1-b)x & \xrightarrow{\mu} & B:\ y^2=x^3+2x^2+bx \\ (x,y) & \longmapsto & \left(-\frac{y^2}{2x^2}, \frac{\sqrt{-2}y(1-b-x^2)}{4x^2}\right). \end{array}$$

This isogeny may be factored into two steps: first twist ${}^\tau B$ by $\sqrt{-2}$, then use the two-isogeny provided by [Si, Ex. III.4.5]. Similarly, we find a formula for the dual $\hat{\mu} : B \to {}^\tau B$.

$$\left(-\frac{y^2}{2x^2}, -\frac{\sqrt{-2}y(b-x^2)}{4x^2}\right) \xleftarrow{\hat{\mu}} (x,y)$$

Thus $B/\mathbb{Q}(\sqrt{-3})$ is a $\mathbb{Q}$-curve, but we note that the isogeny μ is *not* defined over $\mathbb{Q}(\sqrt{-3})$. In fact, we may make the following stronger statement.

Proposition 6.2. *There is no twist of $B/\mathbb{Q}(\sqrt{-3})$ such that the isogeny corresponding to μ is defined over $\mathbb{Q}(\sqrt{-3})$.*

Proof. Any twist of $B/\mathbb{Q}(\sqrt{-3})$ is a twist by $\sqrt{a}$ for some $a \in \mathbb{Q}(\sqrt{-3})$; the resulting curve is $B_a:\ ay^2 = x^3+2x^2+bx$. The isogeny corresponding to μ is

$$\begin{array}{ccc} {}^\tau(B_a) & \longrightarrow & B_a \\ (x,y) & \longmapsto & \left(-\frac{\tau(a)y^2}{2x^2}, \sqrt{\frac{-2\tau(a)}{a}}\frac{y(1-b-x^2)}{4x^2}\right), \end{array}$$

which is not defined over $\mathbb{Q}(\sqrt{-3})$, unless $-2\tau(a)/a$ is a square in $\mathbb{Q}(\sqrt{-3})$. But 2 is a prime in this field, and the valuation at 2 of $-2\tau(a)/a$ must be 1; thus $-2\tau(a)/a$ cannot be a square. □

Now we wish to investigate the two-cocycle $c(g,h) = \mu_g \circ {}^g\mu_h \circ \mu_{gh}^{-1}$. In order to do so, we must extend our base field to $K = \mathbb{Q}(\sqrt{-3}, \sqrt{-2})$. By abusing notation, we now write τ for the non-trivial element of $\mathrm{Gal}(K/\mathbb{Q})$ fixing $\mathbb{Q}(\sqrt{-2})$, so that τ restricts to our previous τ on $\mathbb{Q}(\sqrt{-3})$. Also, we introduce $\sigma \in \mathrm{Gal}(K/\mathbb{Q})$ as the non-trivial automorphism fixing $\mathbb{Q}(\sqrt{-3})$. Thus we have ${}^\sigma B = B$ and ${}^{\sigma\tau}B = {}^\tau B$, and isogenies $\mu_\sigma = \mu_1 = \mathrm{id} : B \to B$, and $\mu_{\sigma\tau} = \mu_\tau = \mu : {}^\tau B \to B$. We remark that by Proposition 5.5, the "degree" $\delta(\mu_g)$ is exactly the usual degree of an isogeny; that is, $\delta(\mu_\sigma) = \delta(\mu_1) = 1$ and $\delta(\mu_{\sigma\tau}) = \delta(\mu_\tau) = 2$.

The following identities are immediate from the formulas for μ and $\hat{\mu}$:

$$^\sigma\mu = -\mu, \qquad {}^\tau\mu = -\hat{\mu}, \qquad {}^{\sigma\tau}\mu = \hat{\mu}, \qquad \text{and, of course, } \mu\hat{\mu} = 2.$$

We can now calculate the values of $c(g,h)$:

		h: 1	σ	τ	$\sigma\tau$
g	1	1	1	1	1
	σ	1	1	−1	−1
	τ	1	1	−2	−2
	$\sigma\tau$	1	1	2	2

$c(g,h)$

We would like to split this cocycle as

$$c(g,h) = \frac{\beta(g)\cdot\beta(h)}{\beta(gh)}$$

for some (locally constant) function $\beta : G_\mathbb{Q} \to \overline{\mathbb{Q}}^*$. Notice that in order to define such an β, we will again have to extend our base field. There is no such $\beta : G_\mathbb{Q} \to \mathrm{Gal}(K/\mathbb{Q}) \to \overline{\mathbb{Q}}^*$ because, for example, $\sigma\tau = \tau\sigma$, but $c(\sigma,\tau) \neq c(\tau,\sigma)$. Therefore, we will have to find an extension of K which is non-abelian over $\mathbb{Q}$.

Let L be the splitting field over $\mathbb{Q}$ of the polynomial $X^4 - 4X^2 + 6$. Figure 6.3 diagrams the subfields of L; in particular, L is a quadratic extension of K.

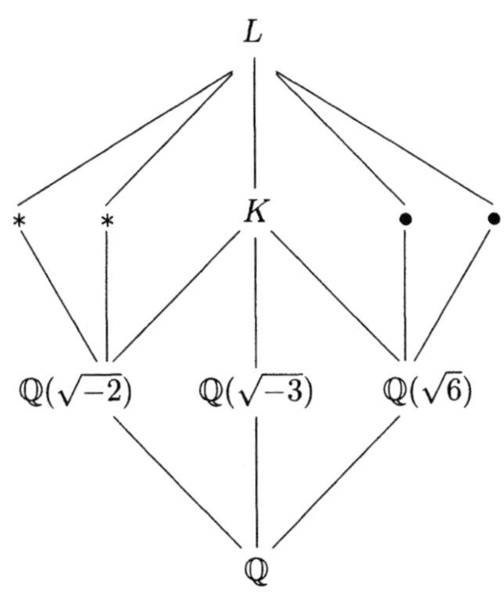

FIGURE 6.3. The subfields of L

Again we redefine our notation: σ is now an element of $\mathrm{Gal}(L/\mathbb{Q})$ generating the subgroup fixing $\mathbb{Q}(\sqrt{-3})$, and τ is now the (non-trivial) element of $\mathrm{Gal}(L/\mathbb{Q})$ fixing one of the $(*)$ fields, so that σ and τ restrict to our previous σ, τ on K. The Galois group $\mathrm{Gal}(L/\mathbb{Q})$ is of course the dihedral group $D_4 = \langle \sigma, \tau : \sigma^4 = \tau^2 = 1,\ \tau\sigma = \sigma^3\tau \rangle$.

We are now able to split $c(g,h)$ by defining $\beta(g)$ as follows:

g	1	σ	σ^2	σ^3	τ	$\sigma\tau$	$\sigma^2\tau$	$\sigma^3\tau$
$\beta(g)$	1	$\sqrt{-1}$	-1	$-\sqrt{-1}$	$\sqrt{-2}$	$\sqrt{2}$	$-\sqrt{-2}$	$-\sqrt{2}$

Let E be the field generated by the values of β, i.e., $E = \mathbb{Q}(\sqrt{-1}, \sqrt{-2})$. Then E is a Galois extension of $\mathbb{Q}$, and we write the elements of $\mathrm{Gal}(E/\mathbb{Q})$ as follows.

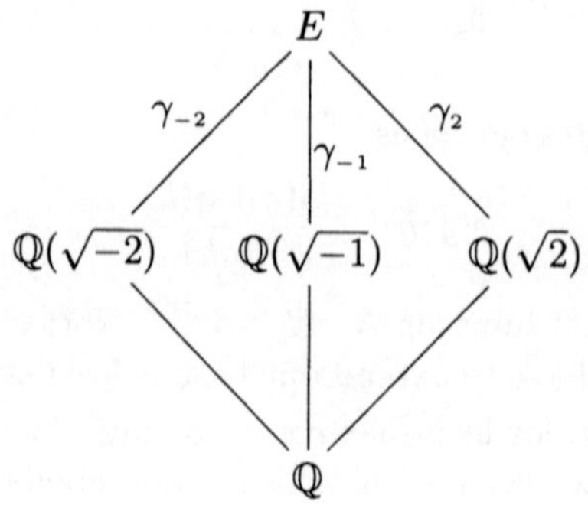

Now we know from Proposition 4.5 that there is an abelian variety $A/\mathbb{Q}$ of $\mathbf{GL}_2$-type with $\mathrm{End}_{\mathbb{Q}}(A) \cong E$, such that $A_{/\overline{\mathbb{Q}}}$ is isogenous to $B \times B \times B \times B$. To compute a specific such A, we will use Weil's "restriction of scalars" functor. We first remind the reader of the salient features of this construction [We, §1.3].

Let $\mathcal{K}/k$ be a finite Galois extension with group G, and let V be an abelian variety over $\mathcal{K}$. The *restriction of scalars* $\mathrm{Res}_{\mathcal{K}/k} V$ from $\mathcal{K}$ to k is an abelian variety W over k, together with a morphism $p : W_{/\mathcal{K}} \to V$, which represents the functor that takes an abelian variety X over k to the abelian group $\mathrm{Hom}_{\mathcal{K}}(X_{/\mathcal{K}}, V)$. The construction of W shows that $W_{/\mathcal{K}}$ is isomorphic to $\prod_{g \in G} {}^{g}V$. Thus, applying the universal property with $X = W$, we find

$$\mathrm{End}_k(W) \cong \mathrm{Hom}_{\mathcal{K}}(W_{/\mathcal{K}}, V) \cong \mathrm{Hom}_{\mathcal{K}}(\textstyle\prod_{g \in G} {}^{g}V,\ V) \cong \bigoplus_{g \in G} \mathrm{Hom}_{\mathcal{K}}({}^{g}V, V).$$

Now we return to the specific $\mathbb{Q}$-curve B, which we view as a curve over the field L. We claim that we may construct A as a quotient of the restriction of scalars $R = \mathrm{Res}_{L/\mathbb{Q}} B$. Indeed, Ribet [Ri 1, Lemma 6.4] has shown that $\mathcal{R} = \mathrm{End}_{\mathbb{Q}}(R)$ is the "twisted group algebra" $\mathbb{Q}[\mathrm{Gal}(L/\mathbb{Q})]^c$, with multiplication defined by $[g][h] = c(g,h)[gh]$ for all $g, h \in \mathrm{Gal}(L/\mathbb{Q})$. The algebra $\mathcal{R}$ maps onto E via $[g] \mapsto \beta(g)$. Moreover, it is easy to see that the kernel of this homomorphism is the ideal $(1 + [\sigma^2])$. Now note that $\frac{1}{2}(1 + [\sigma^2])$ is the central idempotent of the semisimple algebra $\mathcal{R}$ generating this ideal. Therefore, $\frac{1}{2}(1 - [\sigma^2])$ is the complementary central idempotent which projects $\mathcal{R}$ onto E. Thus we may take the abelian variety $A \subset R$ to be the subvariety that is the image of $\frac{1}{2}(1 - [\sigma^2]) \in \mathrm{End}_{\mathbb{Q}}(R)$.

In fact we may describe A more explicitly, by further analysis of the abelian variety R and its endomorphisms over $\mathbb{Q}$. We break down R as $\mathrm{Res}_{K/\mathbb{Q}}(\mathrm{Res}_{L/K} B)$, and first consider $\mathrm{Res}_{L/K} B$, which is an abelian variety over K.

Let B' be the twist of $B_{/K}$ by the quadratic extension L/K. We will continue to write $\mathrm{Gal}(L/K) = \{1, \sigma^2\}$ as a subgroup of $\mathrm{Gal}(L/\mathbb{Q})$.

Lemma 6.4. *The restriction of scalars $\mathrm{Res}_{L/K} B$ is isomorphic over K to $B \times B'$.*

Proof. Recall that by definition, B' is an abelian variety over K which is isomorphic over L (but not over K) to B. Let $\lambda : B \to B'$ be such an isomorphism over L; then $\lambda^{-1} \circ {}^{\sigma^2}\lambda = -1$. Now define $p : B \times B' \to B$ by $p = (\frac{1}{2}, \frac{1}{2}\lambda^{-1})$. We will show that the pair $(B \times B',\ p)$ satisfies the universal property of $\mathrm{Res}_{L/K} B$.

Let X be an abelian variety over K, and $f : X \to B$ a morphism over L. We must demonstrate the existence of a unique morphism $g : X \to B \times B'$ over K such that $p \circ g = f$. Note that if g is to be defined over K, we must also have ${}^{\sigma^2}p \circ g = {}^{\sigma^2}f$. Now using the definition of p, we are able to produce a unique solution to these two equations: $g = (f + {}^{\sigma^2}f,\ \ \lambda \circ f - \lambda \circ {}^{\sigma^2}f)$. Thus $B \times B'$ must be isomorphic to $\mathrm{Res}_{L/K} B$. □

Now on the one hand, $\mathrm{End}_K(B \times B')$ is plainly equal to $\mathrm{End}_K(B) \times \mathrm{End}_K(B') \cong \mathbb{Q} \times \mathbb{Q}$, while on the other hand, we know $\mathrm{End}_K(\mathrm{Res}_{L/K} B)$ is the group algebra $\mathbb{Q}[\mathrm{Gal}(L/K)]$. Using the isomorphism $(p, {}^{\sigma^2}p) : B \times B' \to B \times {}^{\sigma^2}B$ from the above proof, we find that the induced isomorphism $\mathrm{End}_K(B) \times \mathrm{End}_K(B') \xrightarrow{\sim} \mathbb{Q}[\mathrm{Gal}(L/K)]$ is given by $(1, 0) \mapsto \frac{1}{2}(1 + [\sigma^2])$ and $(0, 1) \mapsto \frac{1}{2}(1 - [\sigma^2])$. In other words, $\frac{1}{2}(1+[\sigma^2])$ is the projection $\mathrm{Res}_{L/K} B \to B$, and $\frac{1}{2}(1-[\sigma^2])$ is the projection $\mathrm{Res}_{L/K} B \to B'$.

Proposition 6.5. *The abelian variety $A/\mathbb{Q}$ of $\mathbf{GL}_2$-type is $\mathrm{Res}_{K/\mathbb{Q}} B'$, where B' is the twist of $B_{/K}$ by the quadratic extension L/K.*

Proof. By the universal property of $\mathrm{Res}_{K/\mathbb{Q}}$, there is a map

$$\mathrm{End}_K(\mathrm{Res}_{L/K} B) \to \mathrm{End}_{\mathbb{Q}}(\mathrm{Res}_{K/\mathbb{Q}}(\mathrm{Res}_{L/K} B)).$$

This is, of course, the natural map $\mathbb{Q}[\mathrm{Gal}(L/K)] \to \mathbb{Q}[\mathrm{Gal}(L/\mathbb{Q})]^c$. Now as we have seen, $\frac{1}{2}(1-[\sigma^2])$ is the projection $\mathrm{Res}_{L/K} B \to B'$; therefore as an endomorphism of $\mathrm{Res}_{L/\mathbb{Q}} B$, this element must project $\mathrm{Res}_{K/\mathbb{Q}}(\mathrm{Res}_{L/K} B)$ onto $\mathrm{Res}_{K/\mathbb{Q}} B'$. But we have defined A to be the image of $\frac{1}{2}(1-[\sigma^2])$ in $\mathrm{Res}_{L/\mathbb{Q}} B$. Thus $A = \mathrm{Res}_{K/\mathbb{Q}} B'$. □

REMARK. It is important to note that the other factor $\mathrm{Res}_{K/\mathbb{Q}} B$ of R is *not* isogenous over $\mathbb{Q}$ to A. Indeed, an analysis similar to above shows that over $\mathbb{Q}$,

$$\mathrm{Res}_{K/\mathbb{Q}} B \cong \mathrm{Res}_{\mathbb{Q}(\sqrt{-3})/\mathbb{Q}} B \times \mathrm{Res}_{\mathbb{Q}(\sqrt{-3})/\mathbb{Q}} B_{-2}$$

where B_{-2} is the twist of $B/\mathbb{Q}(\sqrt{-3})$ by $\sqrt{-2}$. Furthermore, recall from our original construction of $\mu : {}^{\tau}B \to B$ that B is isogenous over $\mathbb{Q}(\sqrt{-3})$ to ${}^{\tau}B_{-2}$. Therefore $\mathrm{Res}_{\mathbb{Q}(\sqrt{-3})/\mathbb{Q}} B$ is isogenous over $\mathbb{Q}$ to $\mathrm{Res}_{\mathbb{Q}(\sqrt{-3})/\mathbb{Q}} {}^{\tau}B_{-2} = \mathrm{Res}_{\mathbb{Q}(\sqrt{-3})/\mathbb{Q}} B_{-2}$.

Finally, since we know that ${}^{\tau}B$ is not isogenous to B over $\mathbb{Q}(\sqrt{-3})$, we have $\mathrm{End}_{\mathbb{Q}}(\mathrm{Res}_{\mathbb{Q}(\sqrt{-3})/\mathbb{Q}} B) = \mathbb{Q}$. Thus the endomorphism algebra $\mathrm{End}_{\mathbb{Q}}(\mathrm{Res}_{K/\mathbb{Q}} B)$ is isomorphic to $\mathbf{M}_2(\mathbb{Q})$.

Lastly, we calculate some of the functions we have associated to the abelian variety A of $\mathbf{GL}_2$-type. For example, for each $g \in G_{\mathbb{Q}}$, the automorphism $\varphi \mapsto {}^{g}\varphi$ on $\mathrm{End}(A)$ is given by conjugation by $\alpha(g) = \beta(g)^{-1}$. We now use these $\alpha(g)$ to calculate the characters $\chi_{\gamma} : G_{\mathbb{Q}} \to E^*$ defined by $\chi_{\gamma}(g) = \gamma(\alpha(g))/\alpha(g)$ for $\gamma \in \mathrm{Gal}(E/\mathbb{Q})$. Then writing $\sigma, \tau \in \mathrm{Gal}(L/\mathbb{Q})$ as above, we find

$$\begin{array}{lllllllll} \chi_{\gamma_{-2}}(\sigma) & = & -1 & \chi_{\gamma_{-1}}(\sigma) & = & 1 & \chi_{\gamma_2}(\sigma) & = & -1 \\ \chi_{\gamma_{-2}}(\tau) & = & 1 & \chi_{\gamma_{-1}}(\tau) & = & -1 & \chi_{\gamma_2}(\tau) & = & -1 \end{array}$$

Also, we remind the reader that the character $\varepsilon : G_{\mathbb{Q}} \to E^*$ is equal to the inverse of χ_{γ_2}, because γ_2 acts as complex conjugation on E. Thus $\varepsilon(\sigma) = -1$ and $\varepsilon(\tau) = -1$.

7. Further questions

In conclusion, we would like to discuss a few unanswered questions and suggest some avenues of further exploration regarding the topic of abelian varieties of $\mathbf{GL}_2$-type and their simple components.

The best choice for β?

One question concerns the construction of abelian varieties of $\mathbf{GL}_2$-type from building blocks. Given a building block B, we form the two-cocycle $c(\sigma, \tau)$ on $G_{\mathbb{Q}}$ with values in F^*, which we then split as $c(\sigma, \tau) = \beta(\sigma) \cdot \beta(\tau) \cdot \beta(\sigma\tau)^{-1}$, where $\beta : G_{\mathbb{Q}} \to E^* \subset \overline{F}^*$ is a locally constant function. The question is: what is the "best" choice for β? And what do we mean by "best"? One possible interpretation is to try to minimize the degree $[E : F]$, which would then minimize the dimension of the abelian variety A of $\mathbf{GL}_2$-type that we construct from B. Another "best" choice might be to minimize the order of the character ε attached to A. A related question is: what is the connection between abelian varieties of $\mathbf{GL}_2$-type constructed from the same building block but with different choices of β?

Generalizing the base field to K?

We would like to consider the possibility of extending our results from abelian varieties defined over $\mathbb{Q}$ to abelian varieties defined over an arbitrary field K of characteristic zero. The only trouble seems to be in making the "right" definitions; most of the actual work appears to generalize easily. One complication that arises over K but not over $\mathbb{Q}$ is the possibility of quaternionic multiplication over K. It is not hard to show from our results that if B is a building block over $\mathbb{Q}$, then $\mathrm{End}_{\mathbb{Q}}(B) = \mathrm{End}(B)$ cannot be a quaternion algebra. (There is a more elementary proof of this fact as well, which follows from looking at the action of $\mathrm{End}_{\mathbb{Q}}(B)$ on the Lie algebra $\mathrm{Lie}(B/\mathbb{Q})$.) If K is an arbitrary field, however, then $\mathrm{End}_K(B)$ may be a quaternion algebra.

Now suppose that we try to generalize the definition of building block in the obvious way: a building block is an abelian variety $B/\overline{K}$ such that (i) the endomorphism algebra $\mathrm{End}(B)$ is a central division algebra over a totally real field F with Schur index $t = 1$ or 2, and $t[F:\mathbb{Q}] = \dim B$; (ii) B is a K-abelian variety; i.e., for each $\sigma \in \mathrm{Gal}(\overline{K}/K)$ there exists an isogeny $\mu_\sigma : {}^\sigma B \to B$ that is compatible with $\mathrm{End}(B)$.

Suppose K is a field where a building block and all of its endomorphisms are defined; for instance, consider the case when K is algebraically closed. Then in order to preserve the correspondence between building blocks and abelian varieties of $\mathbf{GL}_2$-type, it is necessary to broaden our definition of $\mathbf{GL}_2$-type. We need: an abelian variety A/K is of $\mathbf{GL}_2$-type if $\mathrm{End}_K(A)$, the algebra of endomorphisms of A defined over K, is a central division algebra over a number field E with Schur index $t = 1$ or 2, and $t[E:\mathbb{Q}] = \dim A$.

These new definitions certainly demonstrate a symmetry that was lacking in the case $K = \mathbb{Q}$. On the other hand, the terminology "of $\mathbf{GL}_2$-type" does not make much sense in the case $2[E:\mathbb{Q}] = \dim A$. Therefore we might want to preserve the original definition of $\mathbf{GL}_2$-type – requiring $\mathrm{End}_K(A) = E$ and $[E:\mathbb{Q}] = \dim A$ – and change the definition of building block accordingly, which would probably require another restriction on the structure of $\mathrm{End}(B)$.

Modularity?

The ultimate goal is to prove that all abelian varieties of $\mathbf{GL}_2$-type are modular, that is, appear as quotients of Jacobians $J_1(N)$. Modularity would provide us with a lot more information about the structure of our abelian varieties of $\mathbf{GL}_2$-type and their building blocks. As mentioned in the introduction, Ribet has shown that this result is implied by Serre's conjecture of [Se 4, (3.2.4$_?$)], but an in-depth discussion of this issue is beyond the scope of this paper.

Appendix A. A non-building block

Recall that in our definition of a building block B, we have required that B be a $\mathbb{Q}$-abelian variety, which we have defined as follows:

Definition. An abelian variety $B/\overline{\mathbb{Q}}$ is a $\mathbb{Q}$-*abelian variety* if for each $\sigma \in \mathrm{Gal}(\overline{\mathbb{Q}}/\mathbb{Q})$, there exists an isogeny $\mu_\sigma : {}^\sigma B \to B$ compatible with the endomorphisms of B; that is, $\mu_\sigma \circ {}^\sigma\varphi = \varphi \circ \mu_\sigma$ for all $\varphi \in \mathrm{End}(B)$.

It is natural to ask whether the compatibility conditions on the μ_σ are necessary. Given an abelian variety B over $\overline{\mathbb{Q}}$ that is isogenous to all of its Galois conjugates, is it perhaps always possible to find a collection of isogenies $\{\mu_\sigma\}$ that respect $\mathrm{End}(B)$? We answer this question in the negative.

We will construct a two-dimensional abelian variety A over $\mathbb{Q}$ such that the full endomorphism algebra $\mathrm{End}(A)$ is isomorphic to $\mathbb{Q}(\sqrt{5})$, but not all of the endomorphisms are defined over $\mathbb{Q}$. Thus, if $\sigma \in G_{\mathbb{Q}}$ acts non-trivially on $\mathrm{End}(A)$, there is no isogeny $\mu_\sigma : A = {}^\sigma A \to A$ such that $\varphi \circ \mu_\sigma = \mu_\sigma \circ {}^\sigma\varphi$ for all $\varphi \in \mathrm{End}(A)$.

The abelian variety A will be the Jacobian of a certain hyperelliptic curve of genus two. A theorem due to Mestre describes how to construct the necessary curve.

Theorem A.1. *Let d be a positive integer. Let E be an elliptic curve over $\overline{\mathbb{Q}}$ equipped with a point S of order $2d+1$ and a point Q such that $2Q$ does not belong to $\langle S\rangle$; let x_E be a function of degree two on E with a double pole at O_E. Define C to be the hyperelliptic curve of genus d described by the equation*

$$y^2 = \prod_{i=0}^{2d}(x - x_E(Q + iS)).$$

Then the Jacobian of C admits an endomorphism whose characteristic polynomial is

$$\prod_{m=1}^{d}\left(X - 2\cos\frac{2m\pi}{2d+1}\right).$$

Proof. [Me, Thm. 1]. □

We remark that we will apply the theorem in the case $d = 2$, when in fact it reduces (via a theorem of Poncelet) to a classical result of Humbert.

NOTE: All the calculations that follow were done on a SparcStation with the GP/PARI Calculator, version 1.37.

Construction of the curve C

In order for the equation of the curve C to be defined over $\mathbb{Q}$, the set of points $\{Q + iS\}_{i=0}^{4}$ must be stable under the action of $G_{\mathbb{Q}}$. We thus look for an elliptic curve over $\mathbb{Q}$ with a rational point Q and a subgroup $\langle S\rangle$ of order five that is $G_{\mathbb{Q}}$-stable. Take E to be the elliptic curve $J_0(11)$, given by the equation

$$E:\ y^2 + y = x^3 - x^2 - 10x - 20.$$

The point $Q = (16, 60)$ is easily seen to be a rational point on E; in fact, it has order five, and it generates all the $\mathbb{Q}$-rational points on E. Now we need to find the rest of the five-torsion subgroup, $E[5]$. According to [Si, III.2.3d], the "duplication formula" for points P on E is

$$x_E(2P) = \frac{x_P^4 + 20x_P^2 + 158x_P + 21}{4x_P^3 - 4x_P^2 - 40x_P - 79} \qquad \text{where } x_P = x_E(P).$$

We use this formula to solve for the points P on E satisfying $x_E(4P) = x_E(-P) = x_E(P)$. We find that the x-coordinates of such points are exactly the roots of the following irreducible polynomials:

$$\begin{aligned}
&x - 16\\
&x - 5\\
&5x^2 + 5x - 29\\
&x^4 + x^3 + 11x^2 + 41x + 101\\
&x^4 + 15x^3 + 120x^2 + 200x + 155\\
&3x^4 - 4x^3 - 60x^2 - 237x - 21.
\end{aligned} \qquad (*)$$

The last polynomial clearly corresponds to the points of order three on E, and we may safely forget about it; the others correspond to $G_{\mathbb{Q}}$-stable sets of points of order five; e.g., $(x - 16)$ corresponds to $\{\pm Q\}$, and $(x - 5)$ to $\{\pm 2Q\}$.

Let S be a point of E such that $x_E(S)$ is a root of the quadratic polynomial $(*)$; hence $x_E(S) \in \mathbb{Q}(\sqrt{5})$. Note that $x_E(2S)$ must also lie in $\mathbb{Q}(\sqrt{5})$ by the duplication formula, and so $x_E(2S)$ must be the other root of $(*)$. Therefore the points of order five corresponding to $(*)$ are $\{\pm S, \pm 2S\}$, and furthermore, these points must be stable under $G_{\mathbb{Q}}$. In other words, the subgroup $\langle S \rangle$ is $G_{\mathbb{Q}}$-stable, as desired. (In fact, $\langle S \rangle$ is the Shimura subgroup of $J_0(11)$.)

It is now clear that the monic quartic polynomials above correspond to the sets $\{Q + iS\}_{i=1}^4$ and $\{2Q + iS\}_{i=1}^4$, but we need to determine which is which. There are several ways to do this; we choose brute force, and use a computer.

First we claim that S (and hence $E[5]$, which is generated by Q and S) is defined over $\mathbb{Q}(\zeta)$, where ζ is a primitive fifth root of unity. Indeed, suppose that $\sigma \in G_{\mathbb{Q}}$ fixes $\mathbb{Q}(\zeta)$. Then, because $\langle S \rangle$ is stable under $G_{\mathbb{Q}}$, we may write ${}^\sigma S = aS$ for some integer a (mod 5). Now let e_5 be the Weil pairing $E[5] \times E[5] \to \boldsymbol{\mu}_5 \subset \mathbb{Q}(\zeta)$. Note that because Q and S generate $E[5]$, $e_5(Q, S) \neq 1$. Now

$$\begin{aligned}
e_5(Q,S) &= {}^\sigma e_5(Q,S) && \text{because } \sigma \text{ fixes } \mathbb{Q}(\zeta).\\
&= e_5({}^\sigma Q, {}^\sigma S) && \text{because the Weil pairing is Galois invariant.}\\
&= e_5(Q, aS) && \text{by definition of } a.\\
&= e_5(Q,S)^a.
\end{aligned}$$

Thus a must be congruent to 1 (mod 5); i.e., σ must fix S, so S is defined over $\mathbb{Q}(\zeta)$.

Now we may write down explicit coordinates in $\mathbb{Q}(\zeta)$ for S. Take S to be the point (α, β), where

$$\alpha = -\tfrac{11}{5}(\zeta + \zeta^{-1}) - \tfrac{8}{5}, \qquad \beta = \tfrac{11}{5}(2\zeta^3 + \zeta^2 + 3\zeta + \tfrac{14}{11}).$$

A quick computer calculation supplies us with

$$\prod_{i=0}^{4}(x - x_E(Q + iS)) = (x - 16)(x^4 + x^3 + 11x^2 + 41x + 101).$$

We now define C to be the hyperelliptic curve of genus two described by the equation

$$C:\ y^2 = (x - 16)(x^4 + x^3 + 11x^2 + 41x + 101),$$

and write A for the Jacobian of C. Then A is a two-dimensional abelian variety over $\mathbb{Q}$, and by Theorem A.1, there is an endomorphism of A with characteristic polynomial $X^2 + X - 1$; thus $\operatorname{End}(A) \supseteq \mathbb{Q}(\sqrt{5})$.

The structure of $\operatorname{End}(A)$

The author would like to thank F. Hazama of Tokyo-Denki University for suggesting the approach used in this section.

Let A be the abelian variety constructed in the previous section. It is well known that if p is a prime of good reduction for A, then the natural map $\operatorname{End}(A) \to \operatorname{End}(A_p)$ is injective, where A_p is the reduction of A mod p. Indeed, let ℓ be a prime not equal to p; write $T_\ell(A)$ for the Tate module and $V_\ell(A)$ for $T_\ell(A)\otimes\mathbb{Q}$. Then use the facts that the reduction map is injective on ℓ-power torsion [Si, Prop. VII.3.1], that $\operatorname{End}(A) \hookrightarrow \operatorname{End} V_\ell(A)$ [W-M, Thm. 3], and that $\operatorname{End}(A_p) \cong \operatorname{End} V_\ell(A_p)$, because A_p is defined over a finite field [W-M, Thm. 6]. The result then follows from the commutative diagram:

$$\begin{array}{ccc} \operatorname{End}(A) & \longrightarrow & \operatorname{End}(A_p) \\ \big\downarrow & & \big\downarrow \cong \\ \operatorname{End} V_\ell(A) & \hookrightarrow & \operatorname{End} V_\ell(A_p)\,. \end{array}$$

Therefore, in order to study $\operatorname{End}(A)$, we will look at $\operatorname{End}(A_p)$ for various choices of p.

To ease calculation, we search for primes p of good reduction with the properties:

(i) $\operatorname{End}_{\mathbb{F}_p}(A_p) \cong \mathbb{Q}(\pi_p)$, where π_p is the Frobenius endomorphism.
(ii) All endomorphisms of A_p are defined over $\mathbb{F}_p$.

We will use [W-M, Thm. 8]:

Theorem A.2. *Let X be a simple abelian variety over the finite field with q elements. Let π_q be the Frobenius endomorphism of X, and let f be the characteristic polynomial of π_q acting on $T_\ell(X)$, where ℓ is a prime not dividing q. Then*

1. *$f = g^e$, where g is a monic irreducible polynomial with integer coefficients.*
2. *$\operatorname{End}_{\mathbb{F}_q}(X)$ is a division algebra whose center is $\mathbb{Q}(\pi_q)$.*
3. *$\dim_{\mathbb{Q}} \operatorname{End}_{\mathbb{F}_q}(X) = e^2[\mathbb{Q}(\pi_q):\mathbb{Q}]$, and $2\dim X = e[\mathbb{Q}(\pi_q):\mathbb{Q}]$.*

Now, in order to prove that a prime p satisfies conditions (i) and (ii), we claim that it is enough to show that the characteristic polynomial of each power of π_p is irreducible of degree four over $\mathbb{Q}$. Indeed, the full endomorphism algebra $\operatorname{End}(A_p)$ is the union $\bigcup_r \operatorname{End}_{\mathbb{F}_{p^r}}(A_p)$, and the Frobenius endomorphism π_{p^r} is simply π_p^r. Therefore, if the characteristic polynomial of π_p^r is irreducible of degree $4 = 2\dim A_p$ for all r, then by (3) we have $\operatorname{End}_{\mathbb{F}_{p^r}}(A_p) \cong \mathbb{Q}(\pi_p^r)$ for all r. Lastly, we see that $\mathbb{Q}(\pi_p^r) = \mathbb{Q}(\pi_p)$ for all r, because $\mathbb{Q}(\pi_p^r)$ is certainly contained in $\mathbb{Q}(\pi_p)$, and by hypothesis both these fields have degree four over $\mathbb{Q}$.

To calculate the characteristic polynomial of π_p, we recall the Frobenius endomorphism π_{C_p} of the curve C_p, which induces the Frobenius endomorphism π_p on A_p. (We remind the reader that for any variety over $\mathbb{F}_q$ with a projective embedding, the "Frobenius" morphism is simply the map on points $(x_0\colon \ldots :x_n) \mapsto (x_0^q\colon \ldots :x_n^q)$.) The map π_{C_p} has a characteristic polynomial obtained by considering the action of π_{C_p} on the étale cohomology group $\mathrm{H}^1_{et}(C_p, \mathbb{Z}_\ell)$, where ℓ is prime to p. But by [Mil, Cor. 9.6],

$$\mathrm{H}^1_{et}(C_p, \mathbb{Z}_\ell) \cong \mathrm{H}^1_{et}(A_p, \mathbb{Z}_\ell) \cong T_\ell(A_p)^\vee,$$

where $^\vee$ denotes the $\mathbb{Z}_\ell$-dual. Hence the characteristic polynomial of π_{C_p} is in fact the same as the characteristic polynomial of π_p acting on $T_\ell(A_p)$, which is what we want. Therefore we will use the curve C_p to calculate this polynomial.

Here we must note that the equation defining C is singular at infinity, so as yet it does not make sense to speak of primes of good reduction for C. However, C has a non-singular model $\widetilde{C} \subset \mathbb{P}^4$. The map $\widetilde{C} \to C$ is an isomorphism over all affine points of C, and has a unique point over the singularity of C [Ca, §2]. Furthermore, this "point at infinity" of $\widetilde{C}$ is a smooth point mod p for each prime p. Hence to study the reduction of $\widetilde{C}$ at p, it suffices to consider the reduction mod p of our affine equation for C.

First, we calculate the discriminant of the fifth degree polynomial appearing in the equation for C; it is $5^5 \cdot 11^{14}$. Thus the primes of bad reduction for this equation are exactly $p = 5$ or $p = 11$, and these we will avoid. Now we need [M-S, Lemma 3]:

Lemma A.3. *Let C be a curve of genus two over $\mathbb{Q}$ with good reduction at the prime p. Then the characteristic polynomial of the Frobenius element at p is given by*

$$X^4 - tX^3 + sX^2 - ptX + p^2, \tag{A.3}$$

where
$$\begin{aligned} t &= p + 1 - N_1, & N_1 &= \#C_p(\mathbb{F}_p),\\ s &= \tfrac{1}{2}(N_1^2 + N_2) + p - N_1 - pN_1, & N_2 &= \#C_p(\mathbb{F}_{p^2}).\end{aligned}$$

Abusing notation, we write π_p for a root of (A.3). Recall that we want all powers of π_p to generate a field of degree four over $\mathbb{Q}$. In particular, we must have $t \neq 0$ for this situation to occur (since otherwise π_p^2 is quadratic over $\mathbb{Q}$). A computer search now finds that $p = 19$ gives $N_1 = 16$ and $N_2 = 390$, so (A.3) becomes

$$X^4 - 4X^3 + 22X^2 - 76X + 361, \tag{A.4}$$

which is irreducible over $\mathbb{Q}$. Note that over $\mathbb{Q}(\sqrt{5})$ the polynomial (A.4) factors as

$$(X^2 + 4\varepsilon X + 19)(X^2 - (4 + 4\varepsilon)X + 19) \qquad \text{where } \varepsilon = (-1 + \sqrt{5})/2.$$

To fix ideas, let π_{19} be a root of the first factor and let $K = \mathbb{Q}(\pi_{19})$. The field K contains the complex conjugate $\overline{\pi}_{19}$, but is not Galois over $\mathbb{Q}$; it is straightforward to check that the other quadratic polynomial defines a different extension of $\mathbb{Q}(\sqrt{5})$. As a consequence, we see that K has only one non-trivial subfield, namely $\mathbb{Q}(\sqrt{5})$.

(Indeed, if K contained another quadratic subfield, it would be biquadratic over $\mathbb{Q}$, and hence Galois.)

We now claim that all powers of π_{19} have degree four over $\mathbb{Q}$. Suppose this were not the case. Then π_{19}^n would lie in $\mathbb{Q}(\sqrt{5})$ for some n, and we would have $\pi_{19}^n = \bar{\pi}_{19}^n$. But then $\pi_{19}/\bar{\pi}_{19}$ would be a root of unity in K, and clearly the only roots of unity in K are ± 1. But $\bar{\pi}_{19}$ cannot equal $\pm\pi_{19}$, since by definition of π_{19} above, $\pi_{19}\bar{\pi}_{19} = 19$ and $\pi_{19} + \bar{\pi}_{19} = -4\varepsilon$. Therefore π_{19}^n generates $\mathbb{Q}(\pi_{19})$ over $\mathbb{Q}$.

We have now shown that $\mathrm{End}(A_{19}) = \mathrm{End}_{\mathbb{F}_{19}}(A_{19}) \cong \mathbb{Q}(\pi_{19})$.

The next prime we investigate is $p = 59$. In this case $N_1 = 76$ and $N_2 = 3550$, so the characteristic polynomial of Frobenius is

$$X^4 + 16X^3 + 162X^2 + 944X + 59^2. \tag{A.5}$$

The polynomial (A.5) is irreducible over $\mathbb{Q}$ and factors over $\mathbb{Q}(\sqrt{5})$ as

$$(X^2 + (6 - 4\varepsilon)X + 59)(X^2 + (10 - 4\varepsilon)X + 59).$$

We let π_{59} be a root of the first factor and then exactly the same analysis as the case $p = 19$ demonstrates that $\mathrm{End}(A_{59}) = \mathrm{End}_{\mathbb{F}_{59}}(A_{59}) \cong \mathbb{Q}(\pi_{59})$, which is again a non-Galois extension of $\mathbb{Q}$ of degree four whose only non-trivial subfield is $\mathbb{Q}(\sqrt{5})$.

Finally, recall that $\mathrm{End}(A)$ injects into both $\mathrm{End}(A_{19})$ and $\mathrm{End}(A_{59})$, and that $\mathrm{End}(A) \supseteq \mathbb{Q}(\sqrt{5})$. Hence to show $\mathrm{End}(A) \cong \mathbb{Q}(\sqrt{5})$, it suffices to show that $\mathbb{Q}(\pi_{19})$ and $\mathbb{Q}(\pi_{59})$ are distinct number fields. But this follows immediately from the calculation of their discriminants over $\mathbb{Q}$, which are $5^2 \cdot 149$ and $2^6 \cdot 5^2 \cdot 281$, respectively.

To complete our analysis of the structure of $\mathrm{End}(A)$, it remains to show that not all endomorphisms of A are defined over $\mathbb{Q}$. Now as above, for any prime p of good reduction we have $\mathrm{End}_{\mathbb{Q}}(A) \hookrightarrow \mathrm{End}_{\mathbb{F}_p}(A_p)$. But we are now interested in primes p such that *not* all endomorphisms of A_p are defined over $\mathbb{F}_p$. For instance, we take $p = 3$ and calculate that $N_1 = 4$ and $N_2 = 10$. Therefore the characteristic polynomial of the Frobenius element at 3 is $X^4 + 9$, which is irreducible over $\mathbb{Q}$, and so $\mathrm{End}_{\mathbb{F}_3}(A_3) \cong \mathbb{Q}(\pi_3)$. But $\mathbb{Q}(\pi_3)$ is the field $\mathbb{Q}(\sqrt{-1}, \sqrt{6})$, which clearly does not contain $\mathbb{Q}(\sqrt{5})$ as a subfield. Hence $\mathrm{End}_{\mathbb{Q}}(A)$ cannot be isomorphic to $\mathbb{Q}(\sqrt{5})$, and we must have $\mathrm{End}_{\mathbb{Q}}(A) \cong \mathbb{Q}$.

Now we see that $A/\mathbb{Q}$ is an abelian variety that has all of the properties of a building block, *except* the compatibility conditions on the μ_σ. Of course, a building block over $\mathbb{Q}$ is exactly an abelian variety of $\mathbf{GL}_2$-type, and we can also see now that A is not of $\mathbf{GL}_2$-type, since $[\mathrm{End}_{\mathbb{Q}}(A) : \mathbb{Q}] = 1 \neq 2 = \dim A$. This demonstrates the importance of requiring in our definition of building block that all of the isogenies $\mu_\sigma : {}^\sigma B \to B$ respect $\mathrm{End}(B)$.

Appendix B. Useful facts concerning central simple algebras

The endomorphism algebra of an abelian variety is a finite-dimensional semi-simple algebra over $\mathbb{Q}$. In this dissertation, we are concerned mostly with abelian varieties whose endomorphism algebras are in fact simple $\mathbb{Q}$-algebras. We list here a collection of definitions and facts describing the structure of these algebras.

NOTATION

$Z(\mathcal{A}) = \{x \in \mathcal{A} : xa = ax \ \forall a \in \mathcal{A}\}$ is the *center* of an algebra $\mathcal{A}$.

$C_{\mathcal{A}}(\mathcal{B}) = \{x \in \mathcal{A} : xb = bx \ \forall b \in \mathcal{B}\}$ is the *centralizer* of a subalgebra $\mathcal{B} \subseteq \mathcal{A}$.

Let $\mathcal{A}$ be a finite-dimensional simple $\mathbb{Q}$-algebra.

- The center $Z(\mathcal{A})$ is a number field F [Pi, Prop. 12.1]. The algebra $\mathcal{A}$ is then a *central simple algebra* over F.
- $\mathcal{A}$ has a unique simple right module M [Pi, Prop. 3.3b], and as right $\mathcal{A}$-modules, $\mathcal{A} \cong \bigoplus^n M$ for some $n \geq 1$.
- (Wedderburn Structure Theorem.) As F-algebras, $\mathcal{A} \cong \mathbf{M}_n(\mathcal{D})$, where $\mathcal{D} = \operatorname{End}_{\mathcal{A}} M$ is a central division algebra over F [Pi, Cor. 3.5a].
- The dimension $\dim_F \mathcal{A}$ is a square [Pi, Prop. 13.1].
- The *Schur index* of $\mathcal{A}$ is $\sqrt{\dim_F \mathcal{D}}$.
  Note that the Schur index of $\mathcal{A}$ depends only on the class of $\mathcal{A}$ in $\operatorname{Br}(F)$.
- (Hasse-Brauer-Noether Theorem.) The Schur index of $\mathcal{A}$ is equal to the order of the class of $\mathcal{A}$ in the Brauer group $\operatorname{Br}(F)$ [Pi, Thm. 18.6]. (Here it is necessary that F be a number field.)
- (Skolem-Noether Theorem.) Let $\mathcal{B}$ be a simple F-algebra. If $f, g : \mathcal{B} \to \mathcal{A}$ are homomorphisms, then there is an inner automorphism $\alpha : \mathcal{A} \to \mathcal{A}$ such that $\alpha \circ f = g$. In particular, any automorphism of $\mathcal{A}$ is inner [Pi, Thm. 12.6].
- (Double Centralizer Theorem.) If $\mathcal{B}$ is a simple subalgebra of $\mathcal{A}$, then [Pi, Thm. 12.7]
  (i) $C_{\mathcal{A}}(\mathcal{B})$ is simple.
  (ii) $(\dim_F \mathcal{B})(\dim_F C_{\mathcal{A}}(\mathcal{B})) = \dim_F \mathcal{A}$.
  (iii) $C_{\mathcal{A}}(C_{\mathcal{A}}(\mathcal{B})) = \mathcal{B}$.
- A *subfield* of $\mathcal{A}$ is a subalgebra E of $\mathcal{A}$ such that E is a field.
  A *maximal subfield* of $\mathcal{A}$ is a subfield which is maximal with respect to inclusion.
- The following are equivalent for a subfield $E \subseteq \mathcal{A}$ [Pi, Prop. 13.1].
  (Here it is necessary that F be a number field.)
  (i) E is a maximal subfield of $\mathcal{A}$.
  (ii) $[E:F] = \sqrt{\dim_F \mathcal{A}}$.
  (iii) $C_{\mathcal{A}}(E) = E$.
- A maximal subfield $E \subseteq \mathcal{A}$ splits $\mathcal{A}$; i.e., $E \otimes_F \mathcal{A} \cong \mathbf{M}_{nt}(E)$ [Pi, Cor. 13.3].

The following diagram summarizes some of the main points.

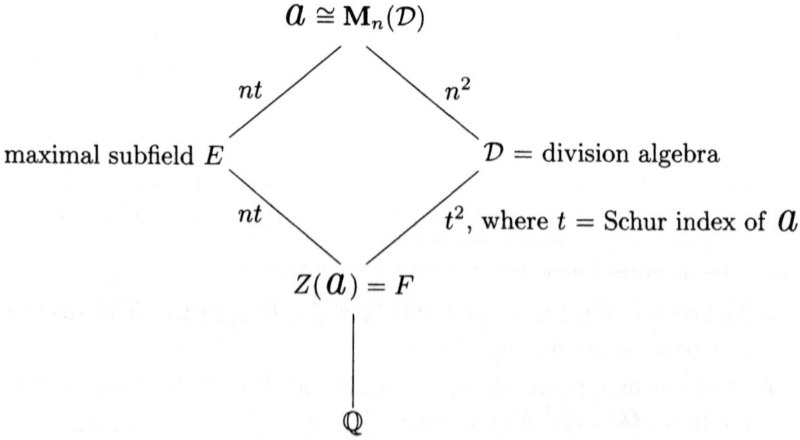

References

[Br] K.S. BROWN. *Cohomology of Groups.* Graduate Texts in Math. **87**, Springer-Verlag, New York, 1982.

[Ca] J.W.S. CASSELS. Arithmetic of curves of genus 2. In *Number Theory and Applications,* pp. 27–35, NATO Adv. Sci. Inst. Ser. C Math. Phys. Sci. **265**, Kluwer Academic Publishers, Boston, 1989.

[El] N. ELKIES. Remarks on elliptic K-curves. Preprint, 1993.

[Ma] B. MAZUR. Number theory as gadfly. *Amer. Math. Monthly* **98** (1991), pp. 593–610.

[M-S] J.R. MERRIMAN AND N.P. SMART. Curves of genus 2 with good reduction away from 2 with a rational Weierstrass point. *Math. Proc. Cambridge Philos. Soc.* **114** (1993), pp. 203–214.

[Me] J.-F. MESTRE. Courbes hyperelliptiques à multiplication réelles. *C. R. Acad. Sci. Paris Sér. I Math.* **307** (1988), pp. 721–724.

[Mil] J.S. MILNE. Jacobian varieties. In G. Cornell and J.H. Silverman (ed.), *Arithmetic Geometry,* pp. 167–212, Springer-Verlag, New York, 1986.

[Mit] B. MITCHELL. *Theory of Categories.* Pure Appl. Math. **17**, Academic Press, New York, 1965.

[Mu] D. MUMFORD. *Abelian Varieties,* Second edition. Tata Inst. Fund. Res. Studies in Math. **5**, Oxford University Press, London, 1974.

[Pi] R.S. PIERCE. *Associative Algebras.* Graduate Texts in Math. **88**, Springer-Verlag, New York, 1982.

[Ri 1] K.A. RIBET. Abelian varieties over $\mathbb{Q}$ and modular forms. In *1992 Proceedings of KAIST Mathematics Workshop,* pp. 53–79, Korea Advanced Institute of Science and Technology, Taejon, 1992.

[Ri 2] K.A. RIBET. Fields of definition of abelian varieties with real multiplication. In *Arithmetic Geometry: Conference on Arithmetic Geometry with an Emphasis on Iwasawa Theory, 1993,* pp. 107–118, Contemp. Math. **174**, American Mathematical Society, Providence, 1994.

[Se 1] J.-P. SERRE. *Cohomologie Galoisienne,* Deuxième édition. Lecture Notes in Math. **5**, Springer-Verlag, Berlin–Göttingen–Heidelberg, 1964.

[Se 2] J.-P. SERRE. *Corps Locaux.* Actualités Sci. Indust. **1296**, Hermann, Paris, 1968.

[Se 3] J.-P. SERRE (prepared in collaboration with C. J. BUSHNELL). Modular forms of weight one and Galois representations. In A. Fröhlich (ed.), *Algebraic Number Fields: L-functions and Galois properties,* pp. 193–268, Academic Press, London, 1977.

[Se 4] J.-P. SERRE. Sur les représentations modulaires de degré 2 de $\mathrm{Gal}(\overline{\mathbb{Q}}/\mathbb{Q})$. *Duke Math. J.* **54** (1987), pp. 179–230.

[Sh 1] G. SHIMURA. Class fields over real quadratic fields and Hecke operators. *Ann. of Math.* (2) **95** (1972), pp. 131–190.

[Sh 2] G. SHIMURA. *Introduction to the Arithmetic Theory of Automorphic Functions.* Publ. Math. Soc. Japan **11**, Princeton University Press, Princeton, 1971.

[Sh 3] G. SHIMURA. On analytic families of polarized abelian varieties and automorphic functions. *Ann. of Math. (2)* **78** (1963), pp. 149–192.

[Sh 4] G. SHIMURA. On the factors of the jacobian variety of a modular function field. *J. Math. Soc. Japan* **25** (1973), pp. 523–544.

[Si] J.H. SILVERMAN. *The Arithmetic of Elliptic Curves.* Graduate Texts in Math. **106**, Springer-Verlag, New York, 1986.

[W-M] W.C. WATERHOUSE AND J.S. MILNE. Abelian varieties over finite fields. In *1969 Number Theory Institute,* pp. 53–64, Proc. Sympos. Pure Math. **20**, American Mathematical Society, Providence, 1971.

[We] A. WEIL. *Adeles and Algebraic Groups.* Progr. Math. **23**, Birkhäuser, Boston, 1982.

Elisabeth E. Pyle
Annapolis
Maryland, USA
e-mail: eepyle@stanfordalumni.org

Progress in Mathematics, Vol. 224, 241–261

Abelian Varieties over Q and Modular Forms

Kenneth A. Ribet

1. Introduction

Let C be an elliptic curve over $\mathbf{Q}$. Let N be the conductor of C. The Taniyama conjecture asserts that there is a non-constant map of algebraic curves $X_o(N) \to C$ which is defined over $\mathbf{Q}$. Here, $X_o(N)$ is the standard modular curve associated with the problem of classifying elliptic curves E together with cyclic subgroups of E having order N.

The Taniyama conjecture may be reformulated in various ways. For example, let $X_1(M)$ be the modular curve associated with the problem of classifying elliptic curves E along with a point of order M on E. One knows that if there is a non-constant map $X_1(M) \to C$ over $\mathbf{Q}$ for some $M \geq 1$, then there is a non-constant map $X_o(N) \to C$. (For a result in this direction, see [11].)

In a recent article [12], B. Mazur introduced another type of reformulation. Mazur says that C possesses a "hyperbolic uniformization of arithmetic type" if there is a non-constant map over the complex field $\pi\colon X_1(M)_{\mathbf{C}} \to C_{\mathbf{C}}$ for some $M \geq 1$. The map π is not required to be defined over $\mathbf{Q}$. In [12, Appendix], Mazur proves: Suppose that there is a non-constant map $\pi\colon X_1(M)_{\mathbf{C}} \to C_{\mathbf{C}}$. Then there is a non-constant map $\pi'\colon X_1(M') \to C$ over $\mathbf{Q}$, where M' is a suitable positive integer (which may be different from M). As mentioned above, it follows easily from the existence of π' that C satisfies Taniyama's conjecture. We thus arrive at the following conclusion:

Theorem 1.1 (Mazur). *Let C be an elliptic curve over $\mathbf{Q}$. Then C satisfies Taniyama's conjecture if and only if C admits a hyperbolic uniformization of arithmetic type over $\mathbf{C}$.*

In connection with [12], Serre asked for a conjectural characterization of elliptic curves over $\mathbf{C}$ which admit a hyperbolic uniformization of arithmetic type. All such curves are defined over $\overline{\mathbf{Q}}$, the algebraic closure of $\mathbf{Q}$ in $\mathbf{C}$. Thus one wishes

---

This text appeared in the KAIST proceedings volume *Algebra and topology 1992*, Korea Adv. Inst. Sci. Tech., Taejŏn, 1992, pp. 53–79, and is reprinted here with permission from the editors of that volume. It was written in conjunction with the author's lectures at the seventh KAIST mathematics workshop in Taejŏn, Korea in 1992. This article describes work that was begun during the author's visits to the Hebrew University of Jerusalem and the Université de Paris XI (Orsay) during the summer of 1992. The author thanks J.-P. Serre for suggesting the problem and for pointing out the relevance of Tate's theorem (Theorem 6.3 below).

to characterize those elliptic curves over $\overline{\mathbf{Q}}$ which are quotients of some modular curve $X_1(M)$. Equivalently, one wishes to study the set of elliptic curves which are quotients of the Jacobian $J_1(N)_{\overline{\mathbf{Q}}}$ of $X_1(N)_{\overline{\mathbf{Q}}}$, for some $N \geq 1$. It is well known that all complex multiplication elliptic curves have this property; this follows from [26, Th. 1]. Hence we are principally interested in elliptic curves $C/\overline{\mathbf{Q}}$ without complex multiplication.

In this article, we introduce the concept of an abelian variety over $\mathbf{Q}$ which is of "$\mathbf{GL}_2$-type." Roughly speaking, this is an abelian variety over $\mathbf{Q}$ whose algebra of $\mathbf{Q}$-endomorphisms is a number field of degree equal to the dimension of the abelian variety. It is easy to see that $J_1(N)$ decomposes up to isogeny over $\mathbf{Q}$ as a product of such abelian varieties. Hence any $C/\overline{\mathbf{Q}}$ which is a quotient of some $J_1(N)$ is a quotient (over $\overline{\mathbf{Q}}$) of an abelian variety of $\mathbf{GL}_2$-type over $\mathbf{Q}$.

We prove in this article the following facts (the third is a simple corollary of the first two):

1. Let C be an elliptic curve over $\overline{\mathbf{Q}}$ without complex multiplication. Then C is a quotient of an abelian variety of $\mathbf{GL}_2$-type over $\mathbf{Q}$ if and only if C is isogenous to each of its Galois conjugates ${}^{\sigma}C$ with $\sigma \in \mathrm{Gal}(\overline{\mathbf{Q}}/\mathbf{Q})$. (If C has this latter property, we say that C is a "$\mathbf{Q}$-curve." The terminology is borrowed from Gross [7].)
2. Assume Serre's conjecture [24, (3.2.4?)] on representations of $\mathrm{Gal}(\overline{\mathbf{Q}}/\mathbf{Q})$. Then every abelian variety of $\mathbf{GL}_2$-type over $\mathbf{Q}$ is a quotient of $J_1(N)$ for some $N \geq 1$.
3. Let $C/\overline{\mathbf{Q}}$ be an elliptic curve without complex multiplication. If C is a quotient of $J_1(N)_{\overline{\mathbf{Q}}}$ for some N, then C is a $\mathbf{Q}$-curve. Conversely, suppose that C is a $\mathbf{Q}$-curve. Then if the conjecture [24, (3.2.4?)] is correct, C is a quotient of $J_1(N)_{\overline{\mathbf{Q}}}$ for some N.

In summary, we arrive at a conjectural characterization of elliptic curves over $\overline{\mathbf{Q}}$ which are quotients of modular curves $X_1(N)$: they are exactly the $\mathbf{Q}$-curves in the sense indicated above. This characterization was predicted by Serre.

2. Abelian varieties over Q of GL$_2$-type

We will be concerned with abelian varieties over $\mathbf{Q}$ which admit actions of number fields that are "as large as possible." We first quantify this concept.

Suppose that A is an abelian variety over $\mathbf{Q}$ and that E is a number field acting on A up to isogeny over $\mathbf{Q}$:

$$E \hookrightarrow \mathbf{Q} \otimes_{\mathbf{Z}} \mathrm{End}_{\mathbf{Q}}(A).$$

By functoriality, E acts on the space of tangent vectors $\mathrm{Lie}(A/\mathbf{Q})$, which is a $\mathbf{Q}$-vector space of dimension $\dim A$. (For an account of the Lie algebra attached to a group scheme over a field, see for example [14, §11].) The dimension of this vector space is therefore a multiple of $[E : \mathbf{Q}]$, so that we have $[E\ \mathbf{Q}] \mid \dim A$. In particular, $[E : \mathbf{Q}] \leq \dim A$.

This observation motivates the study of abelian varieties $A/\mathbf{Q}$ whose endomorphism algebras contain number fields of maximal dimension $\dim A$. If E is a number field of degree $\dim A$ which is contained in $\mathbf{Q} \otimes \mathrm{End}_{\mathbf{Q}}(A)$, then the Tate modules $V_\ell(A)$ associated with A are free of rank two over $E \otimes_{\mathbf{Q}} \mathbf{Q}_\ell$. Accordingly, the action of $\mathrm{Gal}(\overline{\mathbf{Q}}/\mathbf{Q})$ on $V_\ell(A)$ defines a representation with values in $\mathbf{GL}(2, E \otimes \mathbf{Q}_\ell)$. We say that A is of "$\mathbf{GL}_2$-type."

Suppose that B is of $\mathbf{GL}_2$-type, and that $F \subseteq \mathbf{Q} \otimes \mathrm{End}_{\mathbf{Q}}(B)$ is a number field of degree $\dim B$. Let E be a number field containing F, and let $n = [E : F]$. After choosing a basis for E over F, we find an embedding $E \subseteq M(n, F)$ of E into the ring of n by n matrices over F. Since $M(n, F)$ acts naturally up to isogeny on $A := B \times \cdots \times B$ (n factors), we obtain an embedding $E \subseteq \mathbf{Q} \otimes \mathrm{End}_{\mathbf{Q}}(A)$. Hence A is again of $\mathbf{GL}_2$-type. (We can summarize the situation by writing $A = E \otimes_F B$.)

We say that an abelian variety $A/\mathbf{Q}$ of $\mathbf{GL}_2$-type is *primitive* if it is not isogenous over $\mathbf{Q}$ to an abelian variety obtained by this matrix construction, relative to an extension E/F of degree $n > 1$ (cf. [31, §8.2]).

Theorem 2.1. *Let A be an abelian variety of $\mathbf{GL}_2$-type over $\mathbf{Q}$. Then the following conditions are equivalent:* (i) *A is primitive;* (ii) *$A/\mathbf{Q}$ is simple;* (iii) *the endomorphism algebra of $A/\mathbf{Q}$ is a number field whose degree coincides with the dimension of A.*

Proof. Let E be a number field of degree $\dim A$ which is contained in the $\mathbf{Q}$-algebra $\mathcal{X} := \mathbf{Q} \otimes_{\mathbf{Z}} \mathrm{End}_{\mathbf{Q}}(A)$. Let D be the commutant of E in $\mathcal{X}$. We claim that D is a division algebra (cf. [16, Th. 2.3]).

To prove the claim, we must show that each non-zero $\mathbf{Q}$-endomorphism of A which commutes with E is an isogeny. Let λ be such an endomorphism and let B be the image of λ. Then B is a non-zero abelian subvariety of A. The field E operates on B (up to isogeny), and thereby acts by functoriality on the $\mathbf{Q}$-vector space $\mathrm{Lie}(B/\mathbf{Q})$. The dimension of $\mathrm{Lie}(B/\mathbf{Q})$ is accordingly a multiple of $[E : \mathbf{Q}]$; on the other hand, it coincides with $\dim B$. Hence $B = A$, so that λ is an isogeny.

The Lie algebra $\mathrm{Lie}(A/\mathbf{Q})$ may now be viewed as a D-vector space. Because of this, the dimension of $\mathrm{Lie}(A/\mathbf{Q})$ is a multiple of $\dim_{\mathbf{Q}}(D)$. In other words, we have $\dim(D) \mid \dim(E)$. This gives the equality $D = E$; i.e., it shows that E is its own commutant in $\mathcal{X}$.

In particular, the center F of $\mathcal{X}$ is a subfield of E. We have then $\mathcal{X} \approx M(n, Q)$, where Q is a division algebra with center F. If Q has dimension t^2 over F, then $nt = [E : F]$. This follows from the fact that E is a maximal commutative semisimple subalgebra of $\mathcal{X}$, which in turn follows from the statement that E is its own commutant in $\mathcal{X}$.

The structure of $\mathcal{X}$ shows that A is isogenous (over $\mathbf{Q}$) to a product of n copies of an abelian variety B whose algebra of $\mathbf{Q}$-endomorphisms is isomorphic to Q. The same Lie algebra argument we have already used shows that $\dim_{\mathbf{Q}}(Q) \mid \dim(B)$, so that $n \cdot \dim_{\mathbf{Q}}(Q) \mid \dim(A)$. This gives the divisibility $nt^2[F : \mathbf{Q}] \mid [E : \mathbf{Q}]$, which implies $nt^2 \mid nt$. One deduces that $t = 1$, i.e., that $Q = F$, and obtains the equality

$n = [E : F]$. Hence the dimension of B coincides with the degree of F over $\mathbf{Q}$. Also, we have $\mathbf{Q} \otimes \mathrm{End}_{\mathbf{Q}}(A) \approx M(n, F)$.

In other words, A is obtained from B and F by the construction we outlined above. The equivalence of the three statements in the theorem is now clear: each assertion is equivalent to the equality $n = 1$. □

3. ℓ-adic representations attached to primitive abelian varieties over Q of GL$_2$-type

In what follows, we study primitive abelian varieties of $\mathbf{GL}_2$-type over $\mathbf{Q}$. Since we never encounter such abelian varieties which are not primitive in the above sense, we will often drop the word "primitive" and refer simply to abelian varieties over $\mathbf{Q}$ of $\mathbf{GL}_2$-type.

To motivate the study of these varieties, we first allude to the existence of a large class of examples of (primitive) abelian varieties of $\mathbf{GL}_2$-type over $\mathbf{Q}$. Suppose that $f = \sum a_n q^n$ is a normalized cuspidal eigenform of weight two on a subgroup of $\mathbf{SL}(2, \mathbf{Z})$ of the form $\Gamma_1(N)$. Then Shimura [27, Th. 7.14] associates to f an abelian variety $A = A_f$ over $\mathbf{Q}$ together with an action on A of the field $E = \mathbf{Q}(\ldots, a_n, \ldots)$. The variety A_f may be constructed as a quotient of $J_1(N)$, the Jacobian of the standard modular curve $X_1(N)$ [29]. The dimension of A and the degree of E are equal. It is well known (and easy to show) that E is the full algebra of endomorphisms of A which are defined over $\mathbf{Q}$ [19, Cor. 4.2]. Thus, A is of $\mathbf{GL}_2$-type over $\mathbf{Q}$. For each $N \geq 1$, the Jacobian $J_1(N)$ is isogenous to a product of abelian varieties of the form A_f [19, Prop. 2.3].

In §4, we study the converse problem: Suppose that $A/\mathbf{Q}$ is of $\mathbf{GL}_2$-type. Is A isogenous to a quotient of $J_1(N)$, for some $N \geq 1$? We show that an affirmative answer is implied by conjecture (3.2.4?) of Serre's article [24] on modular representations of $\mathrm{Gal}(\overline{\mathbf{Q}}/\mathbf{Q})$.

We now begin the study of Galois representations attached to $\mathbf{GL}_2$-type abelian varieties over $\mathbf{Q}$. Suppose that A is such an abelian variety. Let E be the endomorphism algebra of $A/\mathbf{Q}$. Then E is a number field which is either a totally real number field or a "CM field", since each $\mathbf{Q}$-polarization of A defines a positive involution on E.

Recall that for each prime number ℓ, the Tate module $V_\ell = V_\ell(A)$ is free of rank two over $E \otimes_{\mathbf{Q}} \mathbf{Q}_\ell$. For each prime $\lambda \mid \ell$ of E, let E_λ be the completion of E at λ, and set $V_\lambda := V_\ell \otimes_{E \otimes \mathbf{Q}_\ell} E_\lambda$. Thus V_λ is a two-dimensional vector space over E_λ, and V_ℓ is the direct sum of the V_λ with $\lambda \mid \ell$. The action of $\mathrm{Gal}(\overline{\mathbf{Q}}/\mathbf{Q})$ on V_λ defines a "λ-adic representation" ρ_λ. One knows that the collection (ρ_λ) (as λ ranges over the set of finite primes of E) forms a strictly compatible system of E-rational representations whose exceptional set is the set of prime numbers at which A has bad reduction. (For background on this material, see [25, §11.10] and perhaps [17, Ch. II].) We will prove some facts about the λ-adic representations ρ_λ and their reductions mod λ.

For each λ, let $\delta_\lambda\colon \mathrm{Gal}(\overline{\mathbf{Q}}/\mathbf{Q}) \to E_\lambda^*$ be the determinant of ρ_λ. The δ_λ form a compatible system of E-rational one-dimensional representations of $\mathrm{Gal}(\overline{\mathbf{Q}}/\mathbf{Q})$. In the case where E is totally real, we have $\det \rho_\lambda = \chi_\ell$, where

$$\chi_\ell\colon \mathrm{Gal}(\overline{\mathbf{Q}}/\mathbf{Q}) \to \mathbf{Z}_\ell^*$$

is the ℓ-adic cyclotomic character, and where ℓ is the prime of **Q** lying below λ [17, Lemma 4.5.1]. This formula must be modified slightly in the case where E is allowed to be a CM-field:

Lemma 3.1. *There is a character of finite order* $\epsilon\colon \mathrm{Gal}(\overline{\mathbf{Q}}/\mathbf{Q}) \to E^*$ *such that* $\delta_\lambda = \epsilon\chi_\ell$ *for each finite prime* λ *of* E*. This character is unramified at each prime which is a prime of good reduction for* A*.*

Proof. Since the abelian representations δ_λ arise from an abelian variety, they have the Hodge-Tate property. It follows that they are locally algebraic in the sense of [21, Ch. III]; see [21, p. III-50] and [17, Prop. 1.5.3]. One deduces that the family δ_λ is associated with an E-valued Grossencharacter of type A_o of the field **Q** [17, p. 761]. Since the type-A_o Grossencharacters of **Q** are just products of Dirichlet characters with powers of the "norm" character, there is an integer n and an E-valued Dirichlet character ϵ so that we have $\delta_\lambda = \epsilon\chi_\ell^n$ for each λ. Here, ℓ again denotes the residue characteristic of λ. (We will use the association $\ell \leftrightarrow \lambda$ from time to time without comment.) We have blurred the distinction between characters of $\mathrm{Gal}(\overline{\mathbf{Q}}/\mathbf{Q})$ of finite order and Dirichlet characters: if ϵ is a Dirichlet character, then its Galois-theoretic avatar takes the value $\epsilon(p)$ on a Frobenius element for p in $\mathrm{Gal}(\overline{\mathbf{Q}}/\mathbf{Q})$.

By the criterion of Néron-Ogg-Shafarevich, ρ_λ is ramified at a prime $p \neq \ell$ if and only if A has bad reduction at p. Thus, δ_λ is unramified at a prime $p \neq \ell$ if A has good reduction at p. In other words, if p is a prime of good reduction for A, and if $\ell \neq p$, then $\epsilon\chi_\ell^n$ is unramified at p. Since χ_ℓ is unramified at p, ϵ is unramified at p.

It remains to check that $n = 1$. For this, fix a prime number ℓ. It is well known that the action of $\mathrm{Gal}(\overline{\mathbf{Q}}/\mathbf{Q})$ on $\det_{\mathbf{Q}_\ell}(V_\ell)$ is given by the character $\chi_\ell^{\dim(A)} = \chi_\ell^{[E:\mathbf{Q}]}$. On the other hand, the determinant of the action of $\mathrm{Gal}(\overline{\mathbf{Q}}/\mathbf{Q})$ on V_ℓ is given by the character

$$\prod_{\lambda|\ell} \mathbf{N}_{E_\lambda/\mathbf{Q}_\ell}(\delta_\lambda) = \mathbf{N}_{E/\mathbf{Q}}(\epsilon)\cdot\chi_\ell^{n\cdot[E:\mathbf{Q}]}.$$

(Here, **N** denotes a norm.) Since χ_ℓ has infinite order, we deduce that $\mathbf{N}(\epsilon) = 1$ and that $n = 1$. □

Lemma 3.2. *Each character* δ_λ *is odd in the sense that it takes the value* -1 *on complex conjugations in* $\mathrm{Gal}(\overline{\mathbf{Q}}/\mathbf{Q})$.

[Since χ_ℓ is an odd character, the Lemma may be reformulated as the statement that $\epsilon(-1) = +1$.] For the proof, we use the comparison isomorphism

$$V_\lambda \approx H_1(A(\mathbf{C}), \mathbf{Q}) \otimes_E E_\lambda$$

resulting from an embedding $\overline{\mathbf{Q}} \hookrightarrow \mathbf{C}$. In this view of V_λ, the complex conjugation of $\mathrm{Gal}(\overline{\mathbf{Q}}/\mathbf{Q})$ acts on V_λ as $F_\infty \otimes 1$, where F_∞ is the standard "real Frobenius" on $H_1(A(\mathbf{C}), \mathbf{Q})$ (cf. [4, §0.2]). In particular, δ_λ is odd if and only if we have $\det F_\infty = -1$, where the determinant is taken relative to the E-linear action of F_∞ on $H_1(A(\mathbf{C}), \mathbf{Q})$.

Since F_∞ is an involution, and $H_1(A(\mathbf{C}), \mathbf{Q})$ has dimension two, the indicated determinant is $+1$ if and only if F_∞ acts as a scalar ($= \pm 1$) on $H_1(A(\mathbf{C}), \mathbf{Q})$. To prove that F_∞ does *not* act as a scalar, we recall that $F_\infty \otimes 1$ permutes the two subspaces $H_{0,1}$ and $H_{1,0}$ of $H_1(A(\mathbf{C}), \mathbf{Q}) \otimes_{\mathbf{Q}} \mathbf{C}$ in the Hodge decomposition of $H_1(A(\mathbf{C}), \mathbf{Q}) \otimes_{\mathbf{Q}} \mathbf{C}$. □

Proposition 3.3. *For each λ, ρ_λ is an absolutely irreducible two-dimensional representation of* $\mathrm{Gal}(\overline{\mathbf{Q}}/\mathbf{Q})$ *over E_λ. We have* $\mathrm{End}_{\mathbf{Q}_\ell[\mathrm{Gal}(\overline{\mathbf{Q}}/\mathbf{Q})]} V_\lambda = E_\lambda$.

Proof. Faltings [5] proved that V_ℓ is a semisimple $\mathbf{Q}_\ell[\mathrm{Gal}(\overline{\mathbf{Q}}/\mathbf{Q})]$-module whose commutant is $E \otimes \mathbf{Q}_\ell$. (This is the Tate conjecture for endomorphisms of A.) Since V_ℓ is the product of the V_λ with $\lambda \mid \ell$, each module V_λ is semisimple over $\mathbf{Q}_\ell[\mathrm{Gal}(\overline{\mathbf{Q}}/\mathbf{Q})]$ and satisfies $\mathrm{End}_{\mathbf{Q}_\ell[\mathrm{Gal}(\overline{\mathbf{Q}}/\mathbf{Q})]} V_\lambda = E_\lambda$. This implies that V_λ is simple over E_λ, and that $\mathrm{End}_{E_\lambda} V_\lambda = E_\lambda$. The absolute irreducibility follows. □

For each prime p at which A has good reduction, let a_p be the element of E such that

$$a_p = \mathrm{tr}_{E_\lambda}(\mathrm{Frob}_p \mid V_\lambda)$$

whenever $\ell \neq p$. Let $^-$ denote the canonical involution on E: this involution is the identity if E is totally real, and the "complex conjugation" on E if E is a CM field. The involution $^-$ is the Rosati involution on E induced by every polarization of $A/\mathbf{Q}$.

Proposition 3.4. *We have $a_p = \overline{a}_p \epsilon(p)$ for each prime p of good reduction.*

Proof. Let ℓ be a prime number. Let $\sigma \colon E \hookrightarrow \overline{\mathbf{Q}}_\ell$ be an embedding of fields, and let $\bar{\sigma}$ be the conjugate embedding $x \mapsto \sigma(\overline{x})$. Let $V_\sigma = V_\ell \otimes_{E \otimes \mathbf{Q}_\ell} \overline{\mathbf{Q}}_\ell$, where the tensor product is taken relative to the map of $\mathbf{Q}_\ell$-algebras $E \otimes \mathbf{Q}_\ell \to \overline{\mathbf{Q}}_\ell$ induced by σ. Define $V_{\bar{\sigma}}$ similarly, using $\bar{\sigma}$.

Fix a polarization of A defined over $\mathbf{Q}$. The associated e_{ℓ^ν}-pairings of Weil induce a bilinear map $\langle\, , \rangle \colon V_\ell \times V_\ell \to \mathbf{Q}_\ell(1)$, where $\mathbf{Q}_\ell(1)$ is (as usual) the $\mathbf{Q}_\ell$-adic Tate module attached to the ℓ-power roots of unity in $\overline{\mathbf{Q}}$. We have $\langle ex, y\rangle = \langle x, \overline{e}x\rangle$ for $e \in E$ and $x, y \in V_\ell$. Further, this pairing is $\mathrm{Gal}(\overline{\mathbf{Q}}/\mathbf{Q})$-equivariant in the sense that we have $\langle {}^g x, {}^g y\rangle = {}^g\langle x, y\rangle$ for $g \in \mathrm{Gal}(\overline{\mathbf{Q}}/\mathbf{Q})$. After extending scalars from $\mathbf{Q}_\ell$ to $\overline{\mathbf{Q}}_\ell$, we find an isomorphism of $\overline{\mathbf{Q}}_\ell[\mathrm{Gal}(\overline{\mathbf{Q}}/\mathbf{Q})]$-modules

$$V_{\bar{\sigma}} \approx \mathrm{Hom}(V_\sigma, \overline{\mathbf{Q}}_\ell(1)).$$

Now (3.1) implies that the determinant of V_σ is a one-dimensional $\overline{\mathbf{Q}}_\ell$-vector space on which $\mathrm{Gal}(\overline{\mathbf{Q}}/\mathbf{Q})$ acts by the character ${}^\sigma\epsilon\chi_\ell$. Since V_σ is of dimension two, this gives

$$\mathrm{Hom}(V_\sigma, \overline{\mathbf{Q}}_\ell({}^\sigma\epsilon\chi_\ell)) \approx V_\sigma.$$

In view of the fact that $\mathrm{Hom}(V_\sigma, \overline{\mathbf{Q}}_\ell({}^\sigma\epsilon\chi_\ell))$ is the twist by ${}^\sigma\epsilon$ of $\mathrm{Hom}(V_\sigma, \overline{\mathbf{Q}}_\ell(1))$, we get $V_\sigma \approx V_{\bar\sigma}({}^\sigma\epsilon)$. For $p \neq \ell$ a prime of good reduction, the trace of Frob_p acting on V_σ is $\sigma(a_p)$, and similarly the trace of Frob_p acting on $V_{\bar\sigma}({}^\sigma\epsilon)$ is ${}^\sigma\epsilon(p)\bar\sigma(a_p) = {}^\sigma\epsilon(p)\sigma(\bar a_p)$. This gives $a_p = \epsilon(p)\bar a_p$, as required. □

Proposition 3.5. *Let S be a finite set of prime numbers including the set of primes at which A has bad reduction. Then the field E is generated over $\mathbf{Q}$ by the a_p with $p \notin S$.*

Proof. The Proposition follows from the Tate Conjecture for endomorphisms of A by a simple argument, which we now sketch (see [17, pp. 788–789] for more details). Choose a prime number ℓ, and observe that $\mathrm{End}_{\mathrm{Gal}(\overline{\mathbf{Q}}/\mathbf{Q})} V_\ell = E \otimes \mathbf{Q}_\ell$ because of Faltings's results quoted in the proof of (3.3). Let $\overline{V}_\ell = V_\ell \otimes \overline{\mathbf{Q}}_\ell$; then as a consequence we have $\mathrm{End}_{\overline{\mathbf{Q}}_\ell[\mathrm{Gal}(\overline{\mathbf{Q}}/\mathbf{Q})]} \overline{V}_\ell = E \otimes \overline{\mathbf{Q}}_\ell$. In addition, the semisimplicity of the V_ℓ implies that $\overline{V}_\ell$ is semisimple as a $\overline{\mathbf{Q}}_\ell[\mathrm{Gal}(\overline{\mathbf{Q}}/\mathbf{Q})]$-module. For each $\sigma\colon E \hookrightarrow \overline{\mathbf{Q}}_\ell$, let V_σ be as above. Then the V_σ are simple $\overline{\mathbf{Q}}_\ell[\mathrm{Gal}(\overline{\mathbf{Q}}/\mathbf{Q})]$-modules, and they are pairwise non-isomorphic. Indeed, they are a priori semisimple, but the commutant of their product is $\prod_\sigma \overline{\mathbf{Q}}_\ell$. It follows that their traces are pairwise distinct. Since the trace of Frob_p acting on V_σ is $\sigma(a_p)$ (for $p \notin S \cup \{\ell\}$), the Cebotarev Density Theorem implies that the functions $p \mapsto \sigma(a_p), p \notin S \cup \{\ell\}$ are pairwise distinct. □

For the next result, fix a finite set S as in (3.5), and let F be the subfield of E generated by the numbers $a_p^2/\epsilon(p)$ with $p \notin S$.

Proposition 3.6. *The field F is totally real. The extension E/F is abelian.*

Proof. Let $^-$ again be the canonical complex conjugation of E. For $p \notin S$, we have

$$\frac{\bar a_p^2}{\bar\epsilon(p)} = \frac{a_p^2}{\epsilon(p)^2}\epsilon(p)$$

by (3.4). The first assertion of the Proposition then follows. For the second, let $t_p = a_p^2/\epsilon(p)$. It is clear that E is contained in the extension of F (in an algebraic closure of E) obtained by adjoining to F the square roots of all t_p, and all roots of unity. This gives the second assertion. □

We now consider the reductions of the λ-adic representations ρ_λ. To do this directly, replace A by an abelian variety which is $\mathbf{Q}$-isogenous to A and which has the property that its ring of $\mathbf{Q}$-endomorphisms is the ring of integers $\mathcal{O}$ of E. (This process does not change isomorphism classes of the λ-adic representations ρ_λ.) Write simply A for this new abelian variety. For each λ, consider the kernel $A[\lambda]$: this is the group of $\mathbf{Q}$-valued points of A which are killed by all elements of the maximal ideal λ of $\mathcal{O}$. The action of $\mathcal{O}$ on $A[\lambda]$ makes $A[\lambda]$ into a two-dimensional vector space over the residue field $\mathbf{F}_\lambda$ of λ. The $\mathbf{F}_\lambda$-linear action of $\mathrm{Gal}(\overline{\mathbf{Q}}/\mathbf{Q})$ on $A[\lambda]$ defines a reduction $\overline{\rho}_\lambda$ of ρ_λ (cf. [17, II.2]). (There should be no confusion with the involution $^-$ which appears above.)

Lemma 3.7. *For all but finitely many λ, the representation $\overline{\rho}_\lambda$ is absolutely irreducible.*

Proof. A result of Faltings [6, Theorem 1, page 204] implies that the following holds for almost all λ: $A[\lambda]$ is a semisimple $\mathbf{F}_\ell[\mathrm{Gal}(\overline{\mathbf{Q}}/\mathbf{Q})]$ module whose commuting algebra is $\mathbf{F}_\lambda$. The Lemma follows directly from this statement. □

4. Conjectural connection with modular forms

We continue the discussion of §3, focusing on the possibility of linking A with modular forms, at least conjecturally. According to (3.7), $\overline{\rho}_\lambda$ is absolutely irreducible for almost all λ. For each λ such that $\overline{\rho}_\lambda$ is absolutely irreducible, conjectures of Serre [24, (3.2.3?–3.2.4?)] state that $\overline{\rho}_\lambda$ is "modular" in the sense that it arises from the space of mod ℓ cusp forms of a specific level N_λ, weight k_λ, and character ϵ_λ.

These invariants are essentially constant as functions of λ. Rather than study them for all λ, we will restrict attention to those maximal ideals λ which are odd, prime to the conductor of A, are unramified in E, and have degree one. Let Λ be the set of such ideals λ with the property that $\overline{\rho}_\lambda$ is absolutely irreducible. Lemma 3.7 implies that Λ is an infinite set.

Lemma 4.1. *The levels N_λ are bounded as λ varies in Λ.*

Proof. It follows from the definition of N_λ that this level divides the conductor of the two-dimensional ℓ-adic representation ρ_λ. (To compare the two conductors, one can use the Hilbert Formula of [15, §I].) The conductor of ρ_λ divides the conductor of the full ℓ-adic representation $V_\ell(A)$. According to results of A. Grothendieck [8, Cor. 4.6], this latter conductor is independent of ℓ. (It is by definition the conductor of A.) □

Lemma 4.2. *For all $\lambda \in \Lambda$, we have $k_\lambda = 2$.*

Proof. Take $\lambda \in \Lambda$. The determinant of $\overline{\rho}_\lambda$ is the reduction mod λ of δ_λ. By (3.1), this determinant is the product of the mod ℓ cyclotomic character $\overline{\chi}_\ell$ and the reduction mod λ of ϵ. The definition of Λ shows that ℓ is a prime of good reduction for A, so that ϵ is unramified at ℓ (Lemma 3.1). Hence, if I is an inertia group for ℓ in $\mathrm{Gal}(\overline{\mathbf{Q}}/\mathbf{Q})$, we have $\det \overline{\rho}_\lambda \mid I = \overline{\chi}_\ell$.

Further, suppose that $D \subset \mathrm{Gal}(\overline{\mathbf{Q}}/\mathbf{Q})$ is a decomposition group for ℓ. It is clear that $\overline{\rho}_\lambda \mid D$ is finite at ℓ [24, p. 189], since A has good reduction at ℓ. Indeed, the kernel of multiplication by ℓ on $A_{\mathbf{Q}_\ell}$ extends to a finite flat group scheme $\mathcal{G}$ over $\mathbf{Z}_\ell$ because of this good reduction [8, Cor. 2.2.9]. The Zariski closure of $A[\lambda]$ in $\mathcal{G}$ then prolongs $\overline{\rho}_\lambda$ to a group scheme of type (ℓ, ℓ) over $\mathbf{Z}_\ell$.

By Proposition 4 of [24, §2.8], we find that $k_\lambda = 2$. □

Lemma 4.3. *For all but finitely many $\lambda \in \Lambda$, we have $\epsilon_\lambda = \epsilon$.*

Proof. One checks easily that $\epsilon_\lambda = \epsilon$ whenever ℓ is prime to the order of ϵ. We omit the details, since Lemma 4.3 will not be used below. □

Theorem 4.4. *Let A be an abelian variety over $\mathbf{Q}$ of $\mathbf{GL}_2$-type. Assume Serre's conjecture [24, (3.2.4?)] on representations of* $\mathrm{Gal}(\overline{\mathbf{Q}}/\mathbf{Q})$. *Then A is isogenous to a $\mathbf{Q}$-simple factor of $J_1(N)$, for some $N \geq 1$.*

Proof. Applying [24, (3.2.4?)] to the representations $\overline{\rho}_\lambda$ with $\lambda \in \Lambda$, we find that each $\overline{\rho}_\lambda$ arises from a newform of weight $k_\lambda = 2$ and level dividing N_λ. Since the N_λ's are bounded (Lemma 4.1), there are only a finite number of such newforms.

Hence there is a fixed newform $f = \sum a_n q^n$ which gives rise to an infinite number of the $\overline{\rho}_\lambda$'s. Explicitly, we have the following situation. Let R be the ring of integers of the field $\mathbf{Q}(\ldots, a_n, \ldots)$. For an infinite number of $\lambda \in \Lambda$, there is a ring homomorphism $\varphi_\lambda \colon R \to \overline{\mathbf{F}}_\lambda$ mapping a_p to $\mathrm{tr}(\overline{\rho}_\lambda(\mathrm{Frob}_p))$ for all but finitely many primes p.

Let N be the level of f, and let A_f be the quotient of $J_1(N)$ which is associated to f. Let λ be a prime for which there is a φ_λ as above. By the Cebotarev Density Theorem, we have

$$A_f[\ell] \otimes_{R/\ell R} \overline{\mathbf{F}}_\lambda \approx A[\lambda] \otimes_{\mathbf{F}_\lambda} \overline{\mathbf{F}}_\lambda,$$

where $\overline{\mathbf{F}}_\lambda$ is regarded as an $R/\ell R$-module via φ_λ. Since $A_f[\ell] \otimes_{R/\ell R} \overline{\mathbf{F}}_\lambda$ is a quotient of $A_f[\ell] \otimes_{\mathbf{F}_\ell} \overline{\mathbf{F}}_\lambda$, and since $\mathbf{F}_\lambda = \mathbf{F}_\ell$, we have

$$\mathrm{Hom}_{\overline{\mathbf{F}}_\lambda[\mathrm{Gal}(\overline{\mathbf{Q}}/\mathbf{Q})]}(A_f[\ell] \otimes_{\mathbf{F}_\ell} \overline{\mathbf{F}}_\lambda, A[\lambda] \otimes_{\mathbf{F}_\ell} \overline{\mathbf{F}}_\lambda) \neq 0.$$

It follows that $\mathrm{Hom}_{\mathbf{F}_\ell[\mathrm{Gal}(\overline{\mathbf{Q}}/\mathbf{Q})]}(A_f[\ell], A[\lambda]) \neq 0$.

Hence we have $\mathrm{Hom}_{\mathbf{F}_\ell[\mathrm{Gal}(\overline{\mathbf{Q}}/\mathbf{Q})]}(A_f[\ell], A[\ell]) \neq 0$ for an infinite number of prime numbers ℓ. By the theorem of Faltings quoted above, we get $\mathrm{Hom}_{\mathbf{Q}}(A_f, A) \neq 0$. Since A is a simple abelian variety over $\mathbf{Q}$, A must be a quotient of A_f. Since A_f is, in turn, a quotient of $J_1(N)$, we deduce that A is a quotient of $J_1(N)$. □

5. Decomposition over $\overline{\mathbf{Q}}$

Suppose again that $A/\mathbf{Q}$ is an abelian variety of $\mathbf{GL}_2$-type, and let

$$E = \mathbf{Q} \otimes \mathrm{End}_{\mathbf{Q}}(A).$$

Let $\mathcal{X} = \mathbf{Q} \otimes \mathrm{End}_{\overline{\mathbf{Q}}}(A)$ be the algebra of *all* endomorphisms of A.

Proposition 5.1 (Shimura). *Suppose that $A_{\overline{\mathbf{Q}}}$ has a non-zero abelian subvariety of CM-type. Then $A_{\overline{\mathbf{Q}}}$ is isogenous to a power of a CM elliptic curve.*

This is Proposition 1.5 in [28]. For a generalization of this result, see Proposition 5.2 of [9]. See also the discussion in §4 of [18].

Proposition 5.2. *Suppose that $A_{\overline{\mathbf{Q}}}$ has no non-zero abelian subvariety of CM type. Then the center of $\mathcal{X}$ is a subfield F of E. The algebra $\mathcal{X}$ is isomorphic either to a matrix ring over F, or else to a ring of matrices over a quaternion division algebra over F.*

Proof. We employ the same arguments used to prove (2.1) above and Theorem 2.3 of [16]. Let D be the commutant of E in $\mathcal{X}$. Clearly, D is a division algebra: otherwise, we can make E act on a proper non-zero abelian subvariety of $A_{\overline{\mathbf{Q}}}$, contrary to the hypothesis that no abelian subvariety of $A_{\overline{\mathbf{Q}}}$ is of CM type. Also, E is a subfield of D; it is a maximal commutative subfield because $A_{\overline{\mathbf{Q}}}$ does not have complex multiplication. Hence E is its own commutant in D. Since D is the commutant of E, we get $D = E$. In particular, the center of $\mathcal{X}$ is contained in E, so that the center is a subfield F of E.

Since $\mathcal{X}$ is now a central simple algebra over F, we have $\mathcal{X} \approx M(n, D)$, where D is now a division algebra with center F, and n is a positive integer. Suppose that D has dimension t^2 over F; then $[E : F] = nt$ since E is a maximal commutative subalgebra of $\mathcal{X}$. Up to isogeny, $A_{\overline{\mathbf{Q}}}$ is of the form B^n, where the endomorphism algebra of B contains D. Following an idea of J. Tunnell (cf. [20, Th. 1]), we fix an embedding $\overline{\mathbf{Q}} \hookrightarrow \mathbf{C}$ and form the cohomology group $H^1(B(\mathbf{C}), \mathbf{Q})$. This is a $\mathbf{Q}$-vector space of dimension

$$2\dim(B) = \frac{2}{n}\dim(A) = \frac{2nt}{n}[F : \mathbf{Q}]$$

with a functorial action of D. Hence the $\mathbf{Q}$-dimension $t^2[F : \mathbf{Q}]$ of D divides the dimension over $\mathbf{Q}$ of $H^1(B(\mathbf{C}), \mathbf{Q})$, which is $2t[F : \mathbf{Q}]$. Thus $t \leq 2$, so that either $D = F$, or else D is a quaternion division algebra with center F. □

As in Proposition 3.6, let S be a finite set of primes containing the primes of bad reduction for A. Let F be the center of $\mathcal{X}$. Then we have:

Theorem 5.3. *The field F is generated by the numbers $a_p^2/\epsilon(p)$ with $p \notin S$.*

In other words, the field F which appears in (3.6) is the same as the field F in (5.2).

Proof. To prove (5.3), we let ℓ be a prime which splits completely in E, so that all embeddings $E \hookrightarrow \overline{\mathbf{Q}}_\ell$ take values in $\mathbf{Q}_\ell$. Choose a finite extension K of $\mathbf{Q}$ such that all endomorphisms of A are defined over K, and let H be the corresponding open subgroup of $\mathrm{Gal}(\overline{\mathbf{Q}}/\mathbf{Q})$. Replacing H be a smaller subgroup of $\mathrm{Gal}(\overline{\mathbf{Q}}/\mathbf{Q})$ if necessary, we may assume that H is contained in the kernel of ϵ. We have $\mathcal{X} \otimes \mathbf{Q}_\ell = \mathrm{End}_{\mathbf{Q}_\ell[H]} V_\ell$, by Faltings's results [5]. The center of $\mathcal{X} \otimes \mathbf{Q}_\ell$ is $F \otimes \mathbf{Q}_\ell$.

The Tate module V_ℓ decomposes as a product $\prod_\sigma V_\sigma$, where σ runs over the set Σ of embeddings $\sigma\colon E \to \mathbf{Q}_\ell$, and where $V_\sigma = V_\ell \otimes_{E \otimes \mathbf{Q}_\ell} \mathbf{Q}_\ell$, with $\mathbf{Q}_\ell$ being regarded as an $E \otimes \mathbf{Q}_\ell$ module via σ. (Cf. the proof of (3.4).) Each V_σ is a simple $\mathbf{Q}_\ell[H]$-module because A has no CM subvariety and because the action of H on V_ℓ is semisimple (Faltings). Hence $\mathrm{End}_H V_\sigma = \mathbf{Q}_\ell$ for each σ.

For each prime v of K which is prime to ℓ and the set of bad primes for A, there is a "trace of Frobenius" $t_v \in E$ associated with v. We have $\mathrm{tr}(\mathrm{Frob}_v \mid V_\sigma) = \sigma(t_v)$ for each σ. One knows for $\sigma, \tau \in \Sigma$ that V_σ and V_τ are isomorphic $\mathbf{Q}_\ell[H]$-modules if and only if $\sigma(t_v) = \tau(t_v)$ for all v, i.e., if and only if $\sigma|_L = \tau|_L$, where $L = \mathbf{Q}(\ldots, t_v, \ldots)$ (cf. [17, §IV.4]). This implies that the center of $\mathcal{X} \otimes \mathbf{Q}_\ell$ is $L \otimes \mathbf{Q}_\ell$. Thus $F \otimes \mathbf{Q}_\ell = L \otimes \mathbf{Q}_\ell$ (equality inside $E \otimes \mathbf{Q}_\ell$), which implies that $F = L$.

Suppose now that $\sigma = \tau$ on L, so that $V_\sigma \approx V_\tau$ as representations of H. A simple argument shows that there is a character $\varphi\colon \operatorname{Gal}(\overline{\mathbf{Q}}/\mathbf{Q}) \to \mathbf{Q}_\ell^*$ such that $V_\sigma \approx V_\tau \otimes \varphi$ as $\mathbf{Q}_\ell[\operatorname{Gal}(\overline{\mathbf{Q}}/\mathbf{Q})]$-modules. The character φ is necessarily unramified at all primes $p \neq \ell$ which are primes of good reduction for A. Taking traces, we get $\sigma(a_p) = \varphi(p)\tau(a_p)$ for all such primes. By considering determinants, we get ${}^\sigma\epsilon = \varphi^2 \cdot {}^\tau\epsilon$. These two equations show that σ and τ agree on $a_p^2/\epsilon(p)$ for all good primes $p \neq \ell$.

It follows by the Galois theory that we have $a_p^2/\epsilon(p) \in L$ for all good reduction primes $p \neq \ell$. Hence $a_p^2/\epsilon(p) \in F$ for all such p. By varying ℓ we obtain the inclusion $\mathbf{Q}(\ldots, a_p^2/\epsilon(p), \ldots) \subseteq F$, where p runs over the set of all primes of good reduction for A.

To prove the opposite inclusion, we must show that if $\sigma(a_p^2/\epsilon(p)) = \tau(a_p^2/\epsilon(p))$ for all p, then V_σ and V_τ are H-isomorphic. By the Cebotarev Density Theorem, the hypothesis implies that the functions on $\operatorname{Gal}(\overline{\mathbf{Q}}/\mathbf{Q})$ "$\operatorname{tr}^2 / \det$" are the same for V_σ and V_τ. In particular, we have $\operatorname{tr}(h|V_\sigma) = \pm\operatorname{tr}(h|V_\tau)$ for all $h \in H$. This equality implies that V_σ and V_τ become isomorphic after H is replaced by an open subgroup H_o of H (cf. [22, p. 324]). Since H was already chosen "sufficiently small", we find that V_σ and V_τ are indeed isomorphic as $\mathbf{Q}_\ell[H]$-modules. □

Corollary 5.4. *The center F of $\mathcal{X}$ is a totally real number field. The extension E/F is abelian.*

Proof. The Corollary follows from (5.3) and (3.6). □

Let g be an element of $\operatorname{Gal}(\overline{\mathbf{Q}}/\mathbf{Q})$, and consider the automorphism $x \mapsto {}^g x$ of $\mathcal{X}$ which is induced by g. This automorphism is necessarily inner (Skolem-Noether theorem), and it fixes E; therefore it is given by conjugation by an element $\alpha(g)$ of E^* which is well defined modulo F^*. The map $\alpha\colon \operatorname{Gal}(\overline{\mathbf{Q}}/\mathbf{Q}) \to E^*/F^*$ is a continuous homomorphism (cf. [20, p. 268]). It is a fact that α is unramified at all primes p at which A has semistable reduction [16, Th. 1.1].

Theorem 5.5. *We have $\alpha^2 \equiv \epsilon \pmod{F^*}$. Moreover, suppose that p is a prime of good reduction for A such that $a_p \neq 0$. Then $\alpha(\mathrm{Frob}_p) \equiv a_p \pmod{F^*}$.*

Proof. To prove the first assertion, let ℓ be a prime which splits completely in E. We must prove that $\sigma(\alpha^2(g)/\epsilon(g)) = \tau(\alpha^2(g)/\epsilon(g))$ whenever σ and τ are embeddings $E \hookrightarrow \mathbf{Q}_\ell$ which agree on F.

If σ and τ have this property, then there is a character $\varphi\colon \operatorname{Gal}(\overline{\mathbf{Q}}/\mathbf{Q}) \to \mathbf{Q}_\ell^*$ for which $V_\sigma \approx V_\tau \otimes \varphi$ (the notation is as in the proof of (5.3)). The Galois group $\operatorname{Gal}(\overline{\mathbf{Q}}/\mathbf{Q})$ acts on $\operatorname{Hom}(V_\sigma, V_\tau)$ by multiplication by $\varphi(g)^{-1}$, but also by conjugation by $\alpha(g)$. Since $\alpha(g)$ acts on V_σ and V_τ by ${}^\sigma\alpha(g)$ and ${}^\tau\alpha(g)$ (respectively), $\alpha(g)$ acts on $\operatorname{Hom}(V_\sigma, V_\tau)$ by ${}^\tau\alpha(g)/{}^\sigma\alpha(g)$. Hence, $\varphi(g) = {}^\sigma\alpha(g)/{}^\tau\alpha(g)$ in $\mathbf{Q}_\ell$. On the other hand, as remarked during the proof of (5.3), we have ${}^\sigma\epsilon = \varphi^2 \cdot {}^\tau\epsilon$. The two equalities give the required conclusion

$$\sigma(\alpha^2/\epsilon) = \tau(\alpha^2/\epsilon).$$

For the second assertion, we choose $\ell \neq p$. With the notation as above, $\varphi(p) = {}^{\sigma}\alpha(\mathrm{Frob}_p)/{}^{\tau}\alpha(\mathrm{Frob}_p)$. However, we have $\sigma(a_p) = \varphi(p)\tau(a_p)$, as noted during the proof of (5.3). On comparing two formulas for $\varphi(g)$, we get $\sigma(\alpha(\mathrm{Frob}_p)/a_p) = \tau(\alpha(\mathrm{Frob}_p)/a_p)$. □

Remark. It should be easy to prove that the set of p for which $a_p = 0$ has density 0, since the image of $\mathrm{Gal}(\overline{\mathbf{Q}}/\mathbf{Q})$ in $\mathrm{Aut}\, V_\sigma$ is open in $\mathrm{Aut}\, V_\sigma$, for any σ.

Let $\tilde{\alpha}$ be a lift of α to a function $\mathrm{Gal}(\overline{\mathbf{Q}}/\mathbf{Q}) \to E^*$. The function

$$c\colon (g_1, g_2) \mapsto \frac{\tilde{\alpha}(g_1)\tilde{\alpha}(g_2)}{\tilde{\alpha}(g_1 g_2)}$$

is then a 2-cocycle on $\mathrm{Gal}(\overline{\mathbf{Q}}/\mathbf{Q})$ with values in F^*. (We regard F^* as a $\mathrm{Gal}(\overline{\mathbf{Q}}/\mathbf{Q})$-module with trivial action.) The image $[c]$ of c in $H^2(\mathrm{Gal}(\overline{\mathbf{Q}}/\mathbf{Q}), F^*)$ is independent of the choice of $\tilde{\alpha}$. Since α^2 lifts to the character ϵ, it is clear that $[c]$ has order dividing two.

Consider the map

$$\xi\colon H^2(\mathrm{Gal}(\overline{\mathbf{Q}}/\mathbf{Q}), F^*) \longrightarrow H^2(\mathrm{Gal}(\overline{\mathbf{Q}}/F), \overline{\mathbf{Q}}^*) = \mathrm{Br}\, F$$

obtained from the restriction map $H^2(\mathrm{Gal}(\overline{\mathbf{Q}}/\mathbf{Q}), F^*) \to H^2(\mathrm{Gal}(\overline{\mathbf{Q}}/F), F^*)$ and the map $H^2(\mathrm{Gal}(\overline{\mathbf{Q}}/F), F^*) \to H^2(\mathrm{Gal}(\overline{\mathbf{Q}}/F), \overline{\mathbf{Q}}^*)$ induced by the inclusion of F^* into $\overline{\mathbf{Q}}^*$. (We view $\overline{\mathbf{Q}}^*$ as a $\mathrm{Gal}(\overline{\mathbf{Q}}/F)$-module in the standard way; i.e., we use the Galois action. The notation "$\mathrm{Br}\, F$" indicates the Brauer group of F.)

Theorem 5.6 (Chi). *The class of $\mathcal{X}$ in $\mathrm{Br}\, F$ coincides with $\xi([c])$.*

Proof. According to [3, Th. 3.4], the algebra $\mathcal{X}$ is isomorphic to a twisted matrix algebra $(\mathrm{End}_F\, E)(\alpha)$. Theorem 4.8 of [3] expresses $[(\mathrm{End}_F\, E)(\alpha)] \in \mathrm{Br}\, F$ as the image of a certain two-cocycle whose values are Jacobi sums. By [20, Prop. 1], the class of this Jacobi-sum cocycle coincides with the class of the two-cocycle

$$(g, h) \mapsto \tilde{\alpha}(h)^{g-1} = \frac{{}^{g}\tilde{\alpha}(h)}{\tilde{\alpha}(h)}$$

on $\mathrm{Gal}(\overline{\mathbf{Q}}/F)$. We wish to compare this with the two-cocycle

$$(g, h) \mapsto \frac{\tilde{\alpha}(g)\tilde{\alpha}(h)}{\tilde{\alpha}(gh)}.$$

The product of the two is the map

$$(g, h) \mapsto \frac{{}^{g}\tilde{\alpha}(h)\tilde{\alpha}(g)}{\tilde{\alpha}(gh)},$$

which is a coboundary. Hence the class of $\mathcal{X}$ in $\mathrm{Br}\, F$ is the negative of $\xi([c])$. Since $[c]$ (and $[\mathcal{X}]$) have order two, we get the Theorem as stated. □

A final remark about $\mathcal{X}$ concerns the isogeny class of the simple factors of $A_{\overline{\mathbf{Q}}}$. We have $\mathcal{X} \approx M(n, D)$ for some positive integer n, where D is a division algebra of dimension one or four over F. Accordingly, $A_{\overline{\mathbf{Q}}}$ decomposes up to isogeny as the product of n copies a simple abelian variety B over $\overline{\mathbf{Q}}$ whose endomorphism algebra is isomorphic to D. Since A is defined over $\mathbf{Q}$, we have

$$^{g}B \times \cdots \times {}^{g}B \sim {}^{g}A_{\overline{\mathbf{Q}}} \sim A_{\overline{\mathbf{Q}}} \sim B \times \cdots \times B$$

for each $g \in \mathrm{Gal}(\overline{\mathbf{Q}}/\mathbf{Q})$. (The sign "$\sim$" indicates an isogeny.) By the uniqueness of decomposition up to isogeny, we have ${}^{g}B \sim B$. One says that the isogeny class of B is "defined over $\mathbf{Q}$."

Consider the special case where $n = \dim A$, so that B is of dimension one. The elliptic curve B is then isogenous to each of its conjugates (over $\overline{\mathbf{Q}}$). Borrowing (and bending) a term used by B. H. Gross [7], we say that B is a "$\mathbf{Q}$-curve."

6. Q-curves as factors of abelian varieties of $\mathbf{GL}_2$-type

We have just proved: Suppose that C is an elliptic curve over $\overline{\mathbf{Q}}$ which occurs as a simple factor of a primitive abelian variety of $\mathbf{GL}_2$-type over $\mathbf{Q}$ with no complex multiplication over $\overline{\mathbf{Q}}$. Then C is a $\mathbf{Q}$-curve; it is, more precisely, a $\mathbf{Q}$-curve with no complex multiplication.

In this paragraph, we prove the converse:

Theorem 6.1. *Suppose that C is an elliptic curve over $\overline{\mathbf{Q}}$ with no complex multiplication. Assume that C is isogenous to each of its conjugates ${}^{g}C$ with $g \in \mathrm{Gal}(\overline{\mathbf{Q}}/\mathbf{Q})$. Then there is a primitive abelian variety A of $\mathbf{GL}_2$-type over $\mathbf{Q}$ such that C is a simple factor of A over $\overline{\mathbf{Q}}$.*

Corollary 6.2. *Suppose that C is as in* (6.1). *Assume Serre's conjecture* [24, (3.2.4?)] *on representations of* $\mathrm{Gal}(\overline{\mathbf{Q}}/\mathbf{Q})$. *Then C is a simple factor, over $\overline{\mathbf{Q}}$, of $J_1(N)$ for some $N \geq 1$.*

Proof. The Corollary follows directly from (6.1) and (4.4). □

Proof of (6.1). Let C be as in the statement of the Theorem. We can find a model C_o of C over a number field $K \subset \overline{\mathbf{Q}}$. We may assume that $K/\mathbf{Q}$ is a Galois extension. For each $g \in \mathrm{Gal}(K/\mathbf{Q})$, C_o and ${}^{g}C_o$ are $\overline{\mathbf{Q}}$-isogenous. Enlarging K if necessary, we may assume that there are isogenies $\mu_g\colon {}^{g}C_o \to C_o$ defined over K.

The map

$$c\colon (g,h) \mapsto \mu_g {}^{g}\mu_h \mu_{gh}^{-1}$$

may be regarded as $\mathbf{Q}^*$-valued, since $\mathbf{Q} \otimes \mathrm{End}_K(C_o) = \mathbf{Q}$. A short computation shows that c is a two-cocycle on $\mathrm{Gal}(K/\mathbf{Q})$ with values in $\mathbf{Q}^*$. The class of c in $H^2(\mathrm{Gal}(K/\mathbf{Q}), \mathbf{Q}^*)$ is independent of the choices of the μ_g. By inflation, we may (and will) regard c as a locally constant function on $\mathrm{Gal}(\overline{\mathbf{Q}}/\mathbf{Q})$.

Theorem 6.3 (Tate). *Let M be the discrete* $\mathrm{Gal}(\overline{\mathbf{Q}}/\mathbf{Q})$*-module $\overline{\mathbf{Q}}^*$ with trivial action. Then* $H^2(\mathrm{Gal}(\overline{\mathbf{Q}}/\mathbf{Q}), M) = 0$.

This theorem of Tate is proved as Theorem 4 in Serre's article [23], with $\overline{\mathbf{Q}}$ replaced by $\mathbf{C}$. The proof exposed by Serre begins with the observation that

$$H^2(\mathrm{Gal}(\overline{\mathbf{Q}}/\mathbf{Q}), \mathbf{C}^*) = H^2(\mathrm{Gal}(\overline{\mathbf{Q}}/\mathbf{Q}), W),$$

where W is the torsion subgroup of $\mathbf{C}^*$, i.e., the group of complex roots of unity. This observation follows from the fact that $\mathbf{C}^*/W$ is uniquely divisible. Since $\mathrm{Gal}(\overline{\mathbf{Q}}/\mathbf{Q})$ acts trivially on $\mathbf{C}^*$, W is $\mathrm{Gal}(\overline{\mathbf{Q}}/\mathbf{Q})$-isomorphic to $\mathbf{Q}/\mathbf{Z}$. One proves that $H^2(\mathrm{Gal}(\overline{\mathbf{Q}}/\mathbf{Q}), \mathbf{Q}/\mathbf{Z}) = 0$; in fact, we have $H^2(\mathrm{Gal}(\overline{K}/K), \mathbf{Q}/\mathbf{Z}) = 0$ whenever K is a local or global field [23, §6.5].

Since the quotient of M by its torsion subgroup is uniquely divisible, and since the torsion subgroup of M is isomorphic to $\mathbf{Q}/\mathbf{Z}$, we obtain $H^2(\mathrm{Gal}(\overline{\mathbf{Q}}/\mathbf{Q}), M) = H^2(\mathrm{Gal}(\overline{\mathbf{Q}}/\mathbf{Q}), \mathbf{Q}/\mathbf{Z}) = 0$. □

Because of Tate's theorem, there is a locally constant function

$$\alpha\colon \mathrm{Gal}(\overline{\mathbf{Q}}/\mathbf{Q}) \to \overline{\mathbf{Q}}^*$$

such that we have the identity among functions $G \times G \to \overline{\mathbf{Q}}^*$

$$c(g,h) = \frac{\alpha(g)\alpha(h)}{\alpha(gh)}.$$

After again enlarging K, we may identify α with a function on $\mathrm{Gal}(K/\mathbf{Q})$ and regard g and h as elements of this Galois group.

Let E be the extension of $\mathbf{Q}$ generated by the values of α. The definition of c shows that we have

$$c(g,h)^2 = \frac{\deg \mu_g \deg \mu_h}{\deg \mu_{gh}},$$

where "deg" denotes the degree of an isogeny between elliptic curves (or, more generally, of a non-zero element of $\mathbf{Q} \otimes \mathrm{Hom}(C_1, C_2)$, where C_1 and C_2 are elliptic curves). It follows that the function

$$\epsilon\colon g \mapsto \frac{\alpha^2(g)}{\deg \mu_g}$$

is a Dirichlet character $\mathrm{Gal}(K/\mathbf{Q}) \to E^*$. Because $\alpha^2 \equiv \epsilon \pmod{\mathbf{Q}^*}$, the field E is an abelian extension of $\mathbf{Q}$ (cf. Prop. 3.6).

If one performs the analysis of §3 on the abelian variety A which we construct below, one finds (e.g., in (3.1)) another Dirichlet character called ϵ. It is very likely that the two ϵ's are equal up to "sign", i.e., possible inversion (cf. Lemma 7.1 below).

To simplify notation, let us write simply C for the elliptic curve C_o over K and $C_{\overline{\mathbf{Q}}}$ for the curve originally called C. Let B be the abelian variety $\mathrm{Res}_{K/\mathbf{Q}}\, C$, where "Res" is Weil's "restriction of scalars" functor [33, 13]. Then B is an abelian variety over $\mathbf{Q}$ of dimension $[K:\mathbf{Q}]$. It represents the functor on $\mathbf{Q}$-schemes $S \mapsto C(S_K)$; in particular, we have $\mathrm{Hom}_{\mathbf{Q}}(X, B) = \mathrm{Hom}_K(X_K, C)$ whenever X is an abelian variety over $\mathbf{Q}$.

Applying this formula in the special case where $X = B$, we find $\mathrm{End}_{\mathbf{Q}}(B) = \mathrm{Hom}_K(B_K, C)$. On the other hand [33, p. 5],

$$B_K = \prod_{\sigma\in\mathrm{Gal}(K/\mathbf{Q})} {}^{\sigma}C.$$

Hence we have

$$\mathbf{Q}\otimes\mathrm{End}_Q(B) = \prod_{\sigma} \mathbf{Q}\otimes\mathrm{Hom}_K({}^{\sigma}C, C).$$

Since $\mathbf{Q}\otimes\mathrm{Hom}({}^{\sigma}C, C)$ is the one-dimensional vector space generated by μ_σ, we find that $\mathcal{R} := \mathbf{Q}\otimes\mathrm{End}_{\mathbf{Q}}(B)$ may be written as $\prod \mathbf{Q}\cdot\mu_\sigma$. Let λ_σ be the element of $\mathcal{R}$ which corresponds to $\mu_\sigma\colon {}^{\sigma}C \to C$; then $\mathcal{R}$ is a $\mathbf{Q}$-algebra with vector space basis λ_σ.

Lemma 6.4. *We have* $\lambda_\sigma\lambda_\tau = c(\sigma,\tau)\lambda_{\sigma\tau}$ *in* $\mathcal{R}$ *for* $\sigma,\tau\in\mathrm{Gal}(K/\mathbf{Q})$.

Proof. Each map λ_σ acts on $B_K = \prod_g {}^{g}C$ as a "matrix": it sends the factor ${}^{g\sigma}C$ to ${}^{g}C$ by ${}^{g}\mu_\sigma$ (cf. [7, §15]). A short computation using the identity

$$\mu_\sigma{}^{\sigma}\mu_\tau = c(\sigma,\tau)\mu_{\sigma\tau}$$

gives the desired formula. □

The algebra $\mathcal{R}$ is thus a "twisted group algebra" $\mathbf{Q}[\mathrm{Gal}(K/\mathbf{Q})]$ in which we have the multiplication table $[\sigma][\tau] = c(\sigma,\tau)[\sigma\tau]$. The obvious map of $\mathbf{Q}$-vector spaces

$$\omega\colon \mathcal{R}\to E, \qquad \lambda_\sigma \mapsto \alpha(\sigma)$$

is in fact a surjective homomorphism of $\mathbf{Q}$-algebras because of (6.4).

In [2], $\mathbf{Q}\otimes\mathrm{End}_{\mathbf{Q}}(B)$ is studied in an analogous situation where C has complex multiplication, and where K is the Hilbert class field of the field of complex multiplication. (The curve C is then a "**Q**-curve" in the original sense of the term.) Criteria are given for $\mathbf{Q}\otimes\mathrm{End}_{\mathbf{Q}}(B)$ to be (i) a commutative algebra, i.e., a product of number fields, and (ii) a product of *totally real* fields. It might be interesting to formulate similar criteria in our context.

Let T be the abelian variety $\prod_\sigma C$ over K. We write C_σ for the copy of C in the σth place, so that T becomes the product $\prod C_\sigma$. We have a map $\mathcal{R}\to \mathbf{Q}\otimes\mathrm{End}\,T$ given as follows: For $g\in\mathrm{Gal}(K/\mathbf{Q})$, the element $\lambda_g = [g]$ of $\mathcal{R}$ acts on T by sending C_σ to $C_{g\sigma}$ by the map (of elliptic curves up to isogeny) "multiplication by $c(g,\sigma)$." One checks directly that this is a homomorphism of $\mathbf{Q}$-algebras. The variety T will be interpreted as $\mathcal{R}\otimes_{\mathbf{Q}} C$ by readers who are fond of such tensor-product constructions.

Let $\iota\colon T \xrightarrow{\sim} B_K = \prod {}^{\sigma}C$ be the isomorphism of abelian varieties up to isogeny which takes the factor C_σ of T to the factor ${}^{\sigma^{-1}}C$ of B_K, via the map ${}^{\sigma^{-1}}\mu_\sigma$.

Proposition 6.5. *The map* ι *is* $\mathcal{R}$*-equivariant, for the action of* $\mathcal{R}$ *on* T *just defined and for the structural action of* $\mathcal{R}$ *on* B.

Proof. The proof of this Proposition is an uninteresting computation, which is omitted.

Corollary 6.6. *The Lie algebra* $\mathrm{Lie}(B/\mathbf{Q})$ *is a free* $\mathcal{R}$*-module of rank one.*

Proof. The statement to be proved is true if and only if $\mathrm{Lie}(B_K/K)$ is free of rank one over $R\otimes_{\mathbf{Q}} K$. By (6.5), $\mathrm{Lie}(B_K/K)$ may be identified with $\mathcal{R}\otimes_{\mathbf{Q}}$ $\mathrm{Lie}(C/K)$, with $\mathcal{R}$ operating trivially on the second factor. Hence $\mathrm{Lie}(B_K/K)$ is indeed free of rank one over $\mathcal{R}\otimes_{\mathbf{Q}} K$. □

To complete the proof of (6.1), we let A be the abelian variety $E\otimes_{\mathcal{R}} B$, where E is viewed as a $\mathcal{R}$-module via ω. Explicitly, use the fact that $\mathcal{R}$ is a semisimple $\mathbf{Q}$-algebra to write $\mathcal{R}$ as a direct sum of its quotient E with the kernel of the map ω. Let $\pi\in\mathcal{R}$ be the projector onto E, and let $A\subseteq B$ be the image of π, viewed as an endomorphism of A up to isogeny. (In other words, A is the image of $m\cdot\pi$, where m is a positive integer chosen so that $m\cdot\pi$ is a true endomorphism of B.) Then A is an abelian subvariety of B, defined over $\mathbf{Q}$, whose algebra of $\mathbf{Q}$-endomorphisms is E. By (6.6), E acts without multiplicity on $\mathrm{Lie}(B)$ and therefore, in particular, without multiplicity on $\mathrm{Lie}(A)$. Hence A has dimension equal to $[E:\mathbf{Q}]$, which means that A is of $\mathbf{GL}_2$-type. (It is clear that A is non-zero because π is non-zero.)

Since B_K is isogenous to a product of copies of C, the same holds true for A_K. Thus C is a quotient of B_K.

7. Q-curves over quadratic fields

Suppose that C is a $\mathbf{Q}$-curve as above and that K is a quadratic field. Let σ be the non-trivial automorphism of K over $\mathbf{Q}$. Then, by hypothesis, there is a K-isogeny $\mu=\mu_\sigma\colon {}^\sigma C\to C$. We take the identity map for μ_1, where "1" is the identity automorphism of K. The cocycle c takes the value 1 on all elements of $\mathrm{Gal}(K/\mathbf{Q})\times\mathrm{Gal}(K/\mathbf{Q})$ other than (σ,σ). Its value on that pair is the non-zero integer m such that $\mu\circ{}^\sigma\mu$ is multiplication by m on C. The algebra $\mathcal{R}$ may be written $\mathbf{Q}[X]/(X^2-m)$, where X corresponds to the element of $\mathcal{R}$ we have been calling $[\sigma]$.

Let us split c by defining $\alpha\colon \mathrm{Gal}(K/\mathbf{Q})\to\overline{\mathbf{Q}}^*$ to be the map taking 1 to 1 and σ to a square root of m. The character

$$\theta\colon g\mapsto \frac{\alpha^2(g)}{\deg\mu_g}$$

is then trivial if m is positive and of order two if m is negative. In the case where θ is of order two, it is an isomorphism $\mathrm{Gal}(K/\mathbf{Q})\xrightarrow{\sim}\{\pm1\}$.

If m is a perfect square, then we have $E=\mathbf{Q}$ in the notation of §6. The abelian variety A is then a model of C over $\mathbf{Q}$.

Assume for the rest of this paragraph that m is not a perfect square. Then $\mathcal{R}=E$ is a quadratic number field, and we have $B=A$ in the notation of §6. The field E is real if m is positive and imaginary if m is negative. Thus E is real if and only if θ is trivial.

The λ-adic representations of A define a Dirichlet character ϵ (Lemma 3.1). According to (3.2), ϵ is an even character. Also, ϵ is non-trivial if and only if E is imaginary (3.4). Thus ϵ if non-trivial if and only if θ is non-trivial.

Lemma 7.1. *The characters ϵ and θ are equal.*

Proof. By (6.5), A_K is K-isogenous to the abelian variety "$E \otimes C$", i.e., to the product of two copies of C with E acting through a regular representation $E \hookrightarrow M(2, \mathbf{Q})$. In particular, the λ-adic representations of A_K are just the ℓ-adic representations of C, viewed as taking values in $\mathbf{GL}(2, E_\lambda)$ rather than in $\mathbf{GL}(2, \mathbf{Q}_\ell)$. This implies that the determinants of the $\rho_\lambda|_{\mathrm{Gal}(\overline{\mathbf{Q}}/K)}$ are the cyclotomic characters χ_ℓ. Hence ϵ is trivial on $\mathrm{Gal}(\overline{\mathbf{Q}}/K)$, and therefore may be identified with a character of the group $\mathrm{Gal}(K/\mathbf{Q})$, whose order is two. Since θ is also a character of this latter group, and since the two characters are simultaneously non-trivial, they are equal. □

As mentioned above, it seems very likely that Lemma 7.1 (quite possibly in the form $\epsilon = \theta^{-1}$) generalizes to the situation of §6.

Proposition 7.2 (Serre). *At least one of the two quadratic fields E, K is a real quadratic field.*

Proof. We give two proofs, the first of which was communicated to the author by Serre: Assume that K is a complex quadratic field. After we embed K in $\mathbf{C}$, the automorphism σ of K becomes the restriction to K of $\bar{\ }$, complex conjugation on $\mathbf{C}$. Choose a holomorphic differential ω on C. Then $C_{\mathbf{C}}$ may be identified with the curve $\mathbf{C}/L$, with $L \subset \mathbf{C}$ the period lattice of ω. The curve ${}^\sigma C$ becomes the complex conjugate $\mathbf{C}/\bar{L}$ of C. The isogeny μ is induced by the map "multiplication by γ" on $\mathbf{C}$, for some non-zero complex number γ. The integer m may be identified with $\gamma\bar{\gamma}$, and is therefore positive. This means that E is a real quadratic field.

Another proof can be given as follows. Assume that E is an imaginary quadratic field. Then ϵ is a non-trivial character of $\mathrm{Gal}(K/\mathbf{Q})$. This character is *even* by (3.2). Therefore K is real. □

In connection with Proposition 7.2, it is natural to ask whether each of the three possibilities allowed by the Proposition do, in fact, occur. The case where E is imaginary, so that K is real, was treated in detail by Shimura [28, §10], who constructed a family of numerical examples. This case was later studied by Serre [24, p. 208], who pointed out that Theorem 4.4 holds in this context. A situation where E is a real quadratic field and K is an imaginary quadratic field is described by Koike in [10, §1]. In this example, the abelian variety A is associated to a weight-two newform on $\Gamma_o(81)$. One has $E = \mathbf{Q}(\sqrt{3})$ and $K = \mathbf{Q}(\sqrt{-3})$.

To exhibit an example where both E and K are real, we consider the eight-dimensional complex vector space S of weight-two cusp forms on $\Gamma_o(169)$. According to [1, Table 5], one can find a (normalized) newform $f = \sum a_n q^n \in S$ whose coefficients generate a quadratic field E. Moreover, the only two such forms are f and ${}^\sigma f = \sum {}^\sigma a_n q^n$, where σ is the non-trivial automorphism of E over $\mathbf{Q}$. Since f

is a form with trivial "Nebentypus" character, E is a priori a real quadratic field. In fact, the author has learned from D. Zagier that tables constructed by Cohen and Skoruppa show that $E = \mathbf{Q}(\sqrt{3})$.

Let φ be the quadratic Dirichlet character with conductor 13. Then $f \otimes \varphi := \sum \varphi(n) a_n q^n$ is again a normalized newform in S; its coefficients generate the same quadratic field E as the coefficients of f. Hence we have either $f = \varphi \otimes f$, or else ${}^\sigma f = \varphi \otimes f$. The former possibility is excluded by [18, Th. 4.5], since φ is an even character (i.e., since φ corresponds to a real quadratic field). Hence f is a form with an "extra twist" by φ: we have $\varphi \otimes f = {}^\sigma f$. Moreover, again by Theorem 4.5 of [18], f is not a form with complex multiplication.

The abelian variety $A = A_f$ associated to f satisfies $\mathbf{Q} \otimes \mathrm{End}_{\mathbf{Q}}(A) = E$. According to [19] (see especially Theorem 5.1 of [19]), the full algebra $\mathcal{X} = \mathbf{Q} \otimes \mathrm{End}_{\overline{\mathbf{Q}}}(A)$ of A coincides with its subalgebra $\mathbf{Q} \otimes \mathrm{End}_K(A) = E$, where $K = \mathbf{Q}(\sqrt{13})$ is the extension of $\mathbf{Q}$ cut out by φ. This algebra is a quaternion algebra over $\mathbf{Q}$. Since K may be embedded in $\mathbf{R}$, a well-known theorem of Shimura [30, Th. 0] implies that $\mathcal{X}$ is isomorphic to the matrix algebra $M(2, \mathbf{Q})$. (Once we know that $E = \mathbf{Q}(\sqrt{3})$, we can give an alternative proof that $\mathcal{X}$ is a matrix algebra which is based on the explicit description of $\mathcal{X}$ given in [19].) Hence A becomes isogenous over K to a product $C \times C$, where C is an elliptic curve defined over K. Therefore, we are in the situation described above, with $E = \mathbf{Q}(\sqrt{3})$ and $K = \mathbf{Q}(\sqrt{13})$.

8. Descent of abelian varieties up to isogeny

Suppose that L/K is a Galois extension of fields and that A is an abelian variety over L. A well-known theorem of Weil [32] states that A has a model over K if and only if there are isomorphisms $\mu_\sigma : {}^\sigma A \xrightarrow{\sim} A$ ($\sigma \in \mathrm{Gal}(L/K)$) which satisfy the compatibility condition

$$\mu_\sigma {}^\sigma\mu_\tau = \mu_{\sigma\tau}. \tag{8.1}$$

As an application of the techniques encountered in §6, we will prove an analogous criterion for abelian varieties *up to isogeny*. Namely, suppose that A is isogenous (over L) to an abelian variety B/L which has a model over K. Then one finds isomorphisms $\mu_\sigma : {}^\sigma A \xrightarrow{\sim} A$ of abelian varieties up to isogeny over L which satisfy the compatibility condition (8.1). Conversely, one has

Theorem 8.1. *Suppose that there are isomorphisms of abelian varieties up to isogeny over L, μ: ${}^\sigma A \xrightarrow{\sim} A$, which satisfy* (8.1). *Then there is an abelian variety B over K such that A is L-isogenous to B_L.*

Proof. We can, and do, assume that L is a finite extension of K. Let X be the abelian variety $X = \mathrm{Res}_{L/K} A$. Recalling the discussion of §6, we find a decomposition

$$\mathbf{Q} \otimes \mathrm{End}_K(X) = \prod_{\sigma \in \mathrm{Gal}(L/K)} \mathbf{Q} \otimes \mathrm{Hom}_L({}^\sigma A, A).$$

The homomorphism μ_σ in the "σ^{th} factor" corresponds to an element $[\sigma]$ of $\mathbf{Q} \otimes \operatorname{End}_K(X)$. The element $[g]$ operates on $\prod {}^\sigma A$ as a matrix, sending ${}^{\tau g}A$ to ${}^\tau A$ by ${}^\tau\mu_g$ for each $\tau \in \operatorname{Gal}(L/K)$.

Because the analogue of $c(\sigma, \tau)$ is 1 in this context, we have simply $[\sigma][\tau] = [\sigma\tau]$ for $\sigma, \tau \in \operatorname{Gal}(L/K)$: the identity (8.1) shows that the product of the matrices representing $[\sigma]$ and $[\tau]$ is the matrix representing $[\sigma\tau]$. Hence $\mathcal{R} := \mathbf{Q}[\operatorname{Gal}(L/K)]$ operates on X.

An analogue of Proposition 6.5 shows that we have

$$X_L \approx \mathcal{R} \otimes_{\mathbf{Q}} A$$

in the category of abelian varieties over L up to isogeny. To see this explicitly, we let A_σ be a copy of A indexed by σ and consider the isomorphism of abelian varieties over L up to isogeny

$$\iota\colon \prod_\sigma A_\sigma \xrightarrow{\sim} \prod_\sigma {}^\sigma A = X_L$$

which takes A_σ to ${}^{\sigma^{-1}}A$ via the map ${}^{\sigma^{-1}}\mu_\sigma$. Via this isomorphism, the automorphism $[\sigma]$ of X is transported to the permutation which takes each factor A_g of $\prod_g A_g$ to the factor $A_{\sigma g}$, via the identity map $A \to A$.

Let B be the image of $\eta := \sum_\sigma [\sigma]$, so that B is an abelian subvariety of X which is defined over K. (The sum η need not be a literal endomorphism of X, so that, strictly speaking, one should consider the image of a suitable multiple of η.) The isomorphism ι makes B_L correspond to the diagonal image of A in $\prod A_\sigma = A \times \cdots \times A$. Hence B_L is isogenous to A. □

References

[1] B.J. Birch and W. Kuyk, eds., *Modular functions of one variable IV*, Lecture Notes in Math., vol. 476, Springer-Verlag, Berlin and New York, 1975.

[2] J. Buhler and B.H. Gross, *Arithmetic on elliptic curves with complex multiplication. II*, Invent. Math **79** (1985), 11–29.

[3] W. Chi, *Twists of central simple algebras and endomorphism algebras of some abelian varieties* Math. Ann. **276** (1987), 615–632.

[4] P. Deligne, *Valeurs de fonctions L et périodes d'intégrales*, Proceedings of Symposia in Pure Mathematics **33** (1979), (2) 313–346.

[5] G. Faltings, *Endlichkeitssätze für abelsche Varietäten über Zahlkörpern*, Invent. Math. **73** (1983), 349–366.

[6] G. Faltings, G. Wüstholz et. al., *Rational points*, F. Vieweg & Sohn, Braunschweig-Wiesbaden, 1984.

[7] B.H. Gross, *Arithmetic on elliptic curves with complex multiplication*, Lecture Notes in Math., vol. 776, Springer-Verlag, Berlin and New York, 1980.

[8] A. Grothendieck, *SGA7 I, Exposé IX*, Lecture Notes in Math., vol. 288, Springer-Verlag, Berlin and New York, 1972, pp. 313–523.

[9] H. Hida, *On abelian varieties with complex multiplication as factors of the Jacobians of Shimura curves*, Am. J. Math. **103** (1981), 727–776.

[10] M. Koike, *On certain abelian varieties obtained from new forms of weight* 2 *on* $\Gamma_0(3^4)$ *and* $\Gamma_0(3^5)$, Nagoya Math. J. **62** (1976), 29–39.

[11] D.J. Lorenzini, *On the jacobian of the modular curve* $X_0(N)$, to appear.

[12] B. Mazur, *Number theory as gadfly*, Am. Math. Monthly **98** (1991) 593–610.

[13] J.S. Milne, *On the arithmetic of abelian varieties*, Invent. Math. **17** (1972) 177–190.

[14] D. Mumford, *Abelian varieties*, Second edition, Oxford University Press, London, 1974.

[15] A. Ogg, *Elliptic curves and wild ramification*, Am. J. Math. **89** (1967), 1–21.

[16] K.A. Ribet, *Endomorphisms of semi-stable abelian varieties over number fields*, Annals of Math. **101** (1975), 555–562.

[17] ———, *Galois action on division points on abelian varieties with many real multiplications*, Am. J. Math. **98** (1976), 751–804.

[18] ———, *The ℓ-adic representations attached to an eigenform with Nebentypus: a survey*, Lecture Notes in Math., vol. 601, Springer-Verlag, Berlin and New York, 1977, pp. 17–52.

[19] ———, *Twists of modular forms and endomorphisms of abelian varieties* Math. Ann. **253** (1980), 43–62.

[20] ———, *Endomorphism algebras of Abelian varieties attached to newforms of weight* 2, Progress in Math. **12** (1981), 263–276.

[21] J.-P. Serre, *Abelian ℓ-adic representations and elliptic curves*, Addison-Wesley Publ. Co., Redding, Mass., 1989.

[22] ———, *Propriétés galoisiennes des points d'ordre fini des courbes elliptiques*, Invent. Math. **15** (1972), 259–331.

[23] ———, *Modular forms of weight one and Galois representations*, prepared in collaboration with C. J. Bushnell, Algebraic number fields (*L*-functions and Galois properties), A. Fröhlich, ed., Academic Press, London, New York and San Francisco, 1977, pp. 193–268.

[24] ———, *Sur les représentations modulaires de degré* 2 *de* $\mathrm{Gal}(\overline{\mathbf{Q}}/\mathbf{Q})$, Duke Math. J. **54** (1987), 179–230.

[25] G. Shimura, *Algebraic number fields and symplectic discontinuous groups*, Ann. of Math. **86** (1967), 503–592.

[26] ———, *On elliptic curves with complex multiplication as factors of the Jacobians of modular function fields*, Nagoya Math. J. **43** (1971), 199–208.

[27] ———, *Introduction to the arithmetic theory of automorphic functions*, Princeton University Press, Princeton, 1971.

[28] ———, *Class fields over real quadratic fields and Hecke operators*, Ann. of Math. **95** (1972), 131–190.

[29] ———, *On the factors of the Jacobian variety of a modular function field*, J. Math. Soc. Japan **25** (1973), 523–544.

[30] ———, *On the real points of an arithmetic quotient of a bounded symmetric domain*, Math. Ann. **215** (1975), 135–164.

[31] G. Shimura and Y. Taniyama, *Complex multiplication of abelian varieties and its applications to number theory*, Second printing corrected, Math. Society of Japan, Tokyo, 1975.

[32] A. Weil, *The field of definition of a variety*, Am. J. Math **78** (1956), 509–524.

[33] ———, *Adeles and algebraic groups*, Progress in Math., vol. 23, Birkhäuser, Boston and Basel, 1982.

Kenneth A. Ribet
UC Mathematics Department
Berkeley, CA 94720 USA
e-mail: `ribetmath.berkeley.edu`

Progress in Mathematics, Vol. 224, 263–276

Shimura Curves Embedded in Igusa's Threefold

Victor Rotger

Abstract. Let $\mathcal{O}$ be a maximal order in a totally indefinite quaternion algebra over a totally real number field. In this note we study the locus $\mathcal{Q}_{\mathcal{O}}$ of quaternionic multiplication by $\mathcal{O}$ in the moduli space $\mathcal{A}_g$ of principally polarized abelian varieties of even dimension g with particular emphasis in the two-dimensional case. We describe $\mathcal{Q}_{\mathcal{O}}$ as a union of Atkin-Lehner quotients of Shimura varieties and we compute the number of irreducible components of $\mathcal{Q}_{\mathcal{O}}$ in terms of class numbers of CM-fields.

1. Introduction

Let A be an abelian variety over a fixed algebraic closure $\bar{\mathbb{Q}}$ of the field of rational numbers. By Poincaré's Decomposition Theorem, the algebra of endomorphisms $\mathrm{End}(A) \otimes \mathbb{Q}$ of A decomposes as a direct sum

$$\mathrm{End}(A) \otimes \mathbb{Q} \simeq \oplus_{i=1}^{s} \mathrm{M}_{n_i}(B_i)$$

of matrix algebras over a division algebra B_i.

The ranks $[B_i : \mathbb{Q}]$ are bounded in terms of the dimension of A and, by a classical theorem of Albert, the algebras B_i are isomorphic to either a totally real field, a quaternion algebra over a totally real field or a division algebra over a CM-field (cf. [8], [7]).

We will focus our attention on abelian varieties with totally indefinite quaternionic multiplication. More precisely, let F be a totally real number field of degree $[F : \mathbb{Q}] = n \geq 1$, let R_F be its ring of integers and let ϑ_F denote the different ideal of F over $\mathbb{Q}$. We also let F_+^* denote the group of totally positive elements of F^* and $R_{F+}^* = R_F^* \cap F_+^*$.

Let B be a totally indefinite division quaternion algebra over F, i.e., a division algebra of rank 4 over F such that $B \otimes_{\mathbb{Q}} \mathbb{R} \simeq \oplus_{i=1}^{n} \mathrm{M}_2(\mathbb{R})$, and let $\mathrm{disc}(B)$ denote the reduced discriminant ideal of B. We assume for simplicity that ϑ_F and $\mathrm{disc}(B)$ are coprime ideals of F. Since B is totally indefinite and division, it follows from [16] that $\mathrm{disc}(B) = \wp_1 \cdot \cdots \cdot \wp_{2r}$ for pairwise different prime ideals $\wp_i$ of F and $r \geq 1$. We shall denote by $\mathrm{n} = \mathrm{n}_{B/F}$ and $\mathrm{tr} = \mathrm{tr}_{B/F}$ the reduced norm and reduced trace on B, respectively. For any subset O of B, we shall write

---

Partially supported by a grant FPI from Ministerio de Educación y Ciencia and by Ministerio de Ciencia y Tecnología BFM2000-0627.

$O_0 = \{\beta \in O : \mathrm{tr}(\beta) = 0\}$ for the set of pure quaternions of O. We shall also use the notation $O_+ = \{\beta \in O : \mathrm{n}(\beta) \in F_+^*\}$ and $O^1 = \{\beta \in O : \mathrm{n}(\beta) = 1\}$.

Finally, let $\mathcal{O}$ be a maximal order in B and let $\vartheta(\mathcal{O}) = \{\mu \in \mathcal{O} : \mathrm{disc}(B)|\mathrm{n}(\mu)\}$ denote the reduced different of $\mathcal{O}$ over R_F. This is a two-sided ideal of $\mathcal{O}$ such that $\mathrm{n}(\vartheta(\mathcal{O})) \cdot R_F = \mathrm{disc}(B)$.

Let $\mathcal{A}_g/\mathbb{Q}$ be the moduli space of principally polarized abelian varieties of dimension $g = 2n$. We let $[(A, \mathcal{L})]$ denote the isomorphism class of a principally polarized abelian variety $(A, \mathcal{L})$ regarded as a closed point in $\mathcal{A}_g$.

It is our aim to investigate the nature of the quaternionic locus

$$Q_{\mathcal{O}} \subset \mathcal{A}_g(\mathbb{C})$$

of isomorphism classes of complex principally polarized abelian varieties $[(A, \mathcal{L})]$ of dimension g such that $\mathrm{End}(A) \supseteq \mathcal{O}$.

A reformulation of Proposition 6.2 in [10] yields that the quaternionic locus $Q_{\mathcal{O}}$ is *nonempty* if and only if $\mathrm{disc}(B)$ *is a totally positive principal ideal of* F. In consequence, for the sake of clarity of the exposition, we will assume throughout these notes that the narrow class number of F is $h_+(F) = 1$. This automatically implies the nonemptyness of $Q_{\mathcal{O}}$.

Acknowledgements. I am grateful to A. Arenas for some useful comments and correspondence.

2. Abelian varieties with quaternionic multiplication

As a first step in the study of the quaternionic locus $Q_{\mathcal{O}}$ in the moduli space $\mathcal{A}_g$, it is necessary to understand the geometry of the objects that $Q_{\mathcal{O}}$ parametrizes. Let us review some of the results that were accomplished in [10] in this direction.

Definition 2.1. An abelian variety with quaternionic multiplication by $\mathcal{O}$ over $\bar{\mathbb{Q}}$ is an abelian variety $A/\bar{\mathbb{Q}}$ such that $\mathrm{End}(A) \simeq \mathcal{O}$ and $\dim(A) = 2[F : \mathbb{Q}] = 2n$.

Let A be an abelian variety with quaternionic multiplication by $\mathcal{O}$ over $\bar{\mathbb{Q}}$ and let $\mathrm{NS}(A)$ denote the free $\mathbb{Z}$-module of rank $3n = 3g/2$ of line bundles on A up to algebraic equivalence. We say that two line bundles $\mathcal{L}$, $\mathcal{L}' \in \mathrm{NS}(A)$ are isomorphic if there exists an automorphism $\alpha \in \mathrm{Aut}(A) \simeq \mathcal{O}^*$ such that $\mathcal{L}' = \alpha^*(\mathcal{L})$.

Theorem 2.2. *Let* $A/\bar{\mathbb{Q}}$ *be an abelian variety with quaternion multiplication by* $\mathcal{O}$ *and let* $\iota : \mathcal{O} \simeq \mathrm{End}(A)$ *be a fixed isomorphism of rings. Then, there is an isomorphism of groups*

$$\begin{array}{ccc} \mathrm{NS}(A) & \xrightarrow{\sim} & \vartheta(\mathcal{O})_0 \\ \mathcal{L} & \mapsto & c_1(\mathcal{L}) = \mu \end{array}$$

between the Néron-Severi group of A *and the group of pure quaternions of the reduced different* $\vartheta(\mathcal{O})$ *of* $\mathcal{O}$.

Moreover, for any two non-trivial line bundles $\mathcal{L}$, $\mathcal{L}' \in \mathrm{NS}(A)$, *let* $\mu = c_1(\mathcal{L})$ *and* $\mu' = c_1(\mathcal{L}')$. *Then we have that*

1. $\mathcal{L} \simeq \mathcal{L}'$ *if and only if there exists* $\alpha \in \mathcal{O}^*$ *such that* $\mu' = \bar{\alpha}\mu\alpha$.
2. $\deg(\mathcal{L}) = \deg(\varphi_{\mathcal{L}} : A \to \hat{A})^{1/2} = \mathrm{N}_{F/\mathbb{Q}}(\mathrm{n}(\mu)/D)$.
3. $\mathcal{L}$ *is a polarization if and only if* $\mathrm{n}(\mu) \in F_+^*$ *and* μ *is* ample *(cf.* [10], §5*)*.
4. *The Rosati involution on* $\mathcal{O} \overset{\iota}{\simeq} \mathrm{End}(A)$ *with respect to* $\mathcal{L}$ *is*

$$\circ : \begin{array}{ccc} \mathcal{O} & \to & \mathcal{O} \\ \beta & \mapsto & \mu^{-1}\bar{\beta}\mu \end{array}$$

Let us write the isomorphism c_1 in more explicit terms. Fix an immersion of $\bar{\mathbb{Q}}$ into the field $\mathbb{C}$ of complex numbers and let $A(\mathbb{C}) = V/\Lambda$ for some complex vector space V and a lattice Λ. Upon the choice of an isomorphism $\iota : \mathcal{O} \simeq \mathrm{End}(A)$, the lattice Λ is naturally a left $\mathcal{O}$-module and, since $h(F) = h_+(F) = 1$, by a theorem of Eichler [2] we know that $\Lambda \simeq \mathcal{O}$.

By the Appell-Humbert Theorem, a line bundle $\mathcal{L} \in \mathrm{NS}(A)$ on A can be regarded as a Riemann form $E : \Lambda \times \Lambda \to \mathbb{Z}$ on V. Let us identify $\Lambda = \mathcal{O}$ through a fixed isomorphism and let $t \in F^*$ be any generator of the principal ideal ϑ_F. Then, the inverse isomorphism $c_1^{-1} : \vartheta(\mathcal{O})_0 \overset{\sim}{\to} \mathrm{NS}(A)$ maps a pure quaternion μ to the Riemann form $E_\mu : \mathcal{O} \times \mathcal{O} \to \mathbb{Z}$, $(\beta_1, \beta_2) \mapsto -\mathrm{tr}_{F/\mathbb{Q}}(\mathrm{tr}(\mu\beta_1\bar{\beta}_2)/\mathrm{n}(\mu))$.

Remark 2.3. The isomorphism c_1 is canonical in the sense that it does not depend on the choice of any polarization on A. However, we warn the reader that it does depend on the choice of the isomorphism $\iota : \mathcal{O} \simeq \mathrm{End}(A)$.

Theorem 2.4. *Let A be an abelian variety over $\bar{\mathbb{Q}}$ with quaternionic multiplication by $\mathcal{O}$. Let $D \in F_+^*$ be a totally positive generator of* $\mathrm{disc}(B)$.

Then, A is principally polarizable and the number of isomorphism classes of principal polarizations on A is

$$\pi_0(A) = \frac{1}{2}\sum_S h(S),$$

where S runs among the set of orders in the CM-field $F(\sqrt{-D})$ that contain $R_F[\sqrt{-D}]$ and $h(S)$ denotes its class number.

3. Shimura varieties

As in the preceding sections, let B be a totally indefinite division quaternion algebra over a totally real field F of trivial narrow class number. Let $D \in F_+^*$ be a totally positive generator of $\mathrm{disc}(B)$.

Definition 3.1. A *principally polarized maximal order* of B is a pair $(\mathcal{O}, \mu)$ where $\mathcal{O} \subset B$ is a maximal order and $\mu \in \mathcal{O}$ is a pure quaternion such that $\mu^2 + uD = 0$ for some $u \in R_{F+}^*$.

Attached to a principally polarized maximal order $(\mathcal{O}, \mu)$ there is the following moduli problem over $\mathbb{Q}$: classifying isomorphism classes of triplets $(A, \iota, \mathcal{L})$ given by

- An abelian variety A of dimension $g = 2n$.
- A ring homomorphism $\iota : \mathcal{O} \hookrightarrow \mathrm{End}\,(A)$.
- A principal polarization $\mathcal{L}$ on A such that

$$\iota(\beta)^\circ = \iota(\mu^{-1}\bar{\beta}\mu)$$

for any $\beta \in \mathcal{O}$, where $\circ : \mathrm{End}(A) \to \mathrm{End}(A)$ is the Rosati involution with respect to $\mathcal{L}$.

A triplet $(A, \iota, \mathcal{L})$ will be referred to as a polarized abelian variety with multiplication by $\mathcal{O}$. Two triplets $(A_1, \iota_1, \mathcal{L}_1)$, $(A_2, \iota_2, \mathcal{L}_2)$ are isomorphic if there exists an isomorphism $\alpha \in \mathrm{Hom}(A_1, A_2)$ such that $\alpha\iota_1(\beta) = \iota_2(\beta)\alpha$ for any $\beta \in \mathcal{O}$ and $\alpha^*(\mathcal{L}_2) = \mathcal{L}_1 \in \mathrm{NS}(A_1)$. Note also that, since a priori there is no canonical structure of R_F-algebra on $\mathrm{End}(A)$, the immersion $\iota : \mathcal{O} \hookrightarrow \mathrm{End}(A)$ is just a homomorphism of rings.

As it was proved by Shimura, the corresponding moduli functor is coarsely represented by an irreducible and reduced quasi-projective scheme $\mathfrak{X}_\mu/\mathbb{Q}$ over $\mathbb{Q}$ and of dimension $n = [F : \mathbb{Q}]$. Moreover, since B is division, the Shimura variety $\mathfrak{X}_\mu$ is complete (cf. [14], [15]).

Let $\mathfrak{H} = \{z \in \mathbb{C} : \mathrm{Im}(z) > 0\}$ denote the upper half-plane. Complex analytically, the manifold $\mathfrak{X}_\mu(\mathbb{C})$ can be described independently of the choice of μ as the quotient

$$\mathcal{O}^1 \backslash \mathfrak{H}^n \simeq \mathfrak{X}_\mu(\mathbb{C})$$

of the symmetric space $\mathfrak{H}^n$ by the action of the group $\mathcal{O}^1$ regarded suitably as a discontinuous subgroup of $\mathrm{SL}_2(\mathbb{R})^n$. See [11] for details.

In addition, we are also interested in the Hilbert modular reduced scheme $\mathcal{H}_F/\mathbb{Q}$ that coarsely represents the functor attached to the moduli problem of classifying principally polarized abelian varieties A of dimension g together with an homomorphism $R_F \hookrightarrow \mathrm{End}(A)$. The Hilbert modular variety $\mathcal{H}_F$ has dimension $3n$ and $\mathcal{H}_F(\mathbb{C})$ is the quotient of n copies of the three-dimensional Siegel half-space $\mathfrak{H}_2$ by a suitable discontinuous group (cf. [14], [7]).

Notice that, when $F = \mathbb{Q}$, $\mathcal{H}_F = \mathcal{A}_2$ is Igusa's three-fold, the moduli space of principally polarized abelian surfaces.

There are natural morphisms

$$\begin{array}{rccccc} \pi: & \mathfrak{X}_\mu & \xrightarrow{\pi_F} & \mathcal{H}_F & \longrightarrow & \mathcal{A}_g \\ & (A, \iota, \mathcal{L}) & \mapsto & (A, \iota|_{R_F}, \mathcal{L}) & \mapsto & (A, \mathcal{L}) \end{array}$$

from the Shimura variety $\mathfrak{X}_\mu$ to the Hilbert modular variety $\mathcal{H}_F$ and the moduli space $\mathcal{A}_g$ that consist of gradually *forgetting* the quaternionic endomorphism structure. These morphisms are representable, proper and defined over the field $\mathbb{Q}$ of rational numbers.

As it will be convenient for our purposes in the rest of this paper, we introduce the variety $\widetilde{\mathfrak{X}}_\mu = \pi(\mathfrak{X}_\mu)$ to be the image of the Shimura variety $\mathfrak{X}_\mu$ attached to a pure quaternion $\mu \in \mathcal{O}$ with $\mu^2 + uD = 0$ for some $u \in R^*_{F+}$ in the moduli space $\mathcal{A}_g$ by the forgetful map π. It is important to remark that although the complex analytical structure of $\mathfrak{X}_\mu$ does not depend on the choice of μ, the construction of the forgetful map π and the subvariety $\widetilde{\mathfrak{X}}_\mu$ of $\mathcal{A}_g$ do.

The varieties $\widetilde{\mathfrak{X}}_\mu$ are reduced, irreducible, complete and possibly singular schemes over $\mathbb{Q}$ of dimension n. The set of singularities of $\widetilde{\mathfrak{X}}_\mu$ is a finite set and all the singularities are of *quotient type* (cf. [3] for the terminology).

4. The birational class of the forgetful maps

It is our aim now to describe the forgetful maps $\pi : \mathfrak{X}_\mu \xrightarrow{\pi_F} \mathcal{H}_F \longrightarrow \mathcal{A}_g$ as the projection of the Shimura varieties $\mathfrak{X}_\mu$ onto their quotient by suitable Atkin-Lehner groups up to a birational equivalence. The following groups of Atkin-Lehner involutions were introduced in [12]. We keep the notations and assumptions of the introduction.

Definition 4.1. The *Atkin-Lehner group* W of the maximal order $\mathcal{O}$ is

$$W = \mathrm{Norm}_{B_+^*}(\mathcal{O})/F^* \cdot \mathcal{O}^1.$$

It was shown in [11] that $W \simeq \mathbb{Z}/2\mathbb{Z} \times \overset{2r}{\dots} \times \mathbb{Z}/2\mathbb{Z}$, where $2r$ is the number of ramified prime ideals of B.

Definition 4.2. Let $(\mathcal{O}, \mu)$ be a principally polarized maximal order in B. A *twist* of $(\mathcal{O}, \mu)$ is an element $\chi \in \mathcal{O} \cap \mathrm{Norm}_{B^*}(\mathcal{O})$ such that $\chi^2 + \mathrm{n}(\chi) = 0$ and $\mu\chi = -\chi\mu$.

In other words, a twist of $(\mathcal{O}, \mu)$ is a pure quaternion $\chi \in \mathcal{O} \cap \mathrm{Norm}_{B^*}(\mathcal{O})$ such that

$$B = F + F\mu + F\chi + F\mu\chi = (\frac{-uD, -\mathrm{n}(\chi)}{F}).$$

We say that a principally polarized maximal order $(\mathcal{O}, \mu)$ in B is *twisting* if it admits some twist χ in $\mathcal{O}$. We say that a maximal order $\mathcal{O}$ is *twisting* if there exists $\mu \in \mathcal{O}$ such that $(\mathcal{O}, \mu)$ is a twisting principally polarized order. Finally, we say that B is *twisting* if there exists a twisting maximal order $\mathcal{O}$ in B. Note that B is twisting if and only if $B \simeq (\frac{-uD, m}{F})$ for some $u \in R_{F+}^*$ and $m \in F^*$ such that $m|D$.

Definition 4.3. A *twisting involution* $\omega \in W$ of $(\mathcal{O}, \mu)$ is an Atkin-Lehner involution such that $[\omega] = [\chi] \in W$ is represented in B^* by a twist χ of $(\mathcal{O}, \mu)$.

We let $V_0 = V_0(\mathcal{O}, \mu)$ denote the subgroup of W generated by the twisting involutions of $(\mathcal{O}, \mu)$. For a principally polarized maximal order $(\mathcal{O}, \mu)$, let $R_\mu = F(\mu) \cap \mathcal{O}$ and let $\Omega = \Omega(R_\mu) = \{\xi \in R_\mu : \xi^f = 1, f \geq 1\}$ denote the finite group of roots of unity in the CM-quadratic order R_μ over R_F.

Definition 4.4. The *stable group* of $(\mathcal{O}, \mu)$ is the subgroup

$$W_0 = U_0 \cdot V_0$$

of W generated by

$$U_0 = U_0(\mathcal{O}, \mu) = \mathrm{Norm}_{F(\mu)^*}(\mathcal{O})/F^* \cdot \Omega(R_\mu),$$

and the group V_0 of twisting involutions of $(\mathcal{O}, \mu)$.

As it was also shown in [11], for any principally polarized maximal order $(\mathcal{O}, \mu)$, there are natural monomorphisms of groups $V_0 \subseteq W_0 \subseteq W \subseteq \mathrm{Aut}_{\mathbb{Q}}(\mathfrak{X}_\mu) \subseteq \mathrm{Aut}_{\bar{\mathbb{Q}}}(\mathfrak{X}_\mu \otimes \bar{\mathbb{Q}})$. The question whether the two latter immersions are actually isomorphisms was studied by the author for the case of Shimura curves in [9].

The following was proved in [11].

Theorem 4.5. *Let $(\mathcal{O}, \mu)$ be a principally polarized maximal order in B and let $\mathfrak{X}_\mu$ be the Shimura variety attached to it. Then there is a commutative diagram of finite maps*

$$\begin{array}{ccc} \mathfrak{X}_\mu & \xrightarrow{\pi_F} & \mathcal{H}_F \\ & \searrow \quad \nearrow^{b_F} & \\ & \mathfrak{X}_\mu/W_0, & \end{array}$$

where $\mathfrak{X}_\mu \to \mathfrak{X}_\mu/W_0$ is the natural projection and $b_F : \mathfrak{X}_\mu/W_0 \to \pi_F(\mathfrak{X}_\mu)$ is a birational morphism from $\mathfrak{X}_\mu/W_0$ onto the image of $\mathfrak{X}_B$ in $\mathcal{H}_F$.

The domain of definition of b_F^{-1} is $\pi_F(\mathfrak{X}_\mu) \setminus \mathcal{T}_F$, where $\mathcal{T}_F$ is a finite set (of Heegner points).

5. The quaternionic locus

As above, we let $\mathcal{O}$ be a maximal order in a totally indefinite division quaternion algebra B over a totally real field F of trivial narrow class number and we fix a generator $D \in F_+^*$ of $\mathrm{disc}(B)$.

It is the aim of this section to use the preceding results to study the quaternionic locus $Q_{\mathcal{O}}$ in $\mathcal{A}_g(\mathbb{C})$. As we mentioned in the introduction, since $h_+(F) = 1$, the set $Q_{\mathcal{O}}$ is not empty.

Definition 5.1. A Heegner point in $Q_{\mathcal{O}}$ is an isomorphism class $[(A, \mathcal{L})]$ of a principally polarized abelian variety such that $\mathrm{End}(A) \supsetneq \mathcal{O}$.

According to Definitions 2.1 and 5.1, we note that $Q_{\mathcal{O}}$ is the disjoint union of the set of principally polarized abelian varieties $[(A, \mathcal{L})]$ with quaternionic multiplication by $\mathcal{O}$ and the set of Heegner points. The former is a subset of $Q_{\mathcal{O}}$ whose closure with respect to the analytical topology is $Q_{\mathcal{O}}$ itself. The latter is also a dense but discrete subset of $Q_{\mathcal{O}}$ (cf. [15]).

In order to understand the nature of the locus $Q_{\mathcal{O}}$, we observe that for any principally polarized pair $(\mathcal{O}, \mu)$, the set $\widetilde{\mathfrak{X}}_\mu(\mathbb{C})$ of complex points of the Shimura variety $\widetilde{\mathfrak{X}}_\mu/\mathbb{Q}$ attached to $(\mathcal{O}, \mu)$ sits inside $Q_{\mathcal{O}}$.

Proposition 5.2. *Let μ, $\mu' \in \mathcal{O}_0$ be two pure quaternions such that $\mathrm{n}(\mu) = uD$ and $\mathrm{n}(\mu') = u'D$ for some units u, $u' \in R_{F+}^*$. If $\widetilde{\mathfrak{X}}_\mu(\mathbb{C})$ and $\widetilde{\mathfrak{X}}_{\mu'}(\mathbb{C})$ are different subvarieties of $\mathcal{A}_g(\mathbb{C})$, then $\widetilde{\mathfrak{X}}_\mu(\mathbb{C}) \cap \widetilde{\mathfrak{X}}_{\mu'}(\mathbb{C})$ is a finite set of Heegner points.*

Proof. Assume that the isomorphism class $[A, \mathcal{L}]$ of a principally polarized abelian variety falls at the intersection of $\widetilde{\mathfrak{X}}_\mu$ and $\widetilde{\mathfrak{X}}_{\mu'}$ in $\mathcal{A}_g$. Write $[A, \mathcal{L}] = \pi([A, \iota, \mathcal{L}]) = \pi([A', \iota', \mathcal{L}'])$ as the image by π of points in $\mathfrak{X}_\mu$ and $\mathfrak{X}_{\mu'}$, respectively. Since

$[(A, \mathcal{L})] = [(A', \mathcal{L}')] \in \mathcal{Q}_{\mathcal{O}}$, we can identify the pair $(A, \mathcal{L}) = (A', \mathcal{L}')$ through a fixed isomorphism of polarized abelian varieties.

Let us assume that $[A, \mathcal{L}] = [A', \mathcal{L}']$ was not a Heegner point. Then $\iota : \mathcal{O} \simeq \mathrm{End}(A)$ would be an isomorphism of rings such that $c_1(\mathcal{L}) = \mu$. We then would have by Theorem 2.2, §4, that $c_1(\mathcal{L}) = c_1(\mathcal{L}') = \mu = \mu'$ up to multiplication by elements in F^*. Since $\widetilde{\mathfrak{X}}_\mu = \widetilde{\mathfrak{X}}_{u\mu}$ for all units $u \in R_F^*$, this would contradict the statement. Since the set of Heegner points in $\mathcal{A}_g(\mathbb{C})$ is discrete, we conclude that any two irreducible components of $\mathcal{Q}_{\mathcal{O}}$ meet at a finite set of Heegner points. □

Proposition 5.3.

1. *The locus $Q_{\mathcal{O}}$ is the set of complex points $\mathcal{Q}_{\mathcal{O}}(\mathbb{C})$ of a reduced complete subscheme $\mathcal{Q}_{\mathcal{O}}$ of $\mathcal{A}_g$ defined over $\mathbb{Q}$.*
2. *Let $\rho(\mathcal{O})$ be the number of absolutely irreducible components of $\mathcal{Q}_{\mathcal{O}}$. Then there exist quaternions $\mu_k \in \mathcal{O}_0$ with $\mu_k^2 + u_k D = 0$ for $u_k \in R_{F+}^*$, $1 \leq k \leq \rho(\mathcal{O})$, such that*

$$\mathcal{Q}_{\mathcal{O}} = \bigcup \widetilde{\mathfrak{X}}_{\mu_k}$$

is the decomposition of $\mathcal{Q}_{\mathcal{O}}$ into irreducible components.

Proof. Let $[(A, \mathcal{L})] \in Q_{\mathcal{O}}$ be the isomorphism class of a complex principally polarized abelian variety such that $\mathrm{End}(A) \simeq \mathcal{O}$. Fix an isomorphism $\iota : \mathcal{O} \simeq \mathrm{End}(A)$. By Theorem 2.2, §4, the Rosati involution with respect to $\mathcal{L}$ on $\mathcal{O}$ must be of the form $\varrho_\mu : \mathcal{O} \to \mathcal{O}$, $\beta \mapsto \mu^{-1}\bar{\beta}\mu$ for some $\mu \in \mathcal{O}$ with $\mu^2 + uD = 0$, $u \in R_{F+}^*$. Thus $[(A, \mathcal{L})] = \pi([A, \iota, \mathcal{L}]) \in \widetilde{\mathfrak{X}}_\mu(\mathbb{C})$, namely the set of complex points on a reduced, irreducible, complete and possibly singular scheme over $\mathbb{Q}$ (cf. [14], [15]). Since the set of Heegner points $[(A, \mathcal{L})] \in \widetilde{\mathfrak{X}}_\mu(\mathbb{C})$ is a discrete set which lies on the Zariski closure of its complement, we conclude that $Q_{\mathcal{O}}$ is the union of the Shimura varieties $\widetilde{\mathfrak{X}}_\mu(\mathbb{C})$ as μ varies among pure quaternions satisfying the above properties.

Let us now show that $Q_{\mathcal{O}}$ is actually covered by finitely many pairwise different Shimura varieties. Let $A/\mathbb{C}$ be an arbitrary abelian variety with quaternionic multiplication by $\mathcal{O}$ and fix an isomorphism $\iota : \mathcal{O} \simeq \mathrm{End}(A)$.

Let $(\mathcal{O}, \mu)$ be any principally polarized pair. Since $h_+(F) = 1$, there exists a unit $u \in R_F^*$ such that $u\mu$ is an ample quaternion in the sense of [10], §5. Let $\mathcal{L} \in \mathrm{NS}(A)$ be the line bundle on A such that $c_1(\mathcal{L})^{-1} = u\mu$. From Theorem 2.2, it follows that $\mathcal{L}$ is a principal polarization on A such that the isomorphism class of the triplet $(A, \iota, \mathcal{L})$ corresponds to a closed point in $\mathfrak{X}_\mu(\mathbb{C})$ and hence $[A, \mathcal{L}] \in \widetilde{\mathfrak{X}}_\mu$.

Since, by Proposition 5.2, the intersection points of two different Shimura varieties $\widetilde{\mathfrak{X}}_\mu(\mathbb{C})$ and $\widetilde{\mathfrak{X}}_{\mu'}(\mathbb{C})$ in $\mathcal{A}_g(\mathbb{C})$ are Heegner points, this shows that for every irreducible component of $Q_{\mathcal{O}}$ there exists at least one principal polarization $\mathcal{L}$ on A such that $[A, \mathcal{L}]$ lies on it. Consequently, the number $\pi_0(A)$ of isomorphism classes of principal polarizations on A is an upper bound for the number $\rho(\mathcal{O})$ of irreducible components of $Q_{\mathcal{O}}$. Since, by Theorem 2.4, the number $\pi_0(A)$ is a finite number, this yields the proof of the proposition. □

In view of Proposition 5.3, it is natural to pose the following

Question 5.4. What is the number $\rho(\mathcal{O})$ of irreducible components of $\mathcal{Q}_{\mathcal{O}}$? When is $\mathcal{Q}_{\mathcal{O}}$ irreducible?

6. The distribution of principal polarizations on an abelian variety in $\mathcal{Q}_{\mathcal{O}}$

Let us relate Question 5.4 to the following problem. In Theorem 2.4, we computed the number $\pi_0(A)$ of principal polarizations on an abelian variety A with quaternion multiplication by $\mathcal{O}$ as the finite sum of relative class numbers of suitable orders in the CM-fields $F(\sqrt{-uD})$ for $u \in R^*_{F+}/R^{*2}_F$. This has the following modular interpretation.

Let $\mathcal{L}_1$, ..., $\mathcal{L}_{\pi_0(A)}$ be representatives of the $\pi_0(A)$ distinct principal polarizations on A. Then the pairwise nonisomorphic principally polarized abelian varieties $[(A,\mathcal{L}_1)],\dots,[(A,\mathcal{L}_{\pi_0(A)})]$ correspond to all closed points in $\mathcal{Q}_{\mathcal{O}}$ whose underlying abelian variety is isomorphic to A. We then naturally ask the following

Question 6.1. Let A be an abelian variety with quaternionic multiplication by $\mathcal{O}$. How are the distinct principal polarizations $[(A,\mathcal{L}_j)]$ distributed among the irreducible components $\widetilde{\mathfrak{X}}_{\mu_k}$ of $\mathcal{Q}_{\mathcal{O}}$?

It turns out that the two questions above are related. The linking ingredient is provided by the definition below, which establishes a slightly coarser equivalence relationship on polarizations than the one considered in Theorem 2.2, §1.

Definition 6.2. Let A be an abelian variety with quaternionic multiplication by $\mathcal{O}$ over $\bar{\mathbb{Q}}$.

1. Two polarizations $\mathcal{L}$ and $\mathcal{L}'$ on A are weakly isomorphic if $c_1(\mathcal{L}) \simeq mc_1(\mathcal{L}') \in \mathrm{NS}(A)$ for some $m \in F^*_+$. We shall denote it $\mathcal{L} \simeq_w \mathcal{L}'$.
2. Two principal polarizations $\mathcal{L}$ and $\mathcal{L}'$ on A are *Atkin-Lehner isogenous*, denoted $\mathcal{L} \sim \mathcal{L}'$, if there is an isogeny $\omega \in \mathcal{O} \cap \mathrm{Norm}_{B^*_+}(\mathcal{O})$ of A such that
$$\omega^*(\mathcal{L}) \simeq_w \mathcal{L}'.$$

We note that there is a closed relationship between the above definition and the modular interpretation of the Atkin-Lehner group W given in [11].

Definition 6.3. Let A be an abelian variety with quaternionic multiplication by $\mathcal{O}$ over $\bar{\mathbb{Q}}$. We let $\hat{\Pi}_0(A)$ be the set of principal polarizations on A up to Atkin-Lehner isogeny and we let $\hat{\pi}_0(A) = \#\hat{\Pi}_0(A)$ denote its cardinality.

Theorem 6.4 (Distribution of principal polarizations). *Let A be an abelian variety with quaternionic multiplication by $\mathcal{O}$ over $\bar{\mathbb{Q}}$ and let $\mathcal{L}_1,\dots,\mathcal{L}_{\pi_0(A)}$ be representatives of the $\pi_0(A)$ distinct principal polarizations on A.*

Then, two closed points $[A,\mathcal{L}_i]$ and $[A,\mathcal{L}_j]$ lie on the same irreducible component of $\mathcal{Q}_{\mathcal{O}}$ if and only if the polarizations $\mathcal{L}_i$ and $\mathcal{L}_j$ are Atkin-Lehner isogenous.

Proof. We know from Proposition 5.3 that any irreducible component of $\mathcal{Q}_{\mathcal{O}}$ is $\widetilde{\mathfrak{X}}_\mu$ for some principally polarized pair $(\mathcal{O},\mu)$. We single out and fix one of these.

Let $\mathcal{L}$ be a principal polarization on A such that $[(A,\mathcal{L})]$ lies on $\widetilde{\mathfrak{X}}_\mu$ and let $\mathcal{L}'$ be a second principal polarization on A. We claim that $[A,\mathcal{L}'] \in \widetilde{\mathfrak{X}}_\mu$ if and only if there exists $\omega \in \mathcal{O} \cap \mathrm{Norm}_{B_+^*}(\mathcal{O})$ such that $\mathcal{L}'$ and $\omega^*(\mathcal{L})$ are weakly isomorphic.

Assume first that $\mathcal{L}' \simeq_w \omega^*(\mathcal{L})$ for some $\omega \in \mathcal{O} \cap \mathrm{Norm}_{B_+^*}(\mathcal{O})$. This amounts to saying that $\bar{\omega} c_1(\mathcal{L})\omega = m c_1(\mathcal{L}')$ for some $m \in F^*$. Since both $\omega^*(\mathcal{L})$ and $\mathcal{L}'$ are polarizations, we deduce from Theorem 2.2, §3, that $m \in F_+^*$. Moreover, since $\mathcal{L}$ and $\mathcal{L}'$ are principal, we obtain from Theorem 2.2, §2, that $m = u\mathrm{n}(\omega)$ for some $u \in R_{F+}^*$.

Note that $(A,\iota_\omega,\mathcal{L}')$ is a principally polarized abelian variety with quaternionic multiplication such that the Rosati involution that $\mathcal{L}'$ induces on $\iota_\omega(\mathcal{O})$ is ϱ_μ. Indeed, this follows because $\iota_\omega(\beta)^{\circ_{\mathcal{L}'}} = \iota((\omega^{-1}\beta\omega))^{\circ_{\mathcal{L}'}} = \iota((\omega^{-1}\mu\omega)^{-1}\overline{\omega^{-1}\beta\omega}\text{-}(\omega^{-1}\mu\omega)) = \iota_\omega(\mu^{-1}\bar{\beta}\mu)$. This shows that, if $\mathcal{L}'$ and $\omega^*(\mathcal{L})$ are weakly isomorphic for some $\omega \in \mathcal{O} \cap \mathrm{Norm}_{B_+^*}(\mathcal{O})$, then $[A,\mathcal{L}'] \in \widetilde{\mathfrak{X}}_\mu$.

Conversely, let us assume that $[A,\mathcal{L}'] \in \widetilde{\mathfrak{X}}_\mu$. Let $\iota' : \mathcal{O} \hookrightarrow \mathrm{End}(A)$ be such that $[A,\iota',\mathcal{L}'] \in \mathfrak{X}_{(\mathcal{O},\mathcal{I}_\vartheta,\varrho_\mu)}$. By the Skolem-Noether Theorem, it holds that $\iota' = \omega^{-1}\iota\omega$ for some $\omega \in \mathrm{Norm}_{B^*}(\mathcal{O})$; we can assume that $\omega \in \mathcal{O}$ by suitably scaling it. Since it holds that $\iota_\omega(\beta)^{\circ_{\mathcal{L}'}} = \iota_\omega(\mu^{-1}\bar{\beta}\mu)$ for any $\beta \in \mathcal{O}$, we have that $c_1(\mathcal{L}') = u\omega^{-1}c_1(\mathcal{L})\omega$ for some $u \in R_F^*$ such that $u\mathrm{n}(\omega) \in F_+^*$. Since $\mathrm{n}(\mathcal{O}^*) = R_F^*$, we can find $\alpha \in \mathcal{O}^*$ with reduced norm $\mathrm{n}(\alpha) = u^{-1}$ and thus $\omega\alpha \in B_+^*$.

Let $\mathcal{L}_{\omega\alpha}$ be the polarization on A such that $c_1(\mathcal{L}_{\omega\alpha}) = \frac{u}{\mathrm{n}(\omega)} c_1((\omega\alpha)^*(\mathcal{L}))$. The automorphism $\alpha \in \mathcal{O}^* = \mathrm{Aut}(A)$ induces an isomorphism between the polarizations $\mathcal{L}_{\omega\alpha}$ and $\mathcal{L}'$, since $c_1(\alpha^*(\mathcal{L}')) = \bar{\alpha}(u\omega^{-1}c_1(\mathcal{L})\omega)\alpha = c_1(\mathcal{L}_{\omega\alpha})$. Hence $\mathcal{L}'$ is weakly isomorphic to $(\omega\alpha)^*(\mathcal{L})$. This concludes our claim above and also proves the theorem. □

Corollary 6.5. *The number of irreducible components of* $\mathcal{Q}_{\mathcal{O}}$ *is*

$$\rho(\mathcal{O}) = \hat{\pi}_0(A),$$

independently of the choice of A.

For any irreducible component $\widetilde{\mathfrak{X}}_{\mu_k}$ of $\mathcal{Q}_{\mathcal{O}}$, let $\Pi_0^{(k)}(A) \subset \Pi_0(A)$ denote the set of isomorphism classes of the isogeny class of principal polarizations lying on $\widetilde{\mathfrak{X}}_{\mu_k}$.

As another immediate consequence of Theorem 6.4, the following corollary establishes an internal structure on the set $\Pi_0(A)$. Roughly, it asserts that $\Pi_0(A) = \bigcup_{k=1}^{\rho(\mathcal{O})} \Pi_0^{(k)}(A)$ is the disjoint union of the sets $\Pi_0^{(k)}(A)$, which are equipped with a free and transitive action of a 2-torsion finite abelian group.

Corollary 6.6. *Let A be an abelian variety with quaternionic multiplication by $\mathcal{O}$. Let $\widetilde{\mathfrak{X}}_{\mu_k}$ be an irreducible component of $\mathcal{Q}_{\mathcal{O}}$ and let $W_0^{(k)} \subseteq W$ be the stable subgroup attached to the polarized order $(\mathcal{O},\mu_k)$.*

Then there is a free and transitive action of $W/W_0^{(k)}$ on the Atkin-Lehner isogeny class $\Pi_0^{(k)}(A)$ of principal polarizations lying on $\widetilde{\mathfrak{X}}_{\mu_k}$.

In the case of a non-twisting maximal order $\mathcal{O}$, we have that the stable group $W_0(\mathcal{O},\mu)$ attached to a principally polarized pair $(\mathcal{O},\mu)$ is $U_0(\mathcal{O},\mu)$. The following corollary follows from the proof of Lemma 2.10 in [12].

Corollary 6.7. *Let $\mathcal{O}$ be a non-twisting maximal order in B and assume that, for any $u \in R_{F+}^*$, any primitive root of unity of odd order in the CM-field $F(\sqrt{-uD})$ is contained in the order $R_F[\sqrt{-uD}]$.*

Let A be an abelian variety with quaternionic multiplication by $\mathcal{O}$. Then the distinct isomorphism classes of principally polarized abelian varieties $[(A,\mathcal{L}_1)], \ldots, [(A,\mathcal{L}_{\pi_0(A)})]$ are equidistributed *among the $\rho(\mathcal{O})$ irreducible components of $\mathcal{Q}_{\mathcal{O}}$.*

In particular, it then holds that

$$\pi_0(A) = \frac{|W|}{|W_0|} \cdot \rho(\mathcal{O}).$$

7. Shimura curves embedded in Igusa's threefold

The whole picture becomes particularly neat when we consider the simplest case of quaternion algebras over $\mathbb{Q}$. Let then B be an indefinite quaternion algebra over $\mathbb{Q}$ of discriminant $D = p_1 \cdot \cdots \cdot p_{2r}$ and let $\mathcal{O}$ be a maximal order in B. Since $h(\mathbb{Q}) = 1$, there is a single choice of $\mathcal{O}$ up to conjugation by B^*. Moreover, all left ideals of $\mathcal{O}$ are principal and hence isomorphic to $\mathcal{O}$ as left $\mathcal{O}$-modules.

Let A be a complex abelian surface with quaternionic multiplication by $\mathcal{O}$. By Theorem 2.4, A is principally polarizable and the number of isomorphism classes of principal polarizations on A is

$$\pi_0(A) = \frac{\tilde{h}(-D)}{2},$$

where, for any nonzero squarefree integer d, we write

$$\tilde{h}(d) = \begin{cases} h(4d) + h(d) & \text{if } d \equiv 1 \mod 4, \\ h(4d) & \text{otherwise.} \end{cases}$$

For any integral element $\mu \in \mathcal{O}$ such that $\mu^2 + D = 0$, let now $\mathfrak{X}_\mu$ be the Shimura curve that coarsely represents the functor which classifies principally polarized abelian surfaces $(A, \iota, \mathcal{L})$ with quaternionic multiplication by $\mathcal{O}$ such that the Rosati involution with respect to $\mathcal{L}$ on $\mathcal{O}$ is ϱ_μ. This is an algebraic curve over $\mathbb{Q}$ whose isomorphism class does not actually depend on the choice of the quaternion μ, but only on the discriminant D (cf. [15]). Hence, it is usual to simply denote this isomorphism class as $\mathfrak{X}_D$.

Let $W = \{\omega_m : m|D\} \simeq (\mathbb{Z}/2\mathbb{Z})^{2r}$ be the Atkin-Lehner group attached to $\mathcal{O}$ in Section 4. We know that $W \subseteq \mathrm{Aut}_{\mathbb{Q}}(\mathfrak{X}_D)$ is a subgroup of the group of automorphisms of the Shimura curve $\mathfrak{X}_D$.

Let now $\mathcal{A}_2$ be Igusa's three-fold, the moduli space of principally polarized abelian surfaces. By the work of Igusa (cf. [6]), it is an affine scheme over $\mathbb{Q}$ that contains, as a Zariski open and dense subset, the moduli space $\mathcal{M}_2$ of curves of genus 2, immersed in $\mathcal{A}_2$ via the Torelli embedding.

Sitting in $\mathcal{A}_2$ there is the quaternionic locus $\mathcal{Q}_\mathcal{O}$ of isomorphism classes of principally polarized abelian surfaces $[(A, \mathcal{L})]$ such that $\mathrm{End}(A) \supseteq \mathcal{O}$. Since all maximal orders $\mathcal{O}$ in B are pairwise conjugate, the quaternionic locus $\mathcal{Q}_\mathcal{O}$ does not actually depend on the choice of $\mathcal{O}$ and we may simply denote it by $\mathcal{Q}_\mathcal{O} = \mathcal{Q}_D$.

As explained in Sections 3 and 4, there are forgetful finite morphisms $\pi : \mathfrak{X}_\mu \to \mathcal{Q}_D \subset \mathcal{A}_2$ which map the Shimura curve $\mathfrak{X}_\mu$ onto an irreducible component $\widetilde{\mathfrak{X}}_\mu$ of $\mathcal{Q}_D$. We insist on the fact that the image $\widetilde{\mathfrak{X}}_\mu \subset \mathcal{Q}_D$ *does depend* on the choice of the quaternion μ.

Let us now compare the non-twisting and twisting case, respectively. We first assume that

$$B \not\simeq (\frac{-D, m}{\mathbb{Q}})$$

for all positive divisors $m|D$ of D. Then all principally polarized pairs $(\mathcal{O}, \mu)$ in B are *non-twisting* and the stable subgroup attached to $(\mathcal{O}, \mu)$ is

$$W_0 = U_0 = \langle \omega_D \rangle \subset W,$$

independently of the choice of μ. By Theorem 4.5, we deduce that any irreducible component $\widetilde{\mathfrak{X}}_\mu$ of $\mathcal{Q}_D$ is birationally equivalent to the Atkin-Lehner quotient $\mathfrak{X}_D/\langle \omega_D \rangle$ and thus the quaternionic locus $\mathcal{Q}_D$ in $\mathcal{A}_2$ is the union of pairwise birationally equivalent Shimura curves $\widetilde{\mathfrak{X}}_{\mu_1}, \ldots, \widetilde{\mathfrak{X}}_{\mu_{\rho(\mathcal{O})}}$, meeting at a finite set of Heegner points.

Moreover, for any abelian surface A with quaternionic multiplication by $\mathcal{O}$, it follows from Theorem 6.4 that the closed points $\{[(A, \mathcal{L}_j)]\}_{j=1}^{\pi_0(A)}$ are equidistributed among the $\rho(\mathcal{O})$ irreducible components of $\mathcal{Q}_D$. In addition, Corollary 6.7 ensures that

$$|W/W_0| = 2^{2r-1}|\pi_0(A),$$

as genus theory for binary quadratic forms already predicts.

Finally, we obtain that the number of irreducible components of $\mathcal{Q}_D$ in the non-twisting case is

$$\rho(\mathcal{O}) = \frac{\tilde{h}(-D)}{2^{2r}}.$$

On the other hand, let us assume that

$$B \simeq (\frac{-D, m}{\mathbb{Q}})$$

for some $m|D$. In this case, there can be two different birational classes of irreducible components on $\mathcal{Q}_D$. Indeed, the assumption means that there exist pure quaternions $\mu \in \mathcal{O}$, $\mu^2 + D = 0$, such that $(\mathcal{O}, \mu)$ is a *twisting* principally polarized maximal order. Then

$$W_0(\mathcal{O}, \mu) = \langle \omega_m, \omega_D \rangle$$

and $\widetilde{\mathfrak{X}}_\mu$ is birationally equivalent to $\mathfrak{X}_D/\langle \omega_m, \omega_D \rangle$. We may refer to $\widetilde{\mathfrak{X}}_\mu$ as a twisting irreducible component of $\mathcal{Q}_D$.

In addition to these, there may exist non-twisting polarized orders $(\mathcal{O}, \mu)$ such that the corresponding irreducible components $\widetilde{\mathfrak{X}}_\mu$ of $\mathcal{Q}_D$ are birationally equivalent to $\mathfrak{X}_D/\langle\omega_D\rangle$. We may refer to these as the non-twisting irreducible components of $\mathcal{Q}_D$.

We then have the following lower and upper bounds for the number of irreducible components of $\mathcal{Q}_D$:

$$\frac{\tilde{h}(-D)}{2^{2r}} < \quad \rho(\mathcal{O}) \quad \leq \frac{\tilde{h}(-D)}{2^{2r-1}}.$$

Summing up, we obtain the following

Theorem 7.1. *Let B be an indefinite division quaternion algebra over $\mathbb{Q}$ of discriminant $D = p_1 \cdot \dots \cdot p_{2r}$. Then, the quaternionic locus $\mathcal{Q}_D$ in $\mathcal{A}_2$ is irreducible if and only if*

$$\tilde{h}(-D) = \begin{cases} 2^{2r-1} & \text{if } B \simeq (\frac{-D,m}{\mathbb{Q}}) \text{ for some } m|D, \\ 2^{2r} & \text{otherwise.} \end{cases}$$

Proof. If B is not a twisting quaternion algebra, we already know from the above that the number of irreducible components of $\mathcal{Q}_D$ is $\dfrac{\tilde{h}(-D)}{2^{2r}}$. Hence, in this case, the quaternionic locus of discriminant D in $\mathcal{A}_2$ is irreducible if and only if $\tilde{h}(-D) = 2^{2r}$. If on the other hand B is twisting, it follows from the above inequalities that $\mathcal{Q}_D$ is irreducible if and only if $\tilde{h}(-D) = 2^{2r-1}$. □

In view of Theorem 7.1, there arises a closed relationship between the irreducibility of the quaternionic locus in Igusa's threefold and the genus theory of integral binary quadratic forms and the classical *numeri idonei* studied by Euler, Schinzel and others. We refer the reader to [1] and [13] for the latter.

8. Hashimoto-Murabayashi's families

As the simplest examples to be considered, let B_6 and B_{10} be the rational quaternion algebras of discriminant $D = 2 \cdot 3 = 6$ and $2 \cdot 5 = 10$, respectively. Hashimoto and Murabayashi [4] exhibited two families of principally polarized abelian surfaces with quaternionic multiplication by a maximal order in these quaternion algebras. Namely, let

$$C^{(6)}_{(s,t)} : Y^2 = X(X^4 + PX^3 + QX^2 + RX + 1)$$

be the family of curves with

$$P = 2s + 2t, \quad Q = \frac{(1 + 2t^2)(11 - 28t^2 + 8t^4)}{3(1 - t^2)(1 - 4t^2)}, \quad R = 2s - 2t$$

over the base curve

$$g^{(6)}(t, s) = 4s^2t^2 - s^2 + t^2 + 2 = 0.$$

And let

$$C^{(10)}_{(s,t)} : Y^2 = X(P^2X^4 + P^2(1+R)X^3 + PQX^2 + P(1-R)X + 1)$$

be the family of curves with

$$P = \frac{4(2t+1)(t^2-t-1)}{(t-1)^2}, \quad Q = \frac{(t^2+1)(t^4+8t^3-10t^2-8t+1)}{t(t-1)^2(t+1)^2}$$

and

$$R = \frac{(t-1)s}{t(t+1)(2t+1)}$$

over the base curve

$$g^{(10)}(t,s) = s^2 - t(t-2)(2t+1) = 0.$$

Let $J^{(6)}_{(s,t)} = \mathrm{Jac}(C^{(6)}_{(s,t)})$ and $J^{(10)}_{(s,t)} = \mathrm{Jac}(C^{(10)}_{(s,t)})$ be the Jacobian surfaces of the fibres of the families of curves above respectively. It was proved in [4] that their ring of endomorphisms contain a maximal order in B_6 and B_{10}, respectively.

Both $B_6 = (\frac{-6,2}{\mathbb{Q}})$ and $B_{10} = (\frac{-10,2}{\mathbb{Q}})$ are twisting quaternion algebras. Moreover, it turns out from our formula for $\pi_0(A)$ in the above section that any abelian surface A with quaternionic multiplication by a maximal order in either B_6 or B_{10} admits a *single* isomorphism class of principal polarizations. This implies that $\rho(B_6) = \hat{\pi}_0(A) = \pi_0(A) = 1$ and $\rho(B_{10}) = \hat{\pi}_0(A) = \pi_0(A) = 1$, respectively.

Moreover, the Shimura curves $\mathfrak{X}_6/\mathbb{Q}$ and $\mathfrak{X}_{10}/\mathbb{Q}$ have genus 0, although they are not isomorphic to $\mathbb{P}^1_{\mathbb{Q}}$ because there are no rational points on them. However, it is easily seen that $\mathfrak{X}_6/W_0 = \mathfrak{X}_6/W \simeq \mathbb{P}^1_{\mathbb{Q}}$ and $\mathfrak{X}_{10}/W_0 = \mathfrak{X}_{10}/W \simeq \mathbb{P}^1_{\mathbb{Q}}$, respectively.

As it is observed in [4], the base curves $g^{(6)}$ and $g^{(10)}$ are curves of genus 1 and not of genus 0 as it should be expected. This is explained by the fact that there are obvious isomorphisms between the fibres of the families $C^{(6)}$ and $C^{(10)}$, respectively.

Ibukiyama, Katsura and Oort [5] proved that the supersingular locus in $\mathcal{A}_2/\bar{\mathbb{F}}_p$ is irreducible if and only if $p \leq 11$. As a corollary to their work, Hashimoto and Murabayashi obtained that the reduction mod 3 and 5 of the family of Jacobian surfaces with quaternionic multiplication by B_6 and B_{10} respectively yield the single irreducible component of the supersingular locus in $\mathcal{A}_2/\bar{\mathbb{F}}_3$ and $\mathcal{A}_2/\bar{\mathbb{F}}_5$ respectively. The following statement may be considered as a lift to characteristic 0 of these results.

Theorem 8.1.

1. *The quaternionic locus $\mathcal{Q}_6$ in $\mathcal{A}_2/\mathbb{Q}$ is absolutely irreducible and birationally equivalent to $\mathbb{P}^1_{\mathbb{Q}}$ over $\mathbb{Q}$. A universal family over $\bar{\mathbb{Q}}$ is given by Hashimoto-Murabayashi's family $C^{(6)}$.*
2. *The quaternionic locus $\mathcal{Q}_{10}$ in $\mathcal{A}_2/\mathbb{Q}$ is absolutely irreducible and birationally equivalent to $\mathbb{P}^1_{\mathbb{Q}}$ over $\mathbb{Q}$. A universal family over $\bar{\mathbb{Q}}$ is given by Hashimoto-Murabayashi's family $C^{(10)}$.*

Proof. This follows from Theorem 7.1 and the discussion above. □

In particular, we obtain from Theorem 8.1 that every principally polarized abelian surface $(A, \mathcal{L})$ over $\bar{\mathbb{Q}}$ with quaternionic multiplication by a maximal order of discriminant 6 or 10 is isomorphic over $\bar{\mathbb{Q}}$ to the Jacobian variety of one of the curves $C^{(6)}_{(s,t)}$ or $C^{(10)}_{(s,t)}$, except for finitely many degenerate cases.

References

[1] A. Arenas, Genera and rational equivalence classes of integral quadratic forms, *J. Number Theory* **51** (1995), 210–218.

[2] M. Eichler, Zur Zahlentheorie der Quaternionen-Algebren, *J. reine angew. Math.* **195** (1955), 127–151.

[3] G. van der Geer, *Hilbert modular surfaces*, Ergebn. Math. Grenz. **16**, Springer 1987.

[4] K. Hashimoto, N. Murabayashi, Shimura curves as intersections of Humbert surfaces and defining equations of QM-curves og genus two, *Tôhoku Math. J.* **47** (1995), 271–296.

[5] T. Ibukiyama, T. Katsura, F. Oort, Supersingular curves of genus two and class numbers, *Compos. Math.* **57** (1986), 127–152.

[6] J. Igusa, Arithmetic variety of moduli for genus two, *Ann. Math.* **72** (1960), 612–649.

[7] H. Lange, Ch. Birkenhake, *Complex Abelian Varieties*, Grundl. math. Wiss. **302**, Springer, 1992.

[8] D. Mumford, *Abelian varieties*, Tata Institute of Fundamental Research, Bombay, Oxford University Press (1970).

[9] V. Rotger, On the group of automorphisms of Shimura curves and applications, *Compos. Math.* **132** (2002), 229–241.

[10] V. Rotger, Quaternions, polarizations and class numbers, *J. reine angew. Math.* **561** (2003), 177–197.

[11] V. Rotger, Modular Shimura varieties and forgetful maps, to appear in *Trans. Amer. Math. Soc.*

[12] V. Rotger, On the field of moduli of quaternionic multiplication on abelian varieties, submitted to publication.

[13] A. Schinzel, Sur les sommes de trois carrés, *Bull. Acad. Polonaise Sc.* Vol. VII **6** (1959), 307–310.

[14] G. Shimura, On analytic families of polarized abelian varieties and automorphic functions, *Ann. Math.* **78** (1963), 149–192.

[15] G. Shimura, Construction of class fields and zeta functions of algebraic curves, *Ann. Math.* **85** (1967), 58–159.

[16] M.F. Vignéras, *Arithmétique des algèbres de quaternions*, Lect. Notes Math. **800**, Springer, 1980.

Victor Rotger
Universitat Politècnica de Catalunya
Departament de Matemàtica Aplicada IV (EUPVG)
Av. Victor Balaguer s/n
08800 Vilanova i la Geltrú, Spain
e-mail: `vrotger@mat.upc.es`

Progress in Mathematics, Vol. 224, 277–289

Shafarevich-Tate Groups of Nonsquare Order

William A. Stein

Abstract. Let A denote an abelian variety over $\mathbb{Q}$. We give the first known examples in which $\#\text{Ш}(A/\mathbb{Q})$ is neither a square nor twice a square. For example, let E be the elliptic curve $y^2 + y = x^3 - x$ of conductor 37. We prove that for every odd prime $p < 25000$ (with $p \neq 37$), there is a twist A of $E \times \cdots \times E$ ($p-1$ copies) such that $\#\text{Ш}(A/\mathbb{Q}) = pn^2$ for some integer n. We prove this by showing under certain hypothesis on E and p that there is an exact sequence

$$0 \to E(\mathbb{Q})/pE(\mathbb{Q}) \to \text{Ш}(A/\mathbb{Q})[p^\infty] \to \text{Ш}(E/K)[p^\infty] \to \text{Ш}(E/\mathbb{Q})[p^\infty] \to 0,$$

where K is a certain abelian extension of $\mathbb{Q}$ of degree p.

1. Introduction

The Shafarevich-Tate group of an abelian variety A over a number field F is

$$\text{Ш}(A/F) := \operatorname{Ker}\left(H^1(F, A) \to \bigoplus_{\text{all } v} H^1(F_v, A) \right).$$

What are the possibilities for the group structure of $\text{Ш}(A/F)$? It is conjectured that $\text{Ш}(A/F)$ is finite and this is known in some cases.

Theorem 1.1 (Kato, Kolyvagin, Wiles, et al.). *Suppose A is an elliptic curve over $\mathbb{Q}$.*

(1) *If $\operatorname{ord}_{s=1} L(A, s) \leq 1$, then $\text{Ш}(A/\mathbb{Q})$ is finite.*
(2) *If χ is a character of the Galois group of an abelian extension K of $\mathbb{Q}$ and $L(A, \chi, 1) \neq 0$, then the χ-component of $\text{Ш}(A/K) \otimes_{\mathbb{Z}} \mathbb{Z}[\chi]$ is finite. (Here $\mathbb{Z}[\chi]$ is generated by the image of χ.)*

The Cassels-Tate pairing $\text{Ш}(A/F) \times \text{Ш}(A^\vee/F) \to \mathbb{Q}/\mathbb{Z}$ imposes strong constraints on the structure of $\text{Ш}(A/F)$.

Theorem 1.2 (Tate, Flach). *Let p be a prime and suppose that there is a polarization $\lambda : A \to A^\vee$ of degree coprime to p. If $p = 2$ assume also that λ arises from an F-rational divisor on A (this hypothesis is automatic if A is an elliptic curve, but can fail in general). If $\text{Ш}(A/F)[p^\infty]$ is finite then $\#\text{Ш}(A/F)[p^\infty]$ is a perfect square.*

Proof. If λ is F-rational, the Cassels-Tate pairing on $Ш(A/F)[p^\infty]$ (induced by λ) is nondegenerate and alternating (see [Tat63]), so $\#Ш(A/F)[p^\infty]$ is a perfect square. Even when λ is not F-rational, the Cassels-Tate pairing is nondegenerate and antisymmetric (see [Fla90]), which when p is odd implies that $\#Ш(A/F)[p^\infty]$ is a perfect square. □

It is tempting to conjecture that $\#Ш(A/F)$ is always a perfect square. Perhaps squareness is a fundamental property of Shafarevich-Tate groups? While implementing algorithms based on [PS97] for computing with Jacobians of hyperelliptic curves, M. Stoll was shocked to discover an example of an abelian variety of dimension two such that $\#Ш(A/F)[2^\infty] = 2$. This was surprising because, for example, one finds in the literature [SD67, p. 149] the following statement: "[The group $Ш(A/F)$] is conjectured to be finite, and Tate [26] has shown that if it is finite its order is a perfect square." Stoll and B. Poonen discovered what hid behind this and other similar examples in which $\#Ш(A/F)$ is twice a perfect square.

An algebraic curve X of genus g over a local field k is *deficient* if X has no k-rational divisor of degree $g-1$.

Theorem 1.3 (Poonen-Stoll [PS99]). *Suppose A is the Jacobian of an algebraic curve over F that is deficient at an odd number of places. If $\#Ш(A/F)$ is finite, then $\#Ш(A/F)$ is twice a square.*

For example, they prove that the Jacobian J of the nonsingular projective curve defined by

$$y^2 = -3(x^2+1)(x^2-6x+1)(x^2+6x+1)$$

has Shafarevich-Tate group of order 2 (to see that $\#Ш(J) \mid 2$ they observe that J is isogenous to a product of CM elliptic curves and apply a theorem of Rubin; see [PS99, Prop. 27] for details). Also, Jordan and Livné [JL99] give an infinite family of Atkin-Lehner quotients of Shimura curves which are deficient at an odd number of places.

Though $\#Ш(A/F)$ need not be square, one might still be tempted to conjecture that $Ш(A/F)$ must have order either a square or twice a square. Let p be an odd prime. In this paper, we construct (under certain hypotheses that are satisfied for $p < 25000$) abelian varieties A such that $\#Ш(A/\mathbb{Q}) = pn^2$ for some integer n. For example (see Section 3):

Theorem 1.4. *Let E be the elliptic curve $y^2 + y = x^3 - x$ of conductor 37. For every odd prime $p < 25000$ (with $p \neq 37$), there is a twist A of $E^{\times(p-1)}$ such that $\#Ш(A/\mathbb{Q}) = pn^2$ for some integer n.*

This paper was originally motivated by the problem of relating the conjecture of Birch and Swinnerton-Dyer about the ranks of elliptic curves E to the Birch and Swinnerton-Dyer formula for the orders $\#Ш(A)$ for abelian varieties A of analytic rank 0.

Let p be a prime. Under suitable hypotheses, we construct an abelian variety A and a natural map $E(\mathbb{Q})/pE(\mathbb{Q}) \hookrightarrow Ш(A/\mathbb{Q})$. Thus if $E(\mathbb{Q}) \cong \mathbb{Z}$ then

Ш$(A/\mathbb{Q})$ has a natural subgroup of order p, and no other natural subgroup of order p presents itself. Moreover, when E is defined by $y^2 + y = x^3 - x$, the Birch and Swinnerton-Dyer formula predicts that Ш$(A/\mathbb{Q})[3]$ is of order 3. Further investigation led to the results of this paper.

Acknowledgement: It is a pleasure to thank Kevin Buzzard, Frank Calegari, Sol Friedberg, Benedict Gross, Emmanuel Kowalski, Barry Mazur, Bjorn Poonen, and David Rohrlich for their helpful comments, and in particular Michael Stoll for Lemma 2.10 and Cristian González for carefully reading this paper, making many comments, and sending me a proof of Proposition 2.13.

1.1. Notation

If G is an abelian group and n is an integer, then $G[n]$ denotes the subgroup of elements of order n and $G[n^\infty]$ is the subgroup of elements of order any power of n. We refer to elliptic curves using the notation of [C97].

2. Construction of nonsquare Shafarevich-Tate groups

For the rest of this paper we will work with an elliptic curve E over $\mathbb{Q}$. Aside from the significant use of known cases of the Birch and Swinnerton-Dyer conjecture below, much of the construction should generalize to the situation when E is replaced by a principally polarized abelian variety over a global field.

For the rest of this section, fix an elliptic curve E over $\mathbb{Q}$. By [BCDT01], E is modular so there is a newform $f = \sum_{n=1}^{\infty} a_n q^n$ of level equal to the conductor $N = N_E$ of E such that $L(E,s) = L(f,s)$. For each prime $q \mid N$, the Tamagawa number c_q of E at q is the order of the group of rational components of the special fiber of the Néron model of E at q.

2.1. Twisting by characters of prime order

Let p be a prime number. For any prime $\ell \equiv 1 \pmod{p}$, let

$$\chi_{p,\ell} : (\mathbb{Z}/\ell\mathbb{Z})^* \to \mu_p \subset \mathbb{C}^*$$

be one of the $p-1$ Galois-conjugate Dirichlet characters of order p and conductor ℓ.

Conjecture 2.1. *Suppose p is a prime such that $\rho_{E,p} : \mathrm{Gal}(\overline{\mathbb{Q}}/\mathbb{Q}) \to \mathrm{Aut}(E[p])$ is surjective. Then there exists a prime $\ell \nmid N$ such that $L(E, \chi_{p,\ell}, 1) \neq 0$, $\ell \equiv 1 \pmod{p}$ and $a_\ell \not\equiv \ell + 1 \pmod{p}$.*

Remarks 2.2.

1. Formulas involving modular symbols imply that $L(E, \chi_{p,\ell}, 1) \neq 0$ if and only if $L(E, \chi_{p,\ell}^\sigma, 1) \neq 0$ for any $\mathrm{Gal}(\overline{\mathbb{Q}}/\mathbb{Q})$-conjugate $\chi_{p,\ell}^\sigma$ of $\chi_{p,\ell}$.
2. J. Fearnley proved related nonvanishing results when $L(E,1) \neq 0$ in [Fea01].
3. If E is the elliptic curve $y^2 + y = x^3 - x$ of conductor 37 and rank 1, then $\ell = 41$ is the only $\ell \equiv 1 \pmod{5}$ with $\ell < 1000$ for which $L(E, \chi_{5,\ell}, 1) = 0$.

The following proposition gives evidence for Conjecture 2.1 for the lowest-conductor elliptic curves of ranks 1, 2, and 3.

Proposition 2.3. *Conjecture* 2.1 *is true for the rank* 1 *elliptic curve* **37A** *for every odd* $p < 25000$ *(with* $p \neq 37$*). The conjecture is true for the rank* 2 *curve* **389A** *for every odd* $p < 1000$ *(with* $p \neq 389$*). The conjecture is true for the rank* 3 *curve* **5077A** *for every odd* $p < 1000$.

Proof. Consider the modular symbol

$$e_{p,\ell} = \sum_{a\in(\mathbb{Z}/\ell\mathbb{Z})^*} \chi_{p,\ell}(a)\cdot\left\{0,\ \frac{a}{\ell}\right\} \in H_1(X_0(N),\mathbb{Q}(\zeta_p)).$$

Then $L(E,\chi_{p,\ell},1)\neq 0$ if and only if the image of $e_{p,\ell}$ under

$$H_1(X_0(N),\mathbb{Q}(\zeta_p)) \to H_1(E,\mathbb{Q}(\zeta_p))$$

is nonzero. In any particular case, we can use modular symbols to determine whether or not this image is nonzero.

When p is large, it is difficult to compute in the field $\mathbb{Q}(\zeta_p)$, so instead we compute in the residue class field $\mathbb{F}_\ell = \mathbb{Z}[\zeta_p]/\mathfrak{m} \cong Z/\ell\mathbb{Z}$, where $\mathfrak{m}$ is one of the maximal ideals of $\mathbb{Z}[\zeta_p]$ that lies over ℓ. (Note that ℓ splits completely in $\mathbb{Z}[\zeta_p]$ because $\ell\equiv 1 \pmod p$.) After reducing modulo $\mathfrak{m}$, we compute the image of

$$\overline{e}_{p,\ell} = \sum_{a\in(\mathbb{Z}/\ell\mathbb{Z})^*} a^{(\ell-1)/p}\cdot\left\{0,\ \frac{a}{\ell}\right\} \in H_1(X_0(N),\mathbb{F}_\ell)$$

in $H_1(E,\mathbb{F}_\ell)$. If it is nonzero, then the image of $e_{p,\ell}$ in $H_1(E,\mathbb{Q}(\zeta_p))$ is nonzero.

A big computation (that takes hundreds of hours using MAGMA [BCP97]) shows that the image of $\overline{e}_{p,\ell}$ is nonzero in the cases asserted by the proposition. So the reader can carry out similar computations, we include the following MAGMA V2.10-6 code, which illustrates verification of the proposition for **37A** for $p < 100$:

```
procedure VerifyConjecture(E, p)
   assert Type(E) eq CrvEll;
   assert Type(p) eq RngIntElt and IsPrime(p) and IsOdd(p);
   N := Conductor(E);
   assert N mod p ne 0;
   M := ModularSymbols(E,+1);  // takes a long time if N large!
   ell := 3; t := Cputime();
   printf "p=%o: ", p;
   while true do
      while (ell mod p ne 1)  or  (N mod ell eq 0) or
       TraceOfFrobenius(ChangeRing(E,GF(ell))) mod p eq (ell+1) do
         ell := NextPrime(ell);
      end while;
      k := FiniteField(ell);
      printf "trying ell=%o...",ell;
      psi := DirichletGroup(ell,k).1;
```

```
      eps := psi^(Order(psi) div p);  // order p character
      M_k := BaseExtend(M,k);
      phi := RationalMapping(M_k);
      e := TwistedWindingElement(M_k,1,eps);
      if phi(e) ne 0 then
         printf " success! (%o seconds)\n", Cputime(t);
         return;
      end if;
      printf "failed. ";
      ell := NextPrime(ell);
   end while;
end procedure;

E := EllipticCurve([0,0,1,-1,0]);  // 37A
for p in [q : q in [3..100] | IsPrime(q) and q ne 37] do
   VerifyConjecture(E,p);
end for;
```

The above input results in the following abbreviated output:

```
p=3: trying ell=7... success! (0.021 seconds)
p=5: trying ell=11... success! (0.039 seconds)
p=7: trying ell=29... success! (0.121 seconds)
...
p=89: trying ell=179... success! (0.739 seconds)
p=97: trying ell=389... success! (1.491 seconds)
```

□

2.2. A restriction of scalars exact sequence

As above, E is an elliptic curve over $\mathbb{Q}$. Let p be any prime (note that $p = 2$ is allowed). Suppose $\ell \equiv 1 \pmod{p}$ is another prime and that $\ell \nmid N_E$. Let $K \subset \mathbb{Q}(\mu_\ell)$ be the abelian extension of $\mathbb{Q}$ that corresponds to $\chi_{p,\ell}$ (thus K is the unique subfield of $\mathbb{Q}(\mu_\ell)$ of degree p).

Let $R = \mathrm{Res}_{K/\mathbb{Q}}(E_K)$ be the restriction of scalars down to $\mathbb{Q}$ of E viewed as an elliptic curve over K. Thus R is an abelian variety over $\mathbb{Q}$ of dimension $p = [K : \mathbb{Q}]$. It is characterized by the fact that it represents the following functor on $\mathbb{Q}$-schemes S:

$$S \mapsto E_K(S_K).$$

As a Galois module,

$$R(\overline{\mathbb{Q}}) = E(\overline{\mathbb{Q}}) \otimes_{\mathbb{Z}} \mathbb{Z}[\mathrm{Gal}(K/\mathbb{Q})],$$

where $\tau \in \mathrm{Gal}(\overline{\mathbb{Q}}/\mathbb{Q})$ acts on $\sum P_\sigma \otimes \sigma$ by

$$\tau\left(\sum P_\sigma \otimes \sigma\right) = \sum \tau(P_\sigma) \otimes \tau_{|K} \cdot \sigma,$$

where $\tau_{|K}$ is the image of τ in $\mathrm{Gal}(K/\mathbb{Q})$.

Proposition 2.4. *The identity map induces a closed immerion $\iota : E \hookrightarrow R$, and the trace* $\mathrm{Tr} : K \to \mathbb{Q}$ *induces a surjection* $\mathrm{Tr} : R \to E$ *whose kernel is geometrically connected. Thus we have an exact sequence of abelian varieties*

$$0 \to A \to R \xrightarrow{\mathrm{Tr}} E \to 0. \tag{1}$$

Proof. The existence of ι and Tr follows from Yoneda's lemma. The map ι is induced by the functorial inclusion $E(S) \hookrightarrow E_K(S_K) = R(S)$, so ι is injective. The Tr map is induced by the functorial trace map on points $R(S) = E_K(S_K) \xrightarrow{\mathrm{Tr}} E(S)$.

To verify that Ker(Tr) is geometrically connected, we base extend the exact sequence (1) to $\overline{\mathbb{Q}}$. First, note that there is an isomorphism

$$R_{\overline{\mathbb{Q}}} \cong E_{\overline{\mathbb{Q}}} \times \cdots \times E_{\overline{\mathbb{Q}}}.$$

After base extension, we identify the trace map with the summation map

$$+ : E_{\overline{\mathbb{Q}}} \times \cdots \times E_{\overline{\mathbb{Q}}} \longrightarrow E_{\overline{\mathbb{Q}}}.$$

Let $n = [K : \mathbb{Q}]$. The map defined by

$$(a_1, \ldots, a_{n-1}) \mapsto \left(a_1, a_2, \ldots, a_{n-1}, -\sum_{i=1}^{n-1} a_i \right),$$

is an isomorphism from $E_{\overline{\mathbb{Q}}}^{\times(n-1)}$ to $\mathrm{Ker}(+) = \mathrm{Ker}(\mathrm{Tr}_{\overline{\mathbb{Q}}})$. Thus $\mathrm{Ker}(\mathrm{Tr}_{\overline{\mathbb{Q}}})$ is isomorphic to a product of copies of $E_{\overline{\mathbb{Q}}}$, and hence is connected. □

Corollary 2.5. $\iota(E) \cap \mathrm{Ker}(\mathrm{Tr}) = \iota(E)[p]$.

Proof. The composition $\mathbb{Q} \hookrightarrow K \xrightarrow{\mathrm{Tr}} \mathbb{Q}$ is multiplication by p, so the composition $E \xrightarrow{\iota} R \xrightarrow{\mathrm{Tr}} E$ is also multiplication by p. Since $\iota(E) \cap \mathrm{Ker}(\mathrm{Tr})$ is the kernel of $\mathrm{Tr} \circ \iota = [p]$, it equals $E[p]$. □

Lemma 2.6. *The abelian varieties A_K, R_K, and $(R/\iota(E))_K$ are all isomorphic to a product of copies of E_K.*

Proposition 2.7. *The exact sequence $0 \to A \to R \to E \to 0$ of Proposition 2.4 extends to an exact sequence $0 \to \mathcal{A} \to \mathcal{R} \to \mathcal{E} \to 0$ of Néron models over $\mathbb{Z}$.*

Proof. We use results of [BLR90, Ch. 7] and the fact that formation of Néron models commutes with unramified base change (see [BLR90, §1.2, Prop. 2]) to prove that for every prime q, the complex

$$0 \to \mathcal{A}_{\mathbb{Z}_q} \to \mathcal{R}_{\mathbb{Z}_q} \to \mathcal{E}_{\mathbb{Z}_q} \to 0 \tag{2}$$

is exact.

First suppose that $q \neq \ell$, and let $\mathfrak{q}$ be a prime of K lying over q. We use the fact that formation of Néron models commutes with unramified base extension and check exactness of (2) after base extension to the unramified extension $\mathcal{O}_{K,\mathfrak{q}}$ of $\mathbb{Z}_q$. By Lemma 2.6, the generic fiber of the base extension of (2) to $\mathcal{O}_{K,\mathfrak{q}}$ is

$$0 \to E_{K,\mathfrak{q}}^{\oplus(n-1)} \to E_{K,\mathfrak{q}}^{\oplus n} \xrightarrow{\Sigma} E_{K,\mathfrak{q}} \to 0.$$

Thus the corresponding complex of Néron models over $\mathcal{O}_{K,\mathfrak{q}}$ is

$$0 \to \mathcal{E}_{\mathcal{O}_{K,\mathfrak{q}}}^{\oplus(n-1)} \to \mathcal{E}_{\mathcal{O}_{K,\mathfrak{q}}}^{\oplus n} \xrightarrow{\Sigma} \mathcal{E}_{\mathcal{O}_{K,\mathfrak{q}}} \to 0,$$

which is exact, since it is exact on S-points for *any* ring S.

Suppose that $q = \ell$. Since $p \neq \ell$, [BLR90, Prop. 7.5.3 (a)] asserts that the sequence $0 \to \mathcal{A}_{\mathbb{Z}_q} \to \mathcal{R}_{\mathbb{Z}_q} \to \mathcal{E}_{\mathbb{Z}_q}$ is exact. Since $p \neq q$, the map $[p] : \mathcal{E}_{\mathbb{Z}_q} \to \mathcal{E}_{\mathbb{Z}_q}$ is an étale morphism of smooth schemes. Since E has good reduction at q, we also know that the fibers of $\mathcal{E}_{\mathbb{Z}_q}$ are geometrically connected, so $[p]$ is surjective (for more details, see the proof of [AS02, Lem. 3.2]). It follows that $\mathcal{R}_{\mathbb{Z}_q} \to \mathcal{E}_{\mathbb{Z}_q}$ is surjective. □

2.3. The cokernel of trace

Let ℓ be a prime as in Conjecture 2.1. This section is devoted to computing the cokernel of the trace map $R(\mathbb{Q}) \to E(\mathbb{Q})$. Note that $R(\mathbb{Q}) = E(K)$, so this cokernel is also $E(\mathbb{Q})/\operatorname{Tr}_{K/\mathbb{Q}}(E(K))$.

Lemma 2.8. *Let K_ℓ denote the completion of K at the totally ramified prime of K lying over ℓ. Then $E(K)[p] = E(K_\ell)[p] = 0$.*

Proof. The characteristic polynomial of $\operatorname{Frob}_\ell \in \operatorname{Gal}(\mathbb{Q}_\ell^{\mathrm{ur}}/\mathbb{Q}_\ell)$ on $E[p] = E(\mathbb{Q}_\ell^{\mathrm{ur}})[p]$ is $x^2 - a_\ell x + \ell \in \mathbb{F}_p[x]$. By hypothesis $a_\ell \not\equiv \ell + 1 \pmod p$, so $+1$ is not a root of $x^2 - a_\ell x + \ell$ hence

$$E(\mathbb{Q}_\ell)[p] = E(\mathbb{Q}_\ell^{\mathrm{ur}})[p]^{\operatorname{Frob}_\ell - 1} = 0.$$

Since K is totally ramified at ℓ and E has good reduction at ℓ, $E(K_\ell)[p] = 0$ as well, so $E(K)[p] = 0$, as required. □

Proposition 2.9. $\operatorname{Coker}(R(\mathbb{Q}) \to E(\mathbb{Q})) \cong E(\mathbb{Q})/pE(\mathbb{Q})$.

Proof. By Corollary 2.5 the image of $\iota(E(\mathbb{Q})) \subset R(\mathbb{Q})$ in $E(\mathbb{Q})$ is $pE(\mathbb{Q})$, so the cokernel of $R(\mathbb{Q}) \to E(\mathbb{Q})$ is a quotient of $E(\mathbb{Q})/pE(\mathbb{Q})$. Thus it suffices to prove that $R(\mathbb{Q})/\iota(E(\mathbb{Q}))$ is *finite* of order coprime to p.

We have an exact sequence $0 \to E \to R \to A' \to 0$, with A' an abelian variety that is isogenous to A (in fact, A' is the abelian variety dual of A since R is self dual, but we will not use this fact). The L-series of A' is $\prod_{i=1}^{p-1} L(E, \chi_{p,\ell}^i, s)$, so by hypothesis $L(A', 1) \neq 0$ and it follows from Kato's theorem (see [Rub98, §8.1]) that $A'(\mathbb{Q})$ is finite. Thus $R(\mathbb{Q})/\iota(E(\mathbb{Q}))$ is finite since $R(\mathbb{Q})/\iota(E(\mathbb{Q})) \subset A'(\mathbb{Q})$. By Lemma 2.6, $A'_K \approx E_K^{\times(p-1)}$ and by Lemma 2.8 $E(K)[p] = 0$, so $A'(\mathbb{Q})[p] = 0$, which proves the proposition. □

2.4. Étale cohomology and Shafarevich-Tate groups

Fix an elliptic curve E over $\mathbb{Q}$ and a prime $p \nmid \prod c_{E,q}$. In this section, we use results mostly due to Mazur to relate the Shafarevich-Tate groups of A, R, and E to certain étale cohomology groups. We maintain the notation and assumptions of the previous sections, except that we abuse notation slightly and let $\mathcal{A}$, $\mathcal{R}$, and $\mathcal{E}$ also denote the étale sheaves on $\mathrm{Spec}(\mathbb{Z})$ defined by the Néron models $\mathcal{A}$, $\mathcal{R}$, and $\mathcal{E}$. Let $\mathcal{B}$ be either $\mathcal{A}$, $\mathcal{R}$, or $\mathcal{E}$ and let $B = \mathcal{B}_{\mathbb{Q}}$ be the corresponding abelian variety. Let $H^q(\mathbb{Z}, \mathcal{B})$ be the qth étale cohomology group of $\mathcal{B}$.

Lemma 2.10. *There is an isomorphism* $B(\mathbb{Q}_\ell)[p] \cong \mathcal{B}(\mathbb{F}_\ell)[p]$.

Proof. This follows from [ST68, Lem. 2, p. 495], but we sketch a proof for the convenience of the reader. Let $B^1(\mathbb{Q}_\ell)$ denote the kernel of the natural reduction map $r : B(\mathbb{Q}_\ell) \to \mathcal{B}(\mathbb{F}_\ell)$. Using formal groups and that $p \neq \ell$, one sees that $[p] : B^1(\mathbb{Q}_\ell) \to B^1(\mathbb{Q}_\ell)$ is an isomorphism. Since $\mathcal{B}$ is smooth over $\mathbb{Q}_\ell$, Hensel's lemma (see [BLR90, §2.3 Prop. 5]) implies that the reduction map is surjective, so we obtain an exact sequence

$$0 \to B^1(\mathbb{Q}_\ell) \to B(\mathbb{Q}_\ell) \to \mathcal{B}(\mathbb{F}_\ell) \to 0.$$

The snake lemma applied to the multiplication-by-p diagram attached to this exact sequence yields the exact sequence

$$0 \to B(\mathbb{Q}_\ell)[p] \to \mathcal{B}(\mathbb{F}_\ell)[p] \to 0 \to B(\mathbb{Q}_\ell)/pB(\mathbb{Q}_\ell) \to \mathcal{B}(\mathbb{F}_\ell)/p\mathcal{B}(\mathbb{F}_\ell) \to 0,$$

which proves the lemma. □

The *Tamagawa number* of B at a prime q is $c_{B,q} = \#\Phi_{B,q}(\mathbb{F}_q)$, where $\Phi_{B,q}$ is the component group of the closed fiber of the Néron model of B at q.

Lemma 2.11. $p \nmid c_{B,q}$.

Proof. First suppose $q = \ell$. The cokernel of $\mathcal{B}(\mathbb{F}_\ell) \to \Phi_{B,\ell}(\mathbb{F}_\ell)$ is contained in $H^1(\mathbb{F}_\ell, \mathcal{B}^0)$, which is 0 by Lang's theorem (see [Lan56] or [Ser88, §VI.4]), so if $\Phi_{B,\ell}(\mathbb{F}_\ell)[p] \neq 0$ then $\mathcal{B}(\mathbb{F}_\ell)[p] \neq 0$. But by Lemmas 2.6, 2.8, and 2.10,

$$\mathcal{B}(\mathbb{F}_\ell)[p] \cong \mathcal{B}(\mathbb{Q}_\ell)[p] \subset \mathcal{B}(K_\ell)[p] \cong E(K_\ell)[p] \times \cdots \times E(K_\ell)[p] = 0.$$

Next suppose that $q \neq \ell$. Since formation of Néron models commutes with unramified base extension, we have

$$\Phi_{B,q}(\overline{\mathbb{F}}_q)[p] \cong \Phi_{E,q}(\overline{\mathbb{F}}_q)[p] \times \cdots \times \Phi_{E,q}(\overline{\mathbb{F}}_q)[p] = 0,$$

by our hypotheses on p. □

Following the appendix to [Maz72], let

$$\Sigma(B/\mathbb{Q}) = \ker\left(H^1(\mathbb{Q}, B) \to \bigoplus_{\text{all finite } q} H^1(\mathbb{Q}_q, B) \right),$$

where the sum is over all finite primes q of $\mathbb{Q}$. If p is an odd prime, then $\Sigma(B/\mathbb{Q})[p^\infty] = Ш(B/\mathbb{Q})[p^\infty]$; also one can see easily using Tate cohomology for the cyclic group $\mathrm{Gal}(\mathbb{C}/\mathbb{R})$ that

$$\Sigma(B/\mathbb{Q})[2]/Ш(B/\mathbb{Q})[2] \subset H^1(\mathbb{R}, B(\mathbb{C})) \cong B(\mathbb{R})/B(\mathbb{R})^0,$$

where $B(\mathbb{R})/B(\mathbb{R})^0$ has order 2^e for some $e \leq \dim B$.

Proposition 2.12 (Mazur). *Suppose that $a_\ell \not\equiv \ell + 1 \pmod{p}$. If p is odd, then*

$$H^1(\mathbb{Z}, \mathcal{B})[p^\infty] \cong Ш(B/\mathbb{Q})[p^\infty].$$

Also, $\#H^1(\mathbb{Z}, \mathcal{B})[2^\infty]/Ш(B/\mathbb{Q})[2^\infty]$ divides $\#(B(\mathbb{R})/B(\mathbb{R})^0)$.

Proof. It follows from the appendix to [Maz72] that there is an exact sequence

$$0 \to \Sigma(B)[p^\infty] \to H^1(\mathbb{Z}, \mathcal{B})[p^\infty] \to \bigoplus_{\text{all finite } q} H^1\left(\mathbb{F}_q, \Phi_{B,q}(\overline{\mathbb{F}}_q)\right)[p^\infty], \qquad (3)$$

where $\Phi_{B,q}$ is the component group of the fiber of $\mathcal{B}$ at q. By [Ser79, VIII.4.8],

$$\#H^1(\mathbb{F}_q, \Phi_{B,q}(\overline{\mathbb{F}}_q)) = \#\Phi_{B,q}(\mathbb{F}_q) = c_{B,q},$$

so the proposition follows from Lemma 2.11. □

Proposition 2.13. $H^2(\mathbb{Z}, \mathcal{A})[p] = 0$.

Proof. We apply the lemmas in [Sch83, §III.6]. Note that A has good reduction at p by [Mil72, Prop. 1], and $H^1(\mathbb{Z}, \mathcal{A})[p^\infty]$ is finite by Kato's theorem (see [Rub98, §8.1]) and Proposition 2.12. In the proof of Proposition 2.9, we showed that $A'(\mathbb{Q})$ is finite of order coprime to p, where A' is the abelian variety dual of A. We now use[1] Lemma 7 of [Sch83, §III.6], which because $A'(\mathbb{Q})[p] = 0$ implies that $H^2(\mathbb{Z}, \mathcal{A}[p^\infty]) = 0$. (Schneider uses H^q_{fpqf}, but this is not a problem since étale and fpqf cohomology agree on the smooth scheme $\mathcal{A}$.) It is easy to see (see, e.g., the proof of Lemma 6 of [Sch83, §III.6]) that the natural map $H^q(\mathbb{Z}, \mathcal{A}[p^\infty]) \to H^q(\mathbb{Z}, \mathcal{A})[p^\infty]$ is surjective for any $q > 0$, in particular, for $q = 2$, so $H^2(\mathbb{Z}, \mathcal{A})[p^\infty] = 0$ which proves the proposition. □

2.5. The main theorem

Fix an elliptic curve E over $\mathbb{Q}$ and a prime $p \nmid \prod c_{E,q}$ such that $\rho_{E,p} : G_\mathbb{Q} \to \mathrm{Aut}(E[p])$ is surjective. If $p = 2$ assume also that $E(\mathbb{R})$ is connected. Assume that ℓ is one of the primes whose existence is predicted by Conjecture 2.1.

Theorem 2.14. *There is an exact sequence*

$$0 \to E(\mathbb{Q})/pE(\mathbb{Q}) \to Ш(A/\mathbb{Q})[p^\infty] \to Ш(E/K)[p^\infty] \to Ш(E/\mathbb{Q})[p^\infty] \to 0.$$

In particular, if E has odd rank and $Ш(E/\mathbb{Q})[p^\infty]$ is finite, then $\#Ш(A/\mathbb{Q})[p^\infty]$ is not a perfect square.

---

[1] Note that the proof of Lemma 7 of [Sch83, §III.6] relies on a theorem of Artin and Mazur whose proof they never published; generalizations of this theorem have been published by McCallum [McC86, §5] and Milne [Mil86, §III.3.4], and Mazur assures the author that he and Milne both know the proof of Artin-Mazur duality well.

Proof. By Proposition 2.7 we have an exact sequence of étale sheaves

$$0 \to \mathcal{A} \to \mathcal{R} \to \mathcal{E} \to 0,$$

which gives rise to an exact sequence of étale cohomology groups

$$H^0(\mathbb{Z}, \mathcal{R}) \to H^0(\mathbb{Z}, \mathcal{E}) \to H^1(\mathbb{Z}, \mathcal{A}) \to H^1(\mathbb{Z}, \mathcal{R}) \to H^1(\mathbb{Z}, \mathcal{E}) \to H^2(\mathbb{Z}, \mathcal{A}).$$

We have

$$H^0(\mathbb{Z}, \mathcal{R}) = \mathcal{R}(\mathbb{Z}) = R(\mathbb{Q})$$

and likewise for $\mathcal{E}$, so by Propositions 2.9, 2.12, and 2.13 we obtain an exact sequence

$$0 \to E(\mathbb{Q})/pE(\mathbb{Q}) \to Ш(A/\mathbb{Q})[p^\infty] \to Ш(R/\mathbb{Q})[p^\infty] \to Ш(E/\mathbb{Q})[p^\infty] \to 0.$$

By Shapiro's lemma, there is an isomorphism $Ш(R/\mathbb{Q}) \cong Ш(E/K)$ (see [AS02, §1.3]), which yields the claimed exact sequence.

Kato's theorem ([Rub98, § 8.1] and [Kat, Corollary 14.3]) implies that $Ш(E/K)[p^\infty]$ is finite (for the trivial character use our hypothesis that $Ш(E/\mathbb{Q})[p^\infty]$ is finite, and for the non-trivial characters use our hypothesis that $L(E, \chi_{p,\ell}, 1) \neq 0$). Theorem 1.2 then implies that $\#Ш(E/K)[p^\infty]$ is a perfect square. If $E(\mathbb{Q})$ has odd rank then $\#(E(\mathbb{Q})/pE(\mathbb{Q}))$ is an odd power of p (since $E[p]$ is irreducible), so $\#Ш(A/\mathbb{Q})[p^\infty]$ cannot be a perfect square. □

Remark 2.15. In the language of visibility of Shafarevich-Tate groups (see [CM00]), Theorem 2.14 asserts that the visible subgroup of $Ш(A)$ with respect to the embedding $A \hookrightarrow R$ is canonically isomorphic to the Mordell-Weil quotient $E(\mathbb{Q})/pE(\mathbb{Q})$.

Proposition 2.16. *If $q \neq p$ is a prime, then*

$$Ш(E/K)[q^\infty] \cong Ш(E/\mathbb{Q})[q^\infty] \oplus Ш(A/\mathbb{Q})[q^\infty]. \tag{4}$$

In particular, if $Ш(E/\mathbb{Q})[q^\infty]$ is finite, then $Ш(A/\mathbb{Q})[q^\infty]$ has order a perfect square.

Proof. The intersection of E and A in R is $E[p]$, so the summation map $E \times A \to R$ is an isogeny with kernel $E[p]$. Considering the long exact sequence associated to $0 \to E[p] \to E \times A \to R \to 0$, we see that

$$H^1(\mathbb{Q}, E \times A)[q^\infty] \cong H^1(\mathbb{Q}, R)[q^\infty], \tag{5}$$

and likewise for any completion $\mathbb{Q}_v$ of $\mathbb{Q}$. We then obtain (4) by combining (5) with the fact that cohomology commutes with products and that $H^1(\mathbb{Q}, R) \cong H^1(K, E)$.

If $Ш(E/\mathbb{Q})[q^\infty]$ is finite, then since $Ш(A/\mathbb{Q})[q^\infty]$ is finite (since $L(A, 1) \neq 0$, by construction), it follows from (4) that $Ш(E/K)[q^\infty]$ is finite. We have by Theorem 1.2 that both $Ш(E/K)[q^\infty]$ and $Ш(E/\mathbb{Q})[q^\infty]$ have order a perfect square, so (4) implies that $Ш(A/\mathbb{Q})[q^\infty]$ has order a perfect square. □

3. An example

Combining Proposition 2.3, Theorem 2.14, and Propoposition2.16 yields the following theorem.

Theorem 3.1. *Let E be the elliptic curve $y^2 + y = x^3 - x$ of conductor 37. For every odd prime $p < 25000$ (with $p \neq 37$), there is a twist A of $E^{\times(p-1)}$ such that $\#Ш(A/\mathbb{Q}) = pn^2$ for some integer n.*

Remark 3.2. Using the elliptic curve of conductor 43 in place of E one can construct an abelian variety A with $Ш(A/\mathbb{Q}) = 37n^2$ for some integer n.

Though unnecessary for Theorem 3.1, we prove below that $Ш(E/\mathbb{Q}) = 0$, which removes our dependence on Proposition 2.13. We show that $Ш(E/\mathbb{Q})[p^\infty] = 0$ for all odd p using [Kol90, Thm. A], and we use a 2-descent (with [CrB]) to see that $Ш(E/\mathbb{Q})[2] = 0$.

Theorem 3.3 (Kolyvagin). *Let E be an elliptic curve and let $L = \mathbb{Q}(\sqrt{-D})$ be an imaginary quadratic field of odd discriminant $-D$, where all primes dividing the conductor of E split, and assume that $D \neq 3, 4$. If the Heegner point $y_L \in E(L)$ has infinite order (equivalently, by [GZ86], $L'(E/L, 1) \neq 0$), then $\#Ш(E/L) \mid t \cdot [E(L) : \mathbb{Z}y_L]^2$, where the only primes that divide t are 2 or primes where $\rho_{E,p}$ is not surjective.*

By [C97], E is isolated in its isogeny class, so $\rho : \mathrm{Gal}(\overline{\mathbb{Q}}/\mathbb{Q}) \to \mathrm{Aut}(E[p])$ is surjective for all primes p (see [RS01, §1.4]) hence t is a power of 2. Let $L = \mathbb{Q}(\sqrt{-7})$. To compute $[E(L) : \mathbb{Z}y_L]$ up to a power of 2 we use the Gross-Zagier formula and the fact that $[E(L) : E(\mathbb{Q}) + E^D(\mathbb{Q})]$ is a power of 2. By [GZ86, Thm. 6.3],

$$h(y_L) = \frac{u^2|D|^{\frac{1}{2}}}{\|\omega_f\|} L'(E,1)L(E^D,1),$$

where $D = -7$, $u = 1$, and $\|\omega_f\|$ is the Peterson norm of the newform f corresponding to E. Generators for the period lattice of E are $\omega_1 \sim 2.993459$ and $\omega_2 \sim 2.451389i$, so $\|\omega_f\| \sim 7.338133$. The quadratic twist E^D is the curve **1813B1** in [CrA], and $E^D(\mathbb{Q}) = 0$. From [CrA] we find that $L'(E,1) \sim 0.306000$ and $L(E^D,1) \sim 1.853076$, so $h(y_L) \sim 0.204446$. The height of a generator of $E(\mathbb{Q})$ is $\sim 0.051111 \sim h(y_L)/4$, so $[E(L) : \mathbb{Z}y_L]$ is a power of 2. (As a double check, and to avoid dependence on the Gross-Zagier formula, we wrote a program using [BCP97] to compute Heegner points and found that $y_L = (0,0)$, which is a generator for $E(\mathbb{Q})$.) Thus $\#Ш(E/L)$ is a power of 2.

To connect $Ш(E/L)$ with $Ш(E/\mathbb{Q})$, use the inflation-restriction exact sequence

$$0 \to H^1(L/\mathbb{Q}, E(L)) \to H^1(\mathbb{Q}, E(\overline{\mathbb{Q}})) \to H^1(L, E(\overline{\mathbb{Q}})).$$

Let p be an odd prime. Since $H^1(L/\mathbb{Q}, E(L))$ is a 2-group, the above sequence leads to an injective map

$$H^1(\mathbb{Q}, E(\overline{\mathbb{Q}}))[p] \hookrightarrow H^1(L, E(\overline{\mathbb{Q}}))[p],$$

which induces an inclusion

$$Ш(E/\mathbb{Q})[p] \hookrightarrow Ш(E/L)[p] = 0.$$

References

[AS02] A. Agashe and W.A. Stein, *Visibility of Shafarevich-Tate Groups of Abelian Varieties*, J. of Number Theory, **97** (2002), no. 1, 171–184.

[BCDT01] C. Breuil, B. Conrad, F. Diamond, and R. Taylor, *On the modularity of elliptic curves over* $\mathbb{Q}$*: Wild 3-adic exercises*, J. Amer. Math. Soc. **15** (2001), no. 4, 843–939.

[BCP97] W. Bosma, J. Cannon, and C. Playoust, *The Magma algebra system. I. The user language*, J. Symbolic Comput. **24** (1997), no. 3-4, 235–265, Computational algebra and number theory (London, 1993). MR 1 484 478.

[BLR90] S. Bosch, W. Lütkebohmert, and M. Raynaud, *Néron models*, Springer-Verlag, Berlin, 1990. MR **91i**:14034.

[C97] J.E. Cremona, *Algorithms for modular elliptic curves*, second ed., Cambridge University Press, Cambridge, 1997.

[CrA] ———, *Elliptic Curve Data*, http://www.maths.nott.ac.uk/personal/jec/ftp/data/.

[CrB] ———, `mwrank` *(computer software)*, `http://www.maths.nott.ac.uk/personal/jec/ftp/progs/`.

[CM00] J.E. Cremona and B. Mazur, *Visualizing elements in the Shafarevich-Tate group*, Experiment. Math. **9** (2000), no. 1, 13–28. MR 1 758 797.

[Fea01] J. Fearnley, *Vanishing and Non-Vanishing of L-series of Elliptic Curves Twisted by Dirichlet Characters*, Concordia Ph.D. thesis (2001).

[Fla90] M. Flach, *A generalisation of the Cassels-Tate pairing*, J. Reine Angew. Math. **412** (1990), 113–127. MR **92b**:11037.

[GZ86] B. Gross and D. Zagier, *Heegner points and derivatives of L-series*, Invent. Math. **84** (1986), no. 2, 225–320. MR **87j**:11057.

[JL99] B.W. Jordan and R. Livné, *On Atkin-Lehner quotients of Shimura curves*, Bull. London Math. Soc. **31** (1999), no. 6, 681–685. MR **2000j**:11090.

[Kat] K. Kato, *p-adic Hodge theory and values of zeta functions of modular forms*, Preprint, 244 pages.

[Kol90] V.A. Kolyvagin, *Euler systems*, The Grothendieck Festschrift, Vol. II, Birkhäuser Boston, Boston, MA, 1990, pp. 435–483. MR **92g**:11109.

[Lan56] S. Lang, *Algebraic groups over finite fields*, Amer. J. Math. **78** (1956), 555–563. MR 19,174a.

[Maz72] B. Mazur, *Rational points of abelian varieties with values in towers of number fields*, Invent. Math. **18** (1972), 183–266.

[McC86] W.G. McCallum, *Duality theorems for Néron models*, Duke Math. J. **53** (1986), no. 4, 1093–1124. MR **88c:**14062.

[Mil72] J.S. Milne, *On the arithmetic of abelian varieties*, Invent. Math. **17** (1972), 177–190. MR 48 #8512.

[Mil86] ______, *Arithmetic duality theorems*, Academic Press Inc., Boston, Mass., 1986.

[PS97] B. Poonen and E.F. Schaefer, *Explicit descent for Jacobians of cyclic covers of the projective line*, J. Reine Angew. Math. **488** (1997), 141–188. MR **98k:**11087.

[PS99] B. Poonen and M. Stoll, *The Cassels-Tate pairing on polarized abelian varieties*, Ann. of Math. (2) **150** (1999), no. 3, 1109–1149. MR **2000m:**11048.

[RS01] K.A. Ribet and W.A. Stein, *Lectures on Serre's conjectures*, Arithmetic algebraic geometry (Park City, UT, 1999), IAS/Park City Math. Ser., vol. 9, Amer. Math. Soc., Providence, RI, 2001, pp. 143–232. MR **2002h:**11047.

[Rub98] K. Rubin, *Euler systems and modular elliptic curves*, Galois representations in arithmetic algebraic geometry (Durham, 1996), Cambridge Univ. Press, Cambridge, 1998, pp. 351–367. MR **2001a:**11106.

[Sch83] P. Schneider, *Iwasawa L-functions of varieties over algebraic number fields. A first approach*, Invent. Math. **71** (1983), no. 2, 251–293. MR **85d:**11063.

[SD67] P. Swinnerton-Dyer, *The conjectures of Birch and Swinnerton-Dyer, and of Tate*, Proc. Conf. Local Fields (Driebergen, 1966), Springer, Berlin, 1967, pp. 132–157. MR 37 #6287.

[Ser79] J-P. Serre, *Local fields*, Springer-Verlag, New York, 1979, Translated from the French by Marvin Jay Greenberg.

[Ser88] ______, *Algebraic groups and class fields*, Springer-Verlag, New York, 1988, Translated from the French.

[ST68] J-P. Serre and J.T. Tate, *Good reduction of abelian varieties*, Ann. of Math. (2) **88** (1968), 492–517.

[Tat63] J. Tate, *Duality theorems in Galois cohomology over number fields*, Proc. Internat. Congr. Mathematicians (Stockholm, 1962), Inst. Mittag-Leffler, Djursholm, 1963, pp. 288–295. MR 31 #168.

William A. Stein
515 Science Center
Department of Mathematics
Harvard University
e-mail: `was@math.harvard.edu`

